IGCSE
MATHEMATICS
SECOND EDITION

Ric Pimentel and Terry Wall

PART OF HACHETTE LIVRE UK

Orders: please contact Bookpoint Ltd, 130 Milton Park, Abingdon, Oxon OX14 4SB. Telephone: (44) 01235 827720. Fax: (44) 01235 400454. Lines are open 9.00–5.00, Monday to Saturday, with a 24-hour message answering service. Visit our website at www.hoddereducation.co.uk

First published in 1997
by John Murray (Publishers) Ltd

This second edition first published 2006
by Hodder Murray, an imprint of Hodder Education,
Part of Hachette Livre UK,
338 Euston Road,
London NW1 3BH.

Reprinted 2006, 2007 (twice), 2008

Cover photo: Stone/Getty Images
Typeset in 10/12 pt Times by Pantek Arts Ltd, Maidstone, Kent
Printed and bound in Malta

A catalogue record for this title is available from the British Library.

ISBN: 978 0 340 90813 6

CONTENTS

INTRODUCTION

To the student

This IGCSE mathematics textbook has been written by two experienced class teachers to give you, the student, both enjoyment and a sense of achievement. We also hope it will help you to think in a logical way. We hope that you will see each chapter as a puzzle to be solved. You will be told the rules, given some examples and then be expected to think for yourself and to show others your thought process (your working).

Some of the questions are hard and you may need help from your teacher, but try to solve the questions yourself first! The greatest achievement comes from overcoming difficulty without help.

Students in many countries will be studying from this IGCSE mathematics textbook. It is interesting that, whatever language you speak, mathematics, and algebra in particular, is a universal language, understood in every country on Earth.

The chapters in this book follow closely the order of the IGCSE syllabus. However, it is unlikely that you will work through the book in page order. We have indicated where work is essential revision or at core or extended level but there is some overlap. In fact some core questions are really quite difficult.

There are a number of student assessments at the end of each chapter: some require short answers and some require longer answers, but you should always show your working. The student assessments are to help you to determine how clearly you have understood the work in the chapter. Your teacher will inform you how to use the student assessments but these are best done unaided and in silence.

This new edition also includes a section at the end of many of the chapters where Information and Communication Technology (ICT) can be used to solve problems in a different, and more efficient, way.

We assume that you have already studied maths for many years. Some of the work will be familiar to you, some totally new. We hope that you will approach the work in this book in a positive way expecting both difficulty and enjoyment. We hope that you gain a sense of achievement and that this achievement is reflected in the highest grade in the IGCSE examination of which you are capable.

Ric Pimentel
Terry Wall

NUMBER

■ Essential Revision ■
(i) ORDERING

Order quantities by magnitude and demonstrate familiarity with the symbols $=$, $\neq$, $>$, $<$, $\geqslant$, $\leqslant$.

The following symbols have a specific meaning in mathematics:

$=$ is equal to
$\neq$ is not equal to
$>$ is greater than
$\geqslant$ is greater than or equal to
$<$ is less than
$\leqslant$ is less than or equal to

$x \geqslant 3$ implies that x is greater than or equal to 3, i.e. x can be 3, 4, 4.2, 5, 5.6, etc.
$3 \leqslant x$ implies that 3 is less than or equal to x, i.e. x is still 3, 4, 4.2, 5, 5.6, etc.

Therefore:

$5 > x$ can be rewritten as $x < 5$, i.e. x can be 4, 3.2, 3, 2.8, 2, 1, etc.
$-7 \leqslant x$ can be rewritten as $x \geqslant -7$, i.e. x can be -7, -6, -5, etc.

These inequalities can also be represented on a number line:

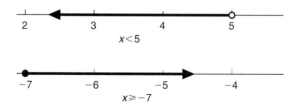

Note that ○→ implies that the number is not included in the solution whilst ●→ implies that the number is included in the solution.

Worked examples   **a)** The maximum number of players from one football team allowed on the pitch at any one time is eleven. Represent this information:
i) as an inequality,
ii) on a number line.

i) Let the number of players be represented by the letter n. n must be less than or equal to 11. Therefore $n \leqslant 11$.
ii)

b) The maximum number of players from one football team allowed on the pitch at any one time is eleven. The minimum allowed is seven players. Represent this information:
i) as an inequality,
ii) on a number line.

i) Let the number of players be represented by the letter n. n must be greater than or equal to 7, but less than or equal to 11.
  Therefore $7 \leqslant n \leqslant 11$.
ii)

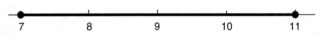

Exercise A   **1.**   Copy each of the following statements, and insert one of the symbols $=, >, <$ into the space to make the statement correct:
a) $7 \times 2 \ldots 8 + 7$ b) $6^2 \ldots 9 \times 4$
c) $5 \times 10 \ldots 7^2$ d) $80 \text{ cm} \ldots 1 \text{ m}$
e) $1000 \text{ litres} \ldots 1 \text{ m}^3$ f) $48 \div 6 \ldots 54 \div 9$

2.   Represent each of the following inequalities on a number line, where x is a real number:
a) $x < 2$ b) $x \geqslant 3$
c) $x \leqslant -4$ d) $x \geqslant -2$
e) $2 < x < 5$ f) $-3 < x < 0$
g) $-2 \leqslant x < 2$ h) $2 \geqslant x \geqslant -1$

3.   Write down the inequalities which correspond to the following number lines:

a)

b)

c)

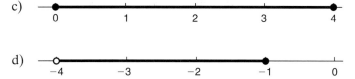

d)

4. Write the following sentences using inequality signs.
   a) The maximum capacity of an athletics stadium is 20 000 people.
   b) In a class the tallest student is 180 cm and the shortest is 135 cm.
   c) Five times a number plus 3 is less than 20.
   d) The maximum temperature in May was 25 °C.
   e) A farmer has between 350 and 400 apples on each tree in his orchard.
   f) In December temperatures in Kenya were between 11 °C and 28 °C.

Exercise B   **1.** Write the following decimals in order of magnitude, starting with the smallest:

6.0   0.6   0.66   0.606   0.06   6.6   6.606

2. Write the following fractions in order of magnitude, starting with the largest:

$$\frac{1}{2} \quad \frac{1}{3} \quad \frac{6}{13} \quad \frac{4}{5} \quad \frac{7}{18} \quad \frac{2}{19}$$

3. Write the following lengths in order of magnitude, starting with the smallest:

2 m   60 cm   800 mm   180 cm   0.75 m

4. Write the following masses in order of magnitude, starting with the largest:

4 kg   3500 g $\frac{3}{4}$ kg   700 g   1 kg

5. Write the following volumes in order of magnitude, starting with the smallest:

1 l   430 ml   800 cm³   120 cl   150 cm³

Student Assessment 1

1. Insert one of the symbols $=, >, <$ into the space to make
    the following statements correct:
    a) $8 \times 5 \ldots 5 \times 8$ b) $9^2 \ldots (100 - 21)$
    c) $45 \text{ cm} \ldots 0.5 \text{ m}$ d) number of days in June ...
    31 days

2. Illustrate the information in each of the following
    statements on a number line.
    a) The greatest number of days in a month is 31.
    b) A month has at least 28 days.
    c) A cake takes between $1\frac{1}{2}$ and $1\frac{3}{4}$ hours to bake.
    d) An aeroplane will land between 2.40 and 2.45 p.m.

3. Write the following sentences using inequalities.
    a) No more than 52 people can be carried on a bus.
    b) The minimum temperature tonight will be 11 °C.
    c) There are between 24 and 38 students in a class.
    d) Three times a number plus 6 is less than 50.
    e) The minimum reaction time for an alarm system is 0.03
        seconds.

4. Illustrate each of the following inequalities on a number
    line, where y is a real number:
    a) $y > 5$ b) $y \leq -3$
    c) $-3 < y < 21$ d) $-2 \leq y < 1$

5. Write the following fractions in order of magnitude,
    starting with the largest:

    $$\frac{1}{6} \quad \frac{2}{3} \quad \frac{7}{12} \quad \frac{13}{18} \quad \frac{6}{7}$$

Student Assessment 2

1. Copy each of the following statements, and insert one of
    the symbols $=, >, <$ into the space to make the statement
    correct:
    a) $4 \times 2 \ldots 2^3$ b) $6^2 \ldots 2^6$
    c) $850 \text{ ml} \ldots 0.5 \text{ litres}$ d) number of days in May ...
    30 days

2. Illustrate the information in each of the following
    statements on a number line.
    a) The temperature during the day reached a maximum of
        35 °C.
    b) There were between 20 and 25 students in a class.
    c) The world record for the 100 m sprint is under 10
        seconds.
    d) Doubling a number and subtracting 4 gives an answer
        greater than 16.

3. Write the information on the following number lines as inequalities:

a)

b)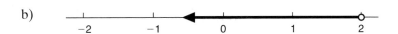

c)

d)

4. Illustrate each of the following inequalities on a number line:

a) $x \geqslant 3$ b) $x < 4$

c) $0 < x < 4$ d) $-3 \leqslant x < 1$

5. Write the following fractions in order of magnitude, starting with the smallest:

$$\frac{4}{7} \quad \frac{3}{14} \quad \frac{9}{10} \quad \frac{1}{2} \quad \frac{2}{5}$$

(ii) STANDARD FORM

Use the standard form $A \times 10^n$ where n is a positive or negative integer, and $1 \leqslant A < 10$.

Standard form is also known as standard index form or sometimes as scientific notation. It involves writing large numbers or very small numbers in terms of powers of 10.

▥ A positive index

$$100 = 1 \times 10^2$$
$$1000 = 1 \times 10^3$$
$$10\,000 = 1 \times 10^4$$
$$3000 = 3 \times 10^3$$

For a number to be in standard form it must take the form $A \times 10^n$ where the index n is a positive or negative integer and A must lie in the range $1 \leqslant A < 10$.

e.g. 3100 can be written in many different ways:

$$3.1 \times 10^3 \quad 31 \times 10^2 \quad 0.31 \times 10^4 \quad \text{etc.}$$

However, only 3.1×10^3 satisfies the above conditions and therefore is the only one which is written in standard form.

Worked examples   **a)** Write 72 000 in standard form.

$$7.2 \times 10^4$$

b) Of the numbers below, ring those which are written in standard form:

$\boxed{4.2 \times 10^3}$ 0.35×10^2 18×10^5 $\boxed{6 \times 10^3}$ 0.01×10^1

c) Multiply the following and write your answer in standard form:

$$600 \times 4000$$
$$= 2\,400\,000$$
$$= 2.4 \times 10^6$$

d) Multiply the following and write your answer in standard form:

$$(2.4 \times 10^4) \times (5 \times 10^7)$$
$$= 12 \times 10^{11}$$
$$= 1.2 \times 10^{12} \text{ when written in standard form}$$

e) Divide the following and write your answer in standard form:

$$(6.4 \times 10^7) \div (1.6 \times 10^3)$$
$$= 4 \times 10^4$$

f) Add the following and write your answer in standard form:

$$(3.8 \times 10^6) + (8.7 \times 10^4)$$

Changing the indices to the same value gives the sum:

$$(380 \times 10^4) + (8.7 \times 10^4)$$
$$= 388.7 \times 10^4$$
$$= 3.887 \times 10^6 \text{ when written in standard form}$$

g) Subtract the following and write your answer in standard form:

$$(6.5 \times 10^7) - (9.2 \times 10^5)$$

Changing the indices to the same value gives

$$(650 \times 10^5) - (9.2 \times 10^5)$$
$$= 640.8 \times 10^5$$
$$= 6.408 \times 10^7 \text{ when written in standard form}$$

Exercise A

1. Which of the following are not in standard form?

  a) 6.2×10^5 b) 7.834×10^{16}
  c) 8.0×10^5 d) 0.46×10^7
  e) 82.3×10^6 f) 6.75×10^1

2. Write the following numbers in standard form:

  a) 600 000   b) 48 000 000
  c) 784 000 000 000   d) 534 000
  e) 7 million   f) 8.5 million

3. Write the following in standard form:

  a) 68×10^5 b) 720×10^6
  c) 8×10^5 d) 0.75×10^8
  e) 0.4×10^{10} f) 50×10^6

4. Multiply the following and write your answers in standard form:

  a) 200×3000 b) 6000×4000
  c) 7 million $\times$ 20   d) 500 $\times$ 6 million
  e) 3 million $\times$ 4 million   f) 4500×4000

5. Light from the Sun takes approximately 8 minutes to reach Earth. If light travels at a speed of 3×10^8 m/s, calculate to three significant figures (s.f.) the distance from the Sun to the Earth.

6. Find the value of the following and write your answers in standard form:

  a) $(4.4 \times 10^3) \times (2 \times 10^5)$ b) $(6.8 \times 10^7) \times (3 \times 10^3)$
  c) $(4 \times 10^5) \times (8.3 \times 10^5)$ d) $(5 \times 10^9) \times (8.4 \times 10^{12})$
  e) $(8.5 \times 10^6) \times (6 \times 10^{15})$ f) $(5.0 \times 10^{12})^2$

7. Find the value of the following and write your answers in standard form:

  a) $(3.8 \times 10^8) \div (1.9 \times 10^6)$ b) $(6.75 \times 10^9) \div (2.25 \times 10^4)$
  c) $(9.6 \times 10^{11}) \div (2.4 \times 10^5)$ d) $(1.8 \times 10^{12}) \div (9.0 \times 10^7)$
  e) $(2.3 \times 10^{11}) \div (9.2 \times 10^4)$ f) $(2.4 \times 10^8) \div (6.0 \times 10^3)$

8. Find the value of the following and write your answers in standard form:
   a) $(3.8 \times 10^5) + (4.6 \times 10^4)$ b) $(7.9 \times 10^9) + (5.8 \times 10^8)$
   c) $(6.3 \times 10^7) + (8.8 \times 10^5)$ d) $(3.15 \times 10^9) + (7.0 \times 10^6)$
   e) $(5.3 \times 10^8) - (8.0 \times 10^7)$ f) $(6.5 \times 10^7) - (4.9 \times 10^6)$
   g) $(8.93 \times 10^{10}) - (7.8 \times 10^9)$ h) $(4.07 \times 10^7) - (5.1 \times 10^6)$

9. The following list shows the distance of the planets of the Solar System from the Sun.

Jupiter	778 million km
Mercury	58 million km
Mars	228 million km
Uranus	2870 million km
Venus	108 million km
Pluto	5920 million km
Earth	150 million km
Saturn	1430 million km

   Write each of the distances in standard form and then arrange them in order of magnitude, starting with the distance of the planet closest to the Sun.

▓ A negative index

A negative index is used when writing a number between 0 and 1 in standard form.

e.g.
$$
\begin{aligned}
100 &= 1 \times 10^2 \\
10 &= 1 \times 10^1 \\
1 &= 1 \times 10^0 \\
0.1 &= 1 \times 10^{-1} \\
0.01 &= 1 \times 10^{-2} \\
0.001 &= 1 \times 10^{-3} \\
0.0001 &= 1 \times 10^{-4}
\end{aligned}
$$

Note that A must still lie within the range $1 \leqslant A < 10$.

Worked examples   a) Write 0.0032 in standard form.

3.2×10^{-3}

b) Write the following numbers in order of magnitude, starting with the largest:

3.6×10^{-3} 5.2×10^{-5} 1×10^{-2} 8.35×10^{-2}
6.08×10^{-8}

8.35×10^{-2} 1×10^{-2} 3.6×10^{-3} 5.2×10^{-5}
6.08×10^{-8}

Exercise B

1. Write the following numbers in standard form:
    a) 0.0006
    b) 0.000 053
    c) 0.000 864
    d) 0.000 000 088
    e) 0.000 000 7
    f) 0.000 414 5

2. Write the following numbers in standard form:
    a) 68×10^{-5}
    b) 750×10^{-9}
    c) 42×10^{-11}
    d) 0.08×10^{-7}
    e) 0.057×10^{-9}
    f) 0.4×10^{-10}

3. Deduce the value of n in each of the following cases:
    a) $0.000\ 25 = 2.5 \times 10^{n}$
    b) $0.00357 = 3.57 \times 10^{n}$
    c) $0.000\ 000\ 06 = 6 \times 10^{n}$
    d) $0.004^{2} = 1.6 \times 10^{n}$
    e) $0.00065^{2} = 4.225 \times 10^{n}$
    f) $0.0002^{n} = 8 \times 10^{-12}$

4. Write these numbers in order of magnitude, starting with the largest:

    $$3.2 \times 10^{-4} \quad 6.8 \times 10^{5} \quad 5.57 \times 10^{-9} \quad 6.2 \times 10^{3}$$
    $$5.8 \times 10^{-7} \quad 6.741 \times 10^{-4} \quad 8.414 \times 10^{2}$$

Student Assessment 1

1. Write the following numbers in standard form:
    a) 8 million
    b) 0.000 72
    c) 75 000 000 000
    d) 0.0004
    e) 4.75 billion
    f) 0.000 000 64

2. Write the following numbers in order of magnitude, starting with the smallest:

    $6.2 \times 10^7 \quad 5.5 \times 10^{-3} \quad 4.21 \times 10^7 \quad 4.9 \times 10^8 \quad 3.6 \times 10^{-5}$
    7.41×10^{-9}

3. Write the following numbers:
    a) in standard form,
    b) in order of magnitude, starting with the largest.

    6 million   820 000   0.0044   0.8   52 000

4. Deduce the value of n in each of the following:
    a) $620 = 6.2 \times 10^n$
    b) $555\,000\,000 = 5.55 \times 10^n$
    c) $0.000\,45 = 4.5 \times 10^n$
    d) $500^2 = 2.5 \times 10^n$
    e) $0.0035^2 = 1.225 \times 10^n$
    f) $0.04^3 = 6.4 \times 10^n$

5. Write the answers to the following calculations in standard form:
    a) $4000 \times 30\,000$
    b) $(2.8 \times 10^5) \times (2.0 \times 10^3)$
    c) $(3.2 \times 10^9) \div (1.6 \times 10^4)$
    d) $(2.4 \times 10^8) \div (9.6 \times 10^2)$

6. The speed of light is 3×10^8 m/s. Venus is 108 million km from the Sun. Calculate the number of minutes it takes for sunlight to reach Venus.

7. A star system is 500 light years away from Earth. If the speed of light is 3×10^5 km/s, calculate the distance the star system is from Earth. Give your answer in kilometres and written in standard form.

Student Assessment 2

1. Write the following numbers in standard form:
    a) 6 million
    b) 0.0045
    c) 3 800 000 000
    d) 0.000 000 361
    e) 460 million
    f) 3

2. Write the following numbers in order of magnitude, starting with the largest:

    $3.6 \times 10^2 \quad 2.1 \times 10^{-3} \quad 9 \times 10^1 \quad 4.05 \times 10^8 \quad 1.5 \times 10^{-2}$
    7.2×10^{-3}

3. Write the following numbers:
    a) in standard form,
    b) in order of magnitude, starting with the smallest.

    15 million   430 000   0.000 435   4.8   0.0085

4. Deduce the value of n in each of the following:
   a) $4750 = 4.75 \times 10^n$
   b) $6\,440\,000\,000 = 6.44 \times 10^n$
   c) $0.0040 = 4.0 \times 10^n$
   d) $1000^2 = 1 \times 10^n$
   e) $0.9^3 = 7.29 \times 10^n$
   f) $800^3 = 5.12 \times 10^n$

5. Write the answers to the following calculations in standard form:
   a) $50\,000 \times 2400$
   b) $(3.7 \times 10^6) \times (4.0 \times 10^4)$
   c) $(5.8 \times 10^7) + (9.3 \times 10^6)$
   d) $(4.7 \times 10^6) - (8.2 \times 10^5)$

6. The speed of light is 3×10^8 m/s. Jupiter is 778 million km from the Sun. Calculate the number of minutes it takes for sunlight to reach Jupiter.

7. A star is 300 light years away from Earth. If the speed of light is 3×10^5 km/s, calculate the distance from the star to Earth. Give your answer in kilometres and written in standard form.

(iii) THE FOUR RULES

Use the four rules for calculations with whole numbers, decimal fractions and vulgar (and mixed) fractions, including correct ordering of operations and use of brackets.

NB: All diagrams are not drawn to scale.

Calculations with whole numbers

Addition, subtraction, multiplication and division are mathematical operations.

Long multiplication

When carrying out long multiplication, it is important to remember place value.

Worked example

$$184 \times 37$$

$$
\begin{array}{rl}
 & 1\ 8\ 4 \\
\times & \ \ 3\ 7 \\
\hline
 & 1\ 2\ 8\ 8 \quad (184 \times 7) \\
 & 5\ 5\ 2\ 0 \quad (184 \times 30) \\
\hline
 & 6\ 8\ 0\ 8 \quad (184 \times 37) \\
\hline
\end{array}
$$

Short division

Worked example

$$453 \div 6$$

$$6\,\overline{)4\ 5\ {}^3 3}\ \ 7\ \ 5\ \text{r}3$$

It is usual, however, to give the final answer in decimal form rather than with a remainder. The division can therefore be continued.

$$453 \div 6$$

$$6\,\overline{)4\ 5\ {}^3 3\ .\ {}^3 0}\ \ 7\ \ 5\ .\ 5$$

Long division

Worked example Calculate $7184 \div 23$ to one decimal place (d.p.).

$$
\begin{array}{r}
3\ 1\ 2.3\ 4 \\
23\,\overline{)7\ 1\ 8\ 4.0\ 0} \\
\underline{6\ 9} \\
2\ 8 \\
\underline{2\ 3} \\
5\ 4 \\
\underline{4\ 6} \\
8\ 0 \\
\underline{6\ 9} \\
1\ 1\ 0 \\
\underline{9\ 2} \\
1\ 8 \\
\end{array}
$$

Therefore $7184 \div 23 = 312.3$ to 1 d.p.

Mixed operations

When a calculation involves a mixture of operations, the order of the operations is important. Multiplications and divisions are done first, whilst additions and subtractions are done afterwards. To override this, brackets need to be used.

Worked examples

a) $3 + 7 \times 2 - 4$
$= 3 + 14 - 4$
$= 13$

b) $(3 + 7) \times 2 - 4$
$= 10 \times 2 - 4$
$= 20 - 4$
$= 16$

c) $3 + 7 \times (2 - 4)$
$= 3 + 7 \times (-2)$
$= 3 - 14$
$= -11$

d) $(3 + 7) \times (2 - 4)$
$= 10 \times (-2)$
$= -20$

Exercise A

1. Evaluate the answer to each of the following:
   a) $3 + 5 \times 2 - 4$
   b) $6 + 4 \times 7 - 12$
   c) $3 \times 2 + 4 \times 6$
   d) $4 \times 5 - 3 \times 6$
   e) $8 \div 2 + 18 \div 6$
   f) $12 \div 8 + 6 \div 4$

2. Copy these and put brackets in the correct places to make the sums correct:
   a) $6 \times 4 + 6 \div 3 = 20$
   b) $6 \times 4 + 6 \div 3 = 36$
   c) $8 + 2 \times 4 - 2 = 12$
   d) $8 + 2 \times 4 - 2 = 20$
   e) $9 - 3 \times 7 + 2 = 44$
   f) $9 - 3 \times 7 + 2 = 54$

3. Without using a calculator, work out the solutions to the following multiplications:
   a) 63×24
   b) 531×64
   c) 785×38
   d) 164×253
   e) 144×144
   f) 170×240

4. Work out the remainders in the following divisions:
   a) $33 \div 7$
   b) $68 \div 5$
   c) $72 \div 7$
   d) $430 \div 9$
   e) $156 \div 5$
   f) $687 \div 10$

5. a) The sum of two numbers is 16, their product is 63. What are the two numbers?
   b) When a number is divided by 7 the result is 14 remainder 2. What is the number?
   c) The difference between two numbers is 5, their product is 176. What are the numbers?
   d) How many 9s can be added to 40 before the total exceeds 100?
   e) A length of rail track is 9 m long. How many complete lengths will be needed to lay 1 km of track?
   f) How many 35 cent stamps can be bought for 10 dollars?

6. Work out the following long divisions to 1 d.p.
   a) $7892 \div 7$
   b) $45\,623 \div 6$
   c) $9452 \div 8$
   d) $4564 \div 4$
   e) $7892 \div 15$
   f) $79\,876 \div 24$

▧ Fractions
Equivalent fractions

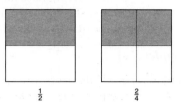
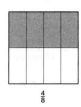

$$\tfrac{1}{2} \qquad \tfrac{2}{4} \qquad \tfrac{4}{8}$$

It should be apparent that $\frac{1}{2}, \frac{2}{4}, \frac{4}{8}$ are equivalent fractions. Similarly, $\frac{1}{3}, \frac{2}{6}, \frac{3}{9}, \frac{4}{12}$ are equivalent, as are $\frac{1}{5}, \frac{10}{50}, \frac{20}{100}$.

Exercise B

1. Copy the following sets of equivalent fractions and fill in the blanks:

   a) $\dfrac{1}{4} = \dfrac{2}{} = \dfrac{}{16} = \dfrac{}{64} = \dfrac{3}{}$

   b) $\dfrac{2}{5} = \dfrac{4}{} = \dfrac{}{20} = \dfrac{}{50} = \dfrac{16}{}$

   c) $\dfrac{3}{8} = \dfrac{6}{} = \dfrac{}{24} = \dfrac{15}{} = \dfrac{}{72}$

   d) $\dfrac{}{7} = \dfrac{8}{14} = \dfrac{12}{} = \dfrac{}{56} = \dfrac{36}{}$

   e) $\dfrac{5}{} = \dfrac{}{27} = \dfrac{20}{36} = \dfrac{}{90} = \dfrac{55}{}$

2. Express the following fractions in their lowest terms, e.g. $\frac{12}{16} = \frac{3}{4}$.

   a) $\frac{5}{10}$ b) $\frac{7}{21}$ c) $\frac{8}{12}$

   d) $\frac{16}{36}$ e) $\frac{75}{100}$ f) $\frac{81}{90}$

3. Write the following improper fractions as mixed numbers, e.g. $\frac{15}{4} = 3\frac{3}{4}$.

   a) $\frac{17}{4}$ b) $\frac{23}{5}$ c) $\frac{8}{3}$

   d) $\frac{19}{3}$ e) $\frac{12}{3}$ f) $\frac{43}{12}$

4. Write the following mixed numbers as improper fractions, e.g. $3\frac{4}{5} = \frac{19}{5}$.

   a) $6\frac{1}{2}$ b) $7\frac{1}{4}$ c) $3\frac{3}{8}$

   d) $11\frac{1}{9}$ e) $6\frac{4}{5}$ f) $8\frac{9}{11}$

Addition and subtraction of fractions

For fractions to be either added or subtracted the denominators need to be the same.

Worked examples

a) $\frac{3}{11} + \frac{5}{11} = \frac{8}{11}$

b) $\frac{7}{8} + \frac{5}{8} = \frac{12}{8}$

c) $\quad \frac{1}{2} + \frac{1}{3}$
$= \frac{3}{6} + \frac{2}{6} = \frac{5}{6}$

d) $\quad \frac{4}{5} - \frac{1}{3}$
$= \frac{12}{15} - \frac{5}{15} = \frac{7}{15}$

When dealing with calculations involving mixed numbers, it is sometimes easier to change them to improper fractions first.

e) $\quad 5\frac{3}{4} - 2\frac{5}{8}$

$= \frac{23}{4} - \frac{21}{8}$

$= \frac{46}{8} - \frac{21}{8}$

$= \frac{25}{8} = 3\frac{1}{8}$

f) $\quad 1\frac{4}{7} + 3\frac{3}{4}$

$= \frac{11}{7} + \frac{15}{4}$

$= \frac{44}{28} + \frac{105}{28}$

$= \frac{149}{28} = 5\frac{9}{28}$

Exercise C

Evaluate each of the following and write the answer as a fraction in its simplest form:

1.
a) $\frac{3}{5} + \frac{4}{5}$
b) $\frac{3}{11} + \frac{7}{11}$
c) $\frac{2}{3} + \frac{1}{4}$
d) $\frac{3}{9} + \frac{4}{9}$
e) $\frac{8}{13} + \frac{2}{5}$
f) $\frac{1}{2} + \frac{2}{3} + \frac{3}{4}$

2.
a) $\frac{3}{7} - \frac{2}{7}$
b) $\frac{4}{5} - \frac{7}{10}$
c) $\frac{8}{9} - \frac{1}{3}$
d) $\frac{7}{12} - \frac{1}{2}$
e) $\frac{5}{8} - \frac{2}{5}$
f) $\frac{3}{4} - \frac{2}{5} + \frac{7}{10}$

3.
a) $2\frac{1}{2} + 3\frac{1}{4}$
b) $3\frac{3}{5} + 1\frac{7}{10}$
c) $6\frac{1}{2} - 3\frac{2}{5}$
d) $8\frac{5}{8} - 2\frac{1}{3}$
e) $5\frac{7}{8} - 4\frac{3}{4}$
f) $3\frac{1}{4} - 2\frac{5}{9}$

Multiplication and division of fractions

Worked examples

a) $\quad \frac{3}{4} \times \frac{2}{3}$

$= \frac{6}{12}$

$= \frac{1}{2}$

b) $\quad 3\frac{1}{2} \times 4\frac{4}{7}$

$= \frac{7}{2} \times \frac{32}{7}$

$= \frac{224}{14}$

$= 16$

Dividing fractions is the same as multiplying by the reciprocal.

c) $\quad \frac{3}{8} \div \frac{3}{4}$

$= \frac{3}{8} \times \frac{4}{3}$

$= \frac{12}{24}$

$= \frac{1}{2}$

d) $\quad 5\frac{1}{2} \div 3\frac{2}{3}$

$= \frac{11}{2} \div \frac{11}{3}$

$= \frac{11}{2} \times \frac{3}{11}$

$= \frac{3}{2}$

Exercise D

1. Write the reciprocal of each of the following:
a) $\frac{3}{4}$
b) $\frac{5}{9}$
c) 7
d) $\frac{1}{9}$
e) $2\frac{3}{4}$
f) $4\frac{5}{8}$

2. Evaluate the following:
a) $\frac{3}{4} \times \frac{4}{5}$
b) $\frac{7}{8} \times \frac{2}{3}$
c) $\frac{3}{4} \times \frac{4}{7} \times \frac{3}{11}$
d) $\left(\frac{4}{5} \div \frac{2}{3}\right) \times \frac{7}{10}$
e) $\frac{1}{2}$ of $\frac{3}{4}$
f) $4\frac{1}{2} \div 3\frac{1}{9}$

3. Evaluate the following:
a) $\left(\frac{3}{8} \times \frac{4}{5}\right) + \left(\frac{1}{2}$ of $\frac{3}{5}\right)$
b) $\left(1\frac{1}{2} \times 3\frac{3}{4}\right) - \left(2\frac{3}{5} \div 1\frac{1}{2}\right)$
c) $\left(\frac{2}{5}$ of $\frac{4}{9}\right) + \left(\frac{4}{9}$ of $\frac{3}{5}\right)$
d) $\left(1\frac{1}{3} \times 2\frac{5}{8}\right)^2$

Changing a fraction to a decimal

To change a fraction to a decimal, divide the numerator by the denominator.

Worked examples **a)** Change $\frac{5}{8}$ to a decimal.

$$\begin{array}{r} 0.6\ \ 2\ \ 5 \\ 8\ \overline{\smash{\big)}\,5.0\ \ ^{2}0\ \ ^{4}0} \end{array}$$

b) Change $2\frac{3}{5}$ to a decimal.
This can be represented as $2 + \frac{3}{5}$.

$$\begin{array}{r} 0.6 \\ 5\ \overline{\smash{\big)}\,3.0} \end{array}$$

therefore $2\frac{3}{5} = 2.6$.

Exercise E **1.** Change the following fractions to decimals:

a) $\frac{3}{4}$ b) $\frac{4}{5}$ c) $\frac{9}{12}$
d) $\frac{17}{60}$ e) $\frac{1}{3}$ f) $\frac{3}{8}$
g) $\frac{7}{16}$ h) $\frac{2}{9}$ i) $\frac{7}{11}$

2. Change the following mixed numbers to decimals:

a) $2\frac{3}{4}$ b) $3\frac{3}{5}$ c) $4\frac{7}{20}$
d) $6\frac{11}{50}$ e) $5\frac{2}{3}$ f) $6\frac{7}{8}$
g) $5\frac{9}{16}$ h) $4\frac{2}{9}$ i) $5\frac{3}{7}$

Changing a decimal to a fraction

Changing a decimal to a fraction is done by knowing the 'value' of each of the numbers in any decimal.

Worked examples **a)** Change 0.45 from a decimal to a fraction.

units	.	tenths	hundredths
0	.	4	5

0.45 is therefore equivalent to 4 tenths and 5 hundredths, which in turn is the same as 45 hundredths.
Therefore $0.45 = \frac{45}{100} = \frac{9}{20}$.

b) Change 2.325 from a decimal to a fraction.

units	.	tenths	hundredths	thousandths
2	.	3	2	5

Therefore $2.325 = 2\frac{325}{1000} = 2\frac{13}{40}$.

Exercise F **1.** Change the following decimals to fractions:

a) 0.5   b) 0.7   c) 0.6
d) 0.75   e) 0.825   f) 0.05
g) 0.050   h) 0.402   i) 0.0002

2. Change the following decimals to mixed numbers:

a) 2.4   b) 6.5   c) 8.2
d) 3.75   e) 10.55   f) 9.204
g) 15.455   h) 30.001   i) 1.0205

Student Assessment 1

1. Evaluate the following:
   a) $5 + 8 \times 3 - 6$ b) $15 + 45 \div 3 - 12$

2. The sum of two numbers is 21 and their product is 90. What are the numbers?

3. How many seconds are there in $2\frac{1}{2}$ hours?

4. Work out 851×27.

5. Work out $6843 \div 19$ giving your answer to 1 d.p.

6. Copy these equivalent fractions and fill in the blanks:

$$\frac{8}{18} = \frac{}{9} = \frac{16}{} = \frac{56}{} = \frac{}{90}$$

7. Evaluate the following:
   a) $3\frac{3}{4} - 1\frac{11}{16}$ b) $4\frac{4}{5} \div \frac{8}{15}$

8. Change the following fractions to decimals:
   a) $\frac{2}{5}$ b) $1\frac{3}{4}$
   c) $\frac{9}{11}$ d) $1\frac{2}{3}$

9. Change the following decimals to fractions. Give each fraction in its simplest form.
   a) 4.2   b) 0.06
   c) 1.85   d) 2.005

Student Assessment 2

1. Evaluate the following:
   a) $6 \times 4 - 3 \times 8$ b) $15 \div 3 + 2 \times 7$

2. The product of two numbers is 72, and their sum is 18. What are the two numbers?

3. How many days are there in 42 weeks?

4. Work out 368×49.

5. Work out $7835 \div 23$ giving your answer to 1 d.p.

6. Copy these equivalent fractions and fill in the blanks:

$$\frac{24}{36} = \frac{}{12} = \frac{4}{30} = \frac{}{} = \frac{60}{}$$

7. Evaluate the following:
   a) $2\frac{1}{2} - \frac{4}{5}$ b) $3\frac{1}{2} \times \frac{4}{7}$

8. Change the following fractions to decimals:
   a) $\frac{7}{8}$ b) $1\frac{2}{5}$
   c) $\frac{8}{9}$ d) $3\frac{2}{7}$

9. Change the following decimals to fractions. Give each fraction in its simplest form.
   a) 6.5   b) 0.04
   c) 3.65   d) 3.008

(iv) *ESTIMATION*

Give approximations to specified numbers of significant figures (s.f.) and decimal places (d.p.) and round off answers to reasonable accuracy in the context of a given problem.

NB: All diagrams are not drawn to scale.

In many instances exact numbers are not necessary or even desirable. In those circumstances approximations are given. The approximations can take several forms. The common types of approximation are dealt with below.

■ Rounding

If 28 617 people attend a gymnastics competition, this figure can be reported to various levels of accuracy.

To the nearest 10 000 this figure would be rounded up to 30 000.
To the nearest 1000 the figure would be rounded up to 29 000.
To the nearest 100 the figure would be rounded down to 28 600.

In this type of situation it is unlikely that the exact number would be reported.

Exercise A

1. Round the following numbers to the nearest 1000:
   a) 68 786   b) 74 245   c) 89 000
   d) 4020   e) 99 500   f) 999 999

2. Round the following numbers to the nearest 100:
   a) 78 540   b) 6858   c) 14 099
   d) 8084   e) 950   f) 2984

3. Round the following numbers to the nearest 10:
   a) 485   b) 692   c) 8847
   d) 83   e) 4   f) 997

Decimal places

A number can also be approximated to a given number of decimal places (d.p.). This refers to the quantity of numbers written after a decimal point.

Worked examples   a) Write 7.864 to 1 d.p.
The answer needs to be written with one number after the decimal point. However, to do this, the second number after the decimal point also needs to be considered. If it is 5 or more then the first number is rounded up.

i.e. 7.864 is written as 7.9 to 1 d.p.

b) Write 5.574 to 2 d.p.
The answer here is to be given with two numbers after the decimal point. In this case the third number after the decimal point needs to be considered. As the third number after the decimal point is less than 5, the second number is not rounded up.

i.e. 5.574 is written as 5.57 to 2 d.p.

Exercise B **1.** Give the following to 1 d.p.

a) 5.58	b) 0.73	c) 11.86
d) 157.39	e) 4.04	f) 15.045
g) 2.95	h) 0.98	i) 12.049

2. Give the following to 2 d.p.

a) 6.473	b) 9.587	c) 16.476
d) 0.088	e) 0.014	f) 9.3048
g) 99.996	h) 0.0048	i) 3.0037

Significant figures

Numbers can also be approximated to a given number of significant figures (s.f.). In the number 43.25 the 4 is the most significant figure as it has a value of 40. In contrast, the 5 is the least significant as it only has a value of 5 hundredths.

Worked examples **a)** Write 43.25 to 3 s.f.
Only the three most significant numbers are written, however the fourth number needs to be considered to see whether the third number is to be rounded up or not.

i.e. 43.25 is written as 43.3 to 3 s.f.

b) Write 0.0043 to 1 s.f.
In this example only two numbers have any significance, the 4 and the 3. The 4 is the most significant and therefore is the only one of the two to be written in the answer.

i.e. 0.0043 is written as 0.004 to 1 s.f.

Exercise C **1.** Write the following to the number of significant figures written in brackets:

a) 48 599 (1 s.f.)	b) 48 599 (3 s.f.)	c) 6841 (1 s.f.)
d) 7538 (2 s.f.)	e) 483.7 (1 s.f.)	f) 2.5728 (3 s.f.)
g) 990 (1 s.f.)	h) 2045 (2 s.f.)	i) 14.952 (3 s.f.)

2. Write the following to the number of significant figures written in brackets:

a) 0.085 62 (1 s.f.)	b) 0.5932 (1 s.f.)	c) 0.942 (2 s.f.)
d) 0.954 (1 s.f.)	e) 0.954 (2 s.f.)	f) 0.003 05 (1 s.f.)
g) 0.003 05 (2 s.f.)	h) 0.009 73 (2 s.f.)	i) 0.009 73 (1 s.f.)

▦ Appropriate accuracy

In many instances calculations carried out using a calculator produce answers which are not whole numbers. A calculator will give the answer to as many decimal places as will fit on its screen. In most cases this degree of accuracy is neither desirable nor necessary. Unless specifically asked for, answers should not be given to more than two decimal places. Indeed, one decimal place is usually sufficient.

Worked example Calculate $4.64 \div 2.3$ giving your answer to an appropriate degree of accuracy.

The calculator will give the answer to $4.64 \div 2.3$ as 2.0173913. However the answer given to 1 d.p. is sufficient. Therefore $4.64 \div 2.3 = 2.0$ (1 d.p.).

▦ Estimating answers to calculations

Even though many calculations can be done quickly and effectively on a calculator, often an estimate for an answer can be a useful check. This is done by rounding each of the numbers in such a way that the calculation becomes relatively straightforward.

Worked examples **a)** Estimate the answer to 57×246.

Here are two possibilities:
i) $60 \times 200 = 12\,000$,
ii) $50 \times 250 = 12\,500$.

b) Estimate the answer to $6386 \div 27$.

$6000 \div 30 = 200$.

Exercise D **1.** Calculate the following, giving your answer to an appropriate degree of accuracy:

a) 23.456×17.89 b) 0.4×12.62 c) 18×9.24
d) $76.24 \div 3.2$ e) 7.6^2 f) 16.42^3

g) $\dfrac{2.3 \times 3.37}{4}$ h) $\dfrac{8.31}{2.02}$ i) $9.2 \div 4^2$

2. Without using a calculator, estimate the answers to the following:

a) 62×19 b) 270×12 c) 55×60
d) 4950×28 e) 0.8×0.95 f) 0.184×475

3. Without using a calculator, estimate the answers to the following:

a) $3946 \div 18$ b) $8287 \div 42$ c) $906 \div 27$
d) $5520 \div 13$ e) $48 \div 0.12$ f) $610 \div 0.22$

4. Without using a calculator, estimate the answers to the following:

a) $78.45 + 51.02$ b) $168.3 - 87.09$ c) 2.93×3.14

d) $84.2 \div 19.5$ e) $\dfrac{4.3 \times 752}{15.6}$ f) $\dfrac{(9.8)^3}{(2.2)^2}$

5. Using estimation, identify which of the following are definitely incorrect. Explain your reasoning clearly.

a) $95 \times 212 = 20\,140$

b) $44 \times 17 = 748$

c) $689 \times 413 = 28\,457$

d) $142\,656 \div 8 = 17\,832$

e) $77.9 \times 22.6 = 2512.54$

f) $\dfrac{84.2 \times 46}{0.2} = 19\,366$

6. Estimate the shaded areas of the following shapes. Do *not* work out an exact answer.

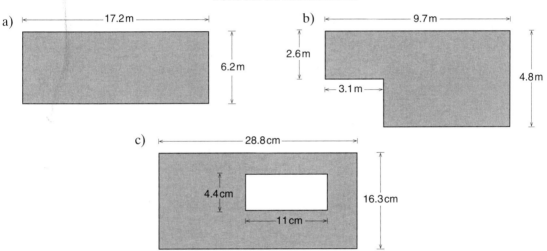

7. Estimate the volume of each of the solids below. Do *not* work out an exact answer.

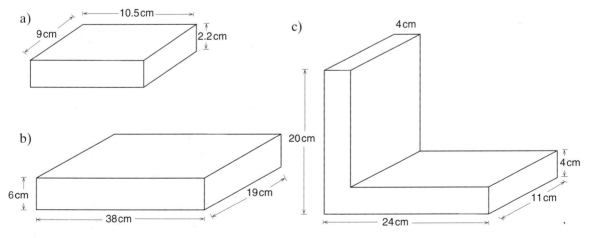

Student Assessment I

NB: All diagrams are not drawn to scale.

1. Round off the following numbers to the degree of accuracy shown in brackets:
    a) 2841 (nearest 100)   b) 7286 (nearest 10)
    c) 48 756 (nearest 1000)   d) 951 (nearest 100)

2. Round off the following numbers to the number of decimal places shown in brackets:
    a) 3.84 (1 d.p.)   b) 6.792 (1 d.p.)
    c) 0.8526 (2 d.p.)   d) 1.5849 (2 d.p.)
    e) 9.954 (1 d.p.)   f) 0.0077 (3 d.p.)

3. Round off the following numbers to the number of significant figures shown in brackets:
    a) 3.84 (1 s.f.)   b) 6.792 (2 s.f.)
    c) 0.7765 (1 s.f.)   d) 9.624 (1 s.f.)
    e) 834.97 (2 s.f.)   f) 0.004 51 (1 s.f.)

4. 1 mile is 1760 yards. Estimate the number of yards in 11.5 miles.

5. Estimate the shaded area of the figure below:

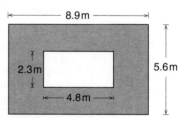

6. Estimate the answers to the following. Do *not* work out an exact answer.

    a) $\dfrac{5.3 \times 11.2}{2.1}$ b) $\dfrac{(9.8)^2}{(4.7)^2}$ c) $\dfrac{18.8 \times (7.1)^2}{(3.1)^2 \times (4.9)^2}$

7. A cuboid's dimensions are given as 12.32 cm by 1.8 cm by 4.16 cm. Calculate its volume, giving your answer to an appropriate degree of accuracy.

Student Assessment 2

NB: All diagrams are not drawn to scale.

1. Round off the following numbers to the degree of accuracy shown in brackets:
    a) 6472 (nearest 10)   b) 88 465 (nearest 100)
    c) 64 785 (nearest 1000)   d) 6.7 (nearest 10)

2. Round off the following numbers to the number of decimal places shown in brackets:
    a) 6.78 (1 d.p.)   b) 4.438 (2 d.p.)
    c) 7.975 (1 d.p.)   d) 63.084 (2 d.p.)
    e) 0.0567 (3 d.p.)   f) 3.95 (2 d.p.)

3. Round off the following numbers to the number of significant figures shown in brackets:
   a) 42.6 (1 s.f.)   b) 5.432 (2 s.f.)
   c) 0.0574 (1 s.f.)   d) 48 572 (2 s.f.)
   e) 687 453 (1 s.f.)   f) 687 453 (3 s.f.)

4. 1 mile is 1760 yards. Estimate the number of yards in 19 miles.

5. Estimate the area of the figure below:

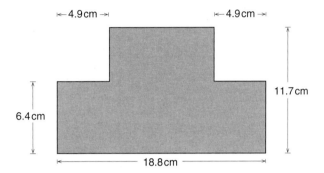

6. Estimate the answers to the following. Do *not* work out an exact answer.

   a) $\dfrac{3.9 \times 26.4}{4.85}$ b) $\dfrac{(3.2)^3}{(5.4)^2}$ c) $\dfrac{2.8 \times (7.3)^2}{(3.2)^2 \times 6.2}$

7. A cuboid's dimensions are given as 3.973 m by 2.4 m by 3.16 m. Calculate its volume, giving your answer to an appropriate degree of accuracy.

1 NUMBER, SET NOTATION AND LANGUAGE

Core Section

Use rational and irrational numbers; continue a given number sequence; recognise patterns in sequences and relationships between different sequences, generalise to simple algebraic statements (including expressions to the *n*th term) relating to such sequences.

NB: All diagrams are not drawn to scale.

▨ Rational and irrational numbers

A **rational number** is any number which can be expressed as a fraction. Examples of some rational numbers and how they can be expressed as a fraction are shown below:

$$0.2 = \tfrac{1}{5} \quad 0.3 = \tfrac{3}{10} \quad 7 = \tfrac{7}{1} \quad 1.53 = \tfrac{153}{100} \quad 0.\dot{2} = \tfrac{2}{9}$$

An **irrational number** cannot be expressed as a fraction. Examples of irrational numbers include:

$$\sqrt{2}, \quad \sqrt{5}, \quad 6 - \sqrt{3}, \quad \pi$$

In summary:
Rational numbers include:

- whole numbers,
- fractions,
- recurring decimals,
- terminating decimals.

Irrational numbers include:

- the square root of any number other than square numbers,
- a decimal which neither repeats nor terminates (e.g. π).

Exercise 1.1

1. For each of the numbers shown below state whether it is rational or irrational:
    a) 1.3
    b) $0.\dot{6}$
    c) $\sqrt{3}$
    d) $-2\tfrac{3}{5}$
    e) $\sqrt{25}$
    f) $\sqrt[3]{8}$
    g) $\sqrt{7}$
    h) 0.625
    i) $0.\dot{1}\dot{1}$

2. For each of the numbers shown below state whether it is rational or irrational:
    a) $\sqrt{2} \times \sqrt{3}$
    b) $\sqrt{2} + \sqrt{3}$
    c) $(\sqrt{2} \times \sqrt{3})^2$
    d) $\dfrac{\sqrt{8}}{\sqrt{2}}$
    e) $\dfrac{2\sqrt{5}}{\sqrt{20}}$
    f) $4 + (\sqrt{9} - 4)$

3. In each of the following decide whether the quantity required is rational or irrational. Give reasons for your answer.

a)
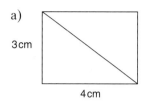
3cm

4cm

The length of the diagonal

b)
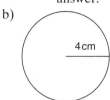
4cm

The circumference of the circle

c)
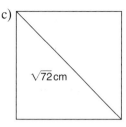
$\sqrt{72}$ cm

The side length of the square

d)
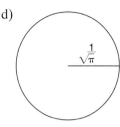
$\frac{1}{\sqrt{\pi}}$

The area of the circle

Sequences

A **sequence** is an ordered set of numbers. Each number in a sequence is known as a **term**. The terms of some sequences form a pattern. For the sequence of numbers

2, 5, 8, 11, 14, 17, ...

the difference between successive terms is +3. The term-to-term rule is therefore +3.

Worked examples **a)** Below is a sequence of numbers.

5, 9, 13, 17, 21, 25, ...

i) What is the term-to-term rule for the sequence?

The term-to-term rule is +4.

ii) Calculate the 10th term of the sequence.
Continuing the pattern gives:

5, 9, 13, 17, 21, 25, 29, 33, 37, 41, ...

Therefore the 10th term is 41.

b) Below is a sequence of numbers.

1, 2, 4, 8, 16, ...

i) What is the term-to-term rule for the sequence?

The term-to-term rule is × 2.

ii) Calculate the 10th term of the sequence.
Continuing the pattern gives:

1, 2, 4, 8, 16, 32, 64, 128, 256, 512, ...

Therefore the 10th term is 512.

Exercise 1.2   For each of the sequences given below:

i)   state a rule to describe the sequence,
ii) calculate the 10th term.
a) 3, 6, 9, 12, 15, ...
b) 8, 18, 28, 38, 48, ...

c) 11, 33, 55, 77, 99, ...
d) 0.7, 0.5, 0.3, 0.1, ...
e) $\frac{1}{2}$, $\frac{1}{3}$, $\frac{1}{4}$, $\frac{1}{5}$, ...
f) $\frac{1}{2}$, $\frac{2}{3}$, $\frac{3}{4}$, $\frac{4}{5}$, ...
g) 1, 4, 9, 16, 25, ...
h) 4, 7, 12, 19, 28, ...
i) 1, 8, 27, 64, ...
j) 5, 25, 125, 625, ...

Sometimes the pattern in a sequence of numbers is not obvious. By looking at the differences between successive terms a pattern is often found.

Worked examples **a)** Calculate the 8th term in the sequence

8, 12, 20, 32, 48, ...

The pattern in this sequence is not immediately obvious, so a row for the differences between successive terms can be constructed.

	8	12	20	32	48
1st differences		4	8	12	16

The pattern in the differences row is +4 and this can be continued to complete the sequence to the 8th term.

8   12   20   32   48   68   92   120
↘ ↗ ↘ ↗ ↘ ↗
1st differences 4   8   12   16   20   24   28

b) Calculate the 8th term in the sequence

3   6   13   28   55
1st differences 3   7   15   27

The row of first differences is not sufficient to spot the pattern, so a row of 2nd differences is constructed.

3   6   13   28   55
1st differences 3   7   15   27
2nd differences 4   8   12

The pattern in the 2nd differences row can be seen to be +4. This can now be used to complete the sequence.

3   6   13   28   55   98   161   248
↗ ↗ ↗
1st differences 3   7   15   27   43   63   87
↗ ↗ ↗
2nd differences 4   8   12   16   20   24

Exercise 1.3 For each of the sequences given below calculate the next two terms:

a) 8, 11, 17, 26, 38
b) 5, 7, 11, 19, 35
c) 9, 3, 3, 9, 21

d) $-2,\ 5,\ 21,\ 51,\ 100$
e) $11,\ 9,\ 10,\ 17,\ 36,\ 79$
f) $4,\ 7,\ 11,\ 19,\ 36,\ 69$
g) $-3,\ 3,\ 8,\ 13,\ 17,\ 21,\ 24$

The *n*th term

So far the method used for generating a sequence relies on knowing the previous term in order to work out the next one. This method works but can be a little cumbersome if the 100th term is needed and only the first five terms are given! A more efficient rule is one which is related to a term's position in a sequence.

Worked examples   **a)** For the sequence shown below give an expression for the *n*th term.

Position	1	2	3	4	5	n
Term	3	6	9	12	15	?

By looking at the sequence it can be seen that the term is always $3 \times$ position.
　　Therefore the *n*th term can be given by the expression $3n$.

b) For the sequence shown below give an expression for the *n*th term.

Position	1	2	3	4	5	n
Term	2	5	8	11	14	?

You will need to spot similarities between sequences. The terms of the above sequence are the same as the terms in example a) above but with 1 subtracted each time.
　　The expression for the *n*th term is therefore $3n - 1$.

Exercise 1.4

For each of the following sequences:

i) write down the next two terms,
ii) give an expression for the *n*th term.
　　a) $5,\ 8,\ 11,\ 14,\ 17,\ \ldots$
　　b) $5,\ 9,\ 13,\ 17,\ 21,\ \ldots$
　　c) $4,\ 9,\ 14,\ 19,\ 24,\ \ldots$
　　d) $8,\ 10,\ 12,\ 14,\ 16,\ \ldots$
　　e) $1,\ 8,\ 15,\ 22,\ 29,\ \ldots$
　　f) $0,\ 4,\ 8,\ 12,\ 16,\ 20,\ \ldots$
　　g) $1,\ 10,\ 19,\ 28,\ 37,\ \ldots$
　　h) $15,\ 25,\ 35,\ 45,\ 55,\ \ldots$
　　i) $9,\ 20,\ 31,\ 42,\ 53,\ \ldots$
　　j) $1.5,\ 3.5,\ 5.5,\ 7.5,\ 9.5,\ 11.5,\ \ldots$
　　k) $0.25,\ 1.25,\ 2.25,\ 3.25,\ 4.25,\ \ldots$
　　l) $0,\ 1,\ 2,\ 3,\ 4,\ 5,\ \ldots$

Exercise 1.5   For each of the following sequences:

  i) write down the next two terms,

  ii) give an expression for the nth term.

    a) 2, 5, 10, 17, 26, 37, ...

    b) 8, 11, 16, 23, 32, ...

    c) 0, 3, 8, 15, 24, 35, ...

    d) 1, 8, 27, 64, 125, ...

    e) 2, 9, 28, 65, 126, ...

    f) 11, 18, 37, 74, 135, ...

    g) −2, 5, 24, 61, 122, ...

    h) 2, 6, 12, 20, 30, 42, ...

Core Section

Student Assessment I

1. State whether the following numbers are rational or irrational:
   a) 1.5
   b) $\sqrt{7}$
   c) $0.\dot{7}$
   d) $0.\dot{7}\dot{3}$
   e) $\sqrt{121}$
   f) π

2. For each of the sequences given below:
   i) calculate the next two terms,
   ii) explain the pattern in words.
   a) 9, 18, 27, 36, ...
   b) 54, 48, 42, 36, ...
   c) 18, 9, 4.5, ...
   d) 12, 6, 0, −6, ...
   e) 216, 125, 64, ...
   f) 1, 3, 9, 27, ...

3. For each of the sequences shown below give an expression for the *n*th term:
   a) 6, 10, 14, 18, 22, ...
   b) 13, 19, 25, 31, ...
   c) 3, 9, 15, 21, 27, ...
   d) 4, 7, 12, 19, 28, ...
   e) 0, 10, 20, 30, 40, ...
   f) 0, 7, 26, 63, 124, ...

Student Assessment 2

1. Show, by expressing them as fractions or whole numbers, that the following numbers are rational:
   a) 0.625
   b) $\sqrt[3]{27}$
   c) 0.44

2. For each of the sequences given below:
   i) calculate the next two terms,
   ii) explain the pattern in words.
   a) 6, 12, 18, 24, ...
   b) 24, 21, 18, 15, ...
   c) 10, 5, 0, ...
   d) 16, 25, 36, 49, ...
   e) 1, 10, 100, ...
   f) 1, $\frac{1}{2}$, $\frac{1}{4}$, $\frac{1}{8}$, ...

3. For each of the sequences shown below give an expression for the *n*th term:
   a) 3, 5, 7, 9, 11, ...
   b) 7, 13, 19, 25, 31, ...
   c) 8, 18, 28, 38, ...
   d) 1, 9, 17, 25, ...
   e) −4, 4, 12, 20, ...
   f) 2, 5, 10, 17, 26, ...

NUMBER, SET NOTATION AND LANGUAGE

Extended Section

Use language, notation and Venn diagrams to describe sets and represent relationships between sets.

▨ Sets

A **set** is a well defined group of objects or symbols. The objects or symbols are called the **elements** of the set. If an element e belongs to a set S, this is represented as $e \in S$. If e does not belong to set S this is represented as $e \notin S$.

Worked examples

a) A particular set consists of the following elements:

{South Africa, Namibia, Egypt, Angola, ...}

i) Describe the set.

The elements of the set are countries of Africa.

ii) Add another two elements to the set.

e.g. Zimbabwe, Ghana

iii) Is the set finite or infinite?

Finite. There is a finite number of countries in Africa.

b) Consider the set

{1, 4, 9, 16, 25, ...}

i) Describe the set.

The elements of the set are square numbers.

ii) Write another two elements of the set.

e.g. 36, 49

iii) Is the set finite or infinite?

Infinite. There is an infinite number of square numbers.

Exercise 1.6

1. In the following questions:

i) describe the set in words,
ii) write down another two elements of the set.

    a) {Asia, Africa, Europe, ...}
    b) {2, 4, 6, 8, ...}
    c) {Sunday, Monday, Tuesday, ...}

    d) {January, March, July, ...}

    e) {1, 3, 6, 10, ...}

    f) {Mehmet, Michael, Mustapha, Matthew, ...}

    g) {11, 13, 17, 19, ...}

    h) {a, e, i, ...}

    i) {Earth, Mars, Venus, ...}

    j) $A = \{x: 3 \leqslant x \leqslant 12\}$

    k) $S = \{y: -5 \leqslant y \leqslant 5\}$

2. The number of elements in a set A is written as $n(A)$.
Give the value of $n(A)$ for the finite sets in Q.1a–k above.

Subsets

If all the elements of one set X are also elements of another set Y, then X is said to be a **subset** of Y.

    This is written as $X \subseteq Y$.

    If a set A is empty (i.e. it has no elements in it), then this is called the empty set and it is represented by the symbol $\emptyset$. Therefore $A = \emptyset$. The empty set is a subset of all sets.

    e.g. Three girls, Winnie, Natalie and Emma form a set A

        $A = \{$Winnie, Natalie, Emma$\}$

        All the possible subsets of A are given below:

        $B = \{$Winnie, Natalie, Emma$\}$

        $C = \{$Winnie, Natalie$\}$

        $D = \{$Winnie, Emma$\}$

        $E = \{$Natalie, Emma$\}$

        $F = \{$Winnie$\}$

        $G = \{$Natalie$\}$

        $H = \{$Emma$\}$

        $I = \emptyset$

Note that the sets B and I above are considered as subsets of A.

    i.e. $B \subseteq A$ and $I \subseteq A$

However, sets C, D, E, F, G and H are considered **proper subsets** of A. This distinction of subset is shown in the notation below.

    $C \subset A$ and $D \subset A$ etc.

Similarly $G \nsubseteq H$ implies that G is not a subset of H

             $G \not\subset H$ implies that G is not a proper subset of H

Worked example $A = \{1, 2, 3, 4, 5, 6, 7, 8, 9, 10\}$

    i) List subset B {even numbers}.

        $B = \{2, 4, 6, 8, 10\}$

    ii) List subset C {prime numbers}.

        $C = \{2, 3, 5, 7\}$

Exercise 1.7

1. $P = \{$whole numbers less than 30$\}$
   a) List the subset Q $\{$even numbers$\}$.
   b) List the subset R $\{$odd numbers$\}$.
   c) List the subset S $\{$prime numbers$\}$.
   d) List the subset T $\{$square numbers$\}$.
   e) List the subset U $\{$triangle numbers$\}$.

2. $A = \{$whole numbers between 50 and 70$\}$
   a) List the subset B $\{$multiples of 5$\}$.
   b) List the subset C $\{$multiples of 3$\}$.
   c) List the subset D $\{$square numbers$\}$.

3. $J = \{$p, q, r$\}$
   a) List all the subsets of J.
   b) List all the proper subsets of J.

4. State whether each of the following statements is true or false:
   a) $\{$Algeria, Mozambique$\} \subseteq \{$countries in Africa$\}$
   b) $\{$mango, banana$\} \subseteq \{$fruit$\}$
   c) $\{1, 2, 3, 4\} \subseteq \{1, 2, 3, 4\}$
   d) $\{1, 2, 3, 4\} \subset \{1, 2, 3, 4\}$
   e) $\{$volleyball, basketball$\} \not\subseteq \{$team sport$\}$
   f) $\{4, 6, 8, 10\} \not\subset \{4, 6, 8, 10\}$
   g) $\{$potatoes, carrots$\} \subseteq \{$vegetables$\}$
   h) $\{12, 13, 14, 15\} \not\subset \{$whole numbers$\}$

The universal set

The **universal set** ($\mathcal{E}$) for any particular problem is the set which contains all the possible elements for that problem.

The **complement** of a set A is the set of elements which are in $\mathcal{E}$ but not in A. The set is identified as A'. Notice that $\mathcal{E}' = \emptyset$ and $\emptyset' = \mathcal{E}$.

Worked examples

a) If $\mathcal{E} = \{1, 2, 3, 4, 5, 6, 7, 8, 9, 10\}$ and $A = \{1, 2, 3, 4, 5\}$ what set is represented by A'?

   A' consists of those elements in $\mathcal{E}$ which are not in A.

   Therefore $A' = \{6, 7, 8, 9, 10\}$.

b) If $\mathcal{E} = \{$all 3D shapes$\}$ and $P = \{$prisms$\}$ what set is represented by P'?

   $P' = \{$all 3D shapes except prisms$\}$.

Venn diagrams

Venn diagrams are the principal way of showing sets diagrammatically. The method consists primarily of entering the elements of a set into a circle or circles.

Some examples of the uses of Venn diagrams are shown below.

$A = \{2, 4, 6, 8, 10\}$ can be represented as:

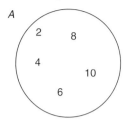

Elements which are in more than one set can also be represented using a Venn diagram.

$P = \{3, 6, 9, 12, 15, 18\}$ and $Q = \{2, 4, 6, 8, 10, 12\}$ can be represented as:

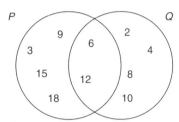

In the diagram above it can be seen that those elements which belong to both sets are placed in the region of overlap of the two circles.

When two sets P and Q overlap as they do above, the notation $P \cap Q$ is used to denote the set of elements in the intersection, i.e. $P \cap Q = \{6, 12\}$.

Note that $6 \in P \cap Q; 8 \notin P \cap Q$.

$J = \{10, 20, 30, 40, 50, 60, 70, 80, 90, 100\}$ and

$K = \{60, 70, 80\}$; as discussed earlier, $K \subset J$ can be represented as shown below:

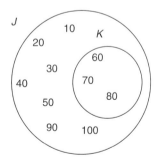

$X = \{1, 3, 6, 7, 14\}$ and $Y = \{3, 9, 13, 14, 18\}$ are represented as:

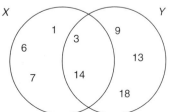

The **union** of two sets is everything which belongs to either or both sets and is represented by the symbol ∪.

Therefore in the example above $X \cup Y = \{1, 3, 6, 7, 9, 13, 14, 18\}$.

Exercise 1.8

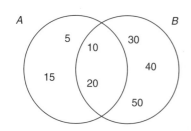

1. Using the Venn diagram (left), indicate whether the following statements are true or false. ∈ means 'is an element of' and ∉ means 'is not an element of'.
   a) $5 \in A$ b) $20 \in B$ c) $20 \notin A$
   d) $50 \in A$ e) $50 \notin B$ f) $A \cap B = \{10, 20\}$

2. Complete the statement $A \cap B = \{\ldots\}$ for each of the Venn diagrams below.

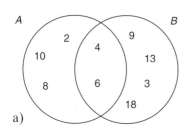

a)

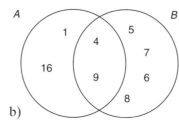

b)

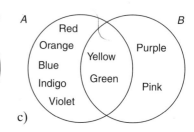

c)

3. Complete the statement $A \cup B = \{\ldots\}$ for each of the Venn diagrams in Q.2 above.

4.

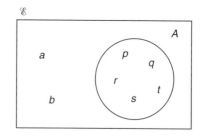

Copy and complete the following statements:
a) $\mathscr{E} = \{\ldots\}$ b) $A' = \{\ldots\}$

5.

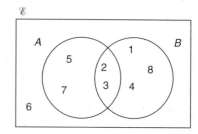

Copy and complete the following statements:
a) $\mathscr{E} = \{\ldots\}$ b) $A' = \{\ldots\}$ c) $A \cap B = \{\ldots\}$
d) $A \cup B = \{\ldots\}$ e) $(A \cap B)' = \{\ldots\}$ f) $A \cap B' = \{\ldots\}$

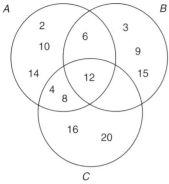

6. a) Describe in words the elements of:
  i) set A ii) set B iii) set C
  b) Copy and complete the following statements:
  i) $A \cap B = \{\ldots\}$ ii) $A \cap C = \{\ldots\}$
  iii) $B \cap C = \{\ldots\}$ iv) $A \cap B \cap C = \{\ldots\}$
  v) $A \cup B = \{\ldots\}$ vi) $C \cup B = \{\ldots\}$

7.

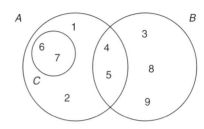

  a) Copy and complete the following statements:
  i) $A = \{\ldots\}$ ii) $B = \{\ldots\}$
  iii) $C' = \{\ldots\}$ iv) $A \cap B = \{\ldots\}$
  v) $A \cup B = \{\ldots\}$ vi) $(A \cap B)' = \{\ldots\}$
  b) State, using set notation, the relationship between C and A.

8.

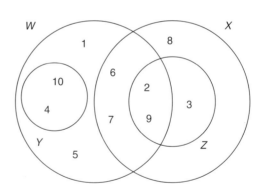

  a) Copy and complete the following statements:
  i) $W = \{\ldots\}$ ii) $X = \{\ldots\}$
  iii) $Z' = \{\ldots\}$ iv) $W \cap Z = \{\ldots\}$
  v) $W \cap X = \{\ldots\}$ vi) $Y \cap Z = \{\ldots\}$
  b) Which of the named sets is a subset of X?

Exercise 1.9   **1.** $A = \{$Egypt, Libya, Morocco, Chad$\}$
    $B = \{$Iran, Iraq, Turkey, Egypt$\}$
    a) Draw a Venn diagram to illustrate the above information.
    b) Copy and complete the following statements:
      i) $A \cap B = \{\ldots\}$ ii) $A \cup B = \{\ldots\}$

2. $P = \{2, 3, 5, 7, 11, 13, 17\}$
    $Q = \{11, 13, 15, 17, 19\}$
    a) Draw a Venn diagram to illustrate the above information.
    b) Copy and complete the following statements:
        i) $P \cap Q = \{\dots\}$ ii) $P \cup Q = \{\dots\}$

3. $B = \{2, 4, 6, 8, 10\}$
    $A \cup B = \{1, 2, 3, 4, 6, 8, 10\}$
    $A \cap B = \{2, 4\}$
    Represent the above information on a Venn diagram.

4. $X = \{a, c, d, e, f, g, l\}$
    $Y = \{b, c, d, e, h, i, k, l, m\}$
    $Z = \{c, f, i, j, m\}$
    Represent the above information on a Venn diagram.

5. $P = \{1, 4, 7, 9, 11, 15\}$
    $Q = \{5, 10, 15\}$
    $R = \{1, 4, 9\}$
    Represent the above information on a Venn diagram.

Problems involving sets

Worked example

In a class of 31 students, some study physics and some study chemistry. If 22 study physics, 20 study chemistry and 5 study neither, calculate the number of students who take both subjects.

The information given above can be entered in a Venn diagram in stages.

The students taking neither physics nor chemistry can be put in first (as shown left).

This leaves 26 students to be entered into the set circles.

If x students take both subjects then

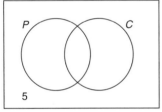

$$n(P) = 22 - x + x$$
$$n(C) = 20 - x + x$$
$$P \cup C = 31 - 5 = 26$$

Therefore $22 - x + x + 20 - x = 26$
$$42 - x = 26$$
$$x = 16$$

Substituting the value of x into the Venn diagram gives:

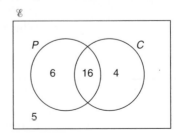

Therefore the number of students taking both physics and chemistry is 16.

Exercise 1.10

1. In a class of 35 students, 19 take Spanish, 18 take French and 3 take neither. Calculate how many take:
    a) both French and Spanish,
    b) just Spanish,
    c) just French.

2. In a year group of 108 students, 60 liked football, 53 liked tennis and 10 liked neither. Calculate the number of students who liked football but not tennis.

3. In a year group of 113 students, 60 liked hockey, 45 liked rugby and 18 liked neither. Calculate the number of students who:
    a) liked both hockey and rugby,
    b) liked only hockey.

4. One year 37 students sat an examination in physics, 48 sat chemistry and 45 sat biology. 15 students sat physics and chemistry, 13 sat chemistry and biology, 7 sat physics and biology and 5 students sat all three.
    a) Draw a Venn diagram to represent this information.
    b) Calculate $n\,(P \cup C \cup B)$.

Extended Section

Student Assessment 1

1. Describe the following sets in words:
   a) {2, 4, 6, 8}
   b) {2, 4, 6, 8, ...}
   c) {1, 4, 9, 16, 25, ...}
   d) {Arctic, Atlantic, Indian, Pacific}

2. Calculate the value of $n(A)$ for each of the sets shown below:
   a) A = {days of the week}
   b) A = {prime numbers between 50 and 60}
   c) A = {x: x is an integer and $5 \leqslant x \leqslant 10$}
   d) A = {days in a leap year}

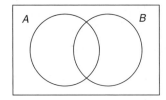

3. Copy out the Venn diagram (left) twice.
   a) On one copy shade and label the region which represents $A \cap B$.
   b) On the other copy shade and label the region which represents $A \cup B$.

4. If A = {a, b} list all the subsets of A.

5. If $\mathscr{E}$ = {m, a, t, h, s} and A = {a, s}, what set is represented by A'?

Student Assessment 2

1. Describe the following sets in words:
   a) {1, 3, 5, 7}
   b) {1, 3, 5, 7, ...}
   c) {1, 3, 6, 10, 15, ...}
   d) {Brazil, Chile, Argentina, Bolivia}

2. Calculate the value of $n(A)$ for each of the sets shown below:
   a) A = {months of the year}
   b) A = {square numbers between 99 and 149}
   c) A = {x: x is an integer and $-9 \leqslant x \leqslant -3$}
   d) A = {students in your class}

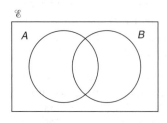

3. Copy out the Venn diagram (left) twice.
   a) On one copy shade and label the region which represents $\mathscr{E}$.
   b) On the other copy shade and label the region which represents $(A \cap B)'$.

4. If A = {w, o, r, k} list all the subsets of A with at least three elements.

5. If $\mathscr{E}$ = {1, 2, 3, 4, 5, 6, 7, 8} and P = {2, 4, 6, 8}, what set is represented by P'?

Student Assessment 3

1. If $A = \{2, 4, 6, 8\}$ write all the proper subsets of A with two or more elements.

2. $J = \{$London, Paris, Rome, Washington, Canberra, Ankara, Cairo$\}$
   $K = \{$Cairo, Maputo, Harare, Ankara$\}$
   a) Draw a Venn diagram to represent the above information.
   b) Copy and complete the statement $J \cap K = \{\ldots\}$.
   c) Copy and complete the statement $J' \cap K = \{\ldots\}$.

3. $M = \{x : x$ is an integer and $2 \leqslant x \leqslant 20\}$
   $N = \{$prime numbers less than 30$\}$
   a) Draw a Venn diagram to illustrate the information above.
   b) Copy and complete the statement $M \cap N = \{\ldots\}$.
   c) Copy and complete the statement $(M \cap N)' = \{\ldots\}$.

4. $\mathscr{E} = \{$natural numbers$\}$, $M = \{$even numbers$\}$ and $N = \{$multiples of 5$\}$.
   a) Draw a Venn diagram and place the numbers 1, 2, 3, 4, 5, 6, 7, 8, 9, 10 in the appropriate places in it.
   b) If $X = M \cap N$, describe set X in words.

5. In a region of mixed farming, farms keep pigs, cattle or sheep. There are 77 farms altogether. 19 farms keep only pigs, 8 keep only cattle and 13 keep only sheep. 13 keep both pigs and cattle, 28 keep both cattle and sheep and 8 keep both pigs and sheep.
   a) Draw a Venn diagram to show the above information.
   b) Calculate $n(P \cap C \cap S)$.

Student Assessment 4

1. $M = \{$a, e, i, o, u$\}$
   a) How many subsets are there of M?
   b) List the subsets of M with four or more elements.

2. $X = \{$lion, tiger, cheetah, leopard, puma, jaguar, cat$\}$
   $Y = \{$elephant, lion, zebra, cheetah, gazelle$\}$
   $Z = \{$anaconda, jaguar, tarantula, mosquito$\}$
   a) Draw a Venn diagram to represent the above information.
   b) Copy and complete the statement $X \cap Y = \{\ldots\}$.
   c) Copy and complete the statement $Y \cap Z = \{\ldots\}$.
   d) Copy and complete the statement $X \cap Y \cap Z = \{\ldots\}$.

3. A group of 40 people were asked whether they like tennis (T) and football (F). The number liking both tennis and football was three times the number liking only tennis. Adding 3 to the number liking only tennis and doubling the answer equals the number of people liking only football. Four said they did not like sport at all.

a) Draw a Venn diagram to represent this information.
b) Calculate $n(T \cap F)$.
c) Calculate $n(T \cap F')$.
d) Calculate $n(T' \cap F)$.

4. The Venn diagram below shows the number of elements in three sets P, Q and R.

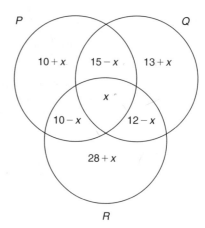

If $n(P \cup Q \cup R) = 93$ calculate:

a) x
b) $n(P)$
c) $n(Q)$
d) $n(R)$
e) $n(P \cap Q)$
f) $n(Q \cap R)$
g) $n(P \cap R)$
h) $n(R \cup Q)$
i) $n(P \cap Q)'$

ICT Section

The spreadsheet below shows a sequence of numbers.

	A	B	C	D	E	F	G	H	I
1	Position	1	2	3	4	5	6	10	20
2	Term	2	4	6	8	10			

1. a) Copy the sequence into a spreadsheet of your own.
   b) By entering formulae into cells G2, H2 and I2, use the spreadsheet to calculate the value of the 6th, 10th and 20th terms.

2. Using a spreadsheet as above, calculate the 6th, 10th and 20th terms of the following sequences:

   a) 3   6   9   12   15
   b) 4   6   8   10   12
   c) 2   5   8   11   14
   d) 0.5   1   1.5   2   2.5
   e) 0   3   8   15   24
   f) 0.5   2   4.5   8   12.5
   g) 1   −2   −5   −8   −11
   h) 2   6   12   20   30

2 LIMITS OF ACCURACY

Core Section

Give appropriate upper and lower bounds for data to a specified accuracy (e.g. measured lengths).

NB: All diagrams are not drawn to scale.

Numbers can be written to different degrees of accuracy. For example 4.5, 4.50 and 4.500, although appearing to represent the same number, do not. This is because they are written to different degrees of accuracy.

4.5 is rounded to one decimal place and therefore any number from 4.45 up to but not including 4.55 would be rounded to 4.5. On a number line this would be represented as:

As an inequality where x represents the number it would be expressed as

$$4.45 \leqslant x < 4.55$$

4.45 is known as the **lower bound** of 4.5, whilst 4.55 is known as the **upper bound**.

4.50 on the other hand is written to two decimal places and therefore only numbers from 4.495 up to but not including 4.505 would be rounded to 4.50. This therefore represents a much smaller range of numbers than that being rounded to 4.5. Similarly the range of numbers being rounded to 4.500 would be even smaller.

Worked example A girl's height is given as 162 cm to the nearest centimetre.

i) Work out the lower and upper bounds within which her height can lie.

      Lower bound = 161.5 cm
      Upper bound = 162.5 cm

ii) Represent this range of numbers on a number line.

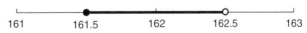

iii) If the girl's height is h cm, express this range as an inequality.

$$161.5 \leqslant h < 162.5$$

Exercise 2.1 **1.** Each of the following numbers is expressed to the nearest whole number.
    i)   Give the upper and lower bounds of each.
    ii) Using x as the number, express the range in which the number lies as an inequality.

a) 6   b) 83   c) 152
d) 1000   e) 0   f) −4

2. Each of the following numbers is correct to one decimal place.
   i) Give the upper and lower bounds of each.
   ii) Using x as the number, express the range in which the number lies as an inequality.
   a) 3.8   b) 15.6   c) 1.0
   d) 10.0   e) 0.3   f) −0.2

3. Each of the following numbers is correct to two significant figures.
   i) Give the upper and lower bounds of each.
   ii) Using x as the number, express the range in which the number lies as an inequality.
   a) 4.2   b) 0.84   c) 420
   d) 5000   e) 0.045   f) 25 000

4. The mass of a sack of vegetables is given as 5.4 kg.
   a) Illustrate the lower and upper bounds of the mass on a number line.
   b) Using M kg for the mass, express the range of values in which M must lie, as an inequality.

5. At a school sports day, the winning time for the 100 m race was given as 11.8 seconds.
   a) Illustrate the lower and upper bounds of the time on a number line.
   b) Using T seconds for the time, express the range of values in which T must lie, as an inequality.

6. The capacity of a swimming pool is given as 620 m³ correct to two significant figures.
   a) Calculate the lower and upper bounds of the pool's capacity.
   b) Using x cubic metres for the capacity, express the range of values in which x must lie, as an inequality.

7. A farmer measures the dimensions of his rectangular field to the nearest 10 m. The length is recorded as 630 m and the width is recorded as 400 m.
   a) Calculate the lower and upper bounds of the length.
   b) Using W metres for the width, express the range of values in which W must lie, as an inequality.

Core Section

Student Assessment I

1. The following numbers are expressed to the nearest whole number. Illustrate on a number line the range in which each must lie.
   a) 7
   b) 40
   c) 0
   d) −200

2. The following numbers are expressed correct to two significant figures. Representing each number by the letter x, express the range in which each must lie, using an inequality.
   a) 210
   b) 64
   c) 3.0
   d) 0.88

3. A school measures the dimensions of its rectangular playing field to the nearest metre. The length was recorded as 350 m and the width as 200 m. Express the range in which the length and width lie using inequalities.

4. A boy's mass was measured to the nearest 0.1 kg. If his mass was recorded as 58.9 kg, illustrate on a number line the range within which it must lie.

5. An electronic clock is accurate to $\frac{1}{1000}$ of a second. The duration of a flash from a camera is timed at 0.004 seconds. Express the upper and lower bounds of the duration of the flash using inequalities.

6. The following numbers are rounded to the degree of accuracy shown in brackets. Express the lower and upper bounds of these numbers as an inequality.
   a) $x = 4.83$ (2 d.p.)
   b) $y = 5.05$ (2 d.p.)
   c) $z = 10.0$ (1 d.p.)
   d) $p = -100.00$ (2 d.p.)

Student Assessment 2

1. The following numbers are rounded to the nearest 100. Illustrate on a number line the range in which they must lie.
   a) 500
   b) 7000
   c) 0
   d) −32 000

2. The following numbers are expressed correct to three significant figures. Represent the limits of each number using inequalities.
   a) 254
   b) 40.5
   c) 0.410
   d) 100

3. The dimensions of a rectangular courtyard are given to the nearest 0.5 m. The length is recorded as 20.5 m and the width as 10.0 m. Represent the limits of these dimensions using inequalities.

4. The circumference c of a tree is measured to the nearest 2 mm. If its circumference is measured as 245.6 cm, illustrate on a number line the range within which it must lie.

5. The time it takes Earth to rotate around the Sun is given as 365.25 days to five significant figures. What are the upper and lower bounds of this time?

6. The following numbers are rounded to the degree of accuracy shown in brackets. Express the lower and upper bounds of these numbers as an inequality.
   a) 10.90 (2 d.p.)   b) 3.00 (2 d.p.)
   c) 0.5 (1 d.p.)   d) −175.00 (2 d.p.)

LIMITS OF ACCURACY

Extended Section

Obtain appropriate upper and lower bounds to solutions of simple problems (e.g. the calculation of the perimeter or the area of a rectangle) given data to a specified accuracy.

NB: All diagrams are not drawn to scale.

When numbers are written to a specific degree of accuracy, calculations involving those numbers also give a range of possible answers.

Worked examples **a)** Calculate the upper and lower bounds for the following calculation, given that each number is given to the nearest whole number.

$$34 \times 65$$

34 lies in the range $33.5 \leqslant x < 34.5$.
65 lies in the range $64.5 \leqslant x < 65.5$.

The lower bound of the calculation is obtained by multiplying together the two lower bounds. Therefore the minimum product is 33.5×64.5, i.e. 2160.75.

The upper bound of the calculation is obtained by multiplying together the two upper bounds. Therefore the maximum product is 34.5×65.5, i.e. 2259.75.

b) Calculate the upper and lower bounds to $\dfrac{33.5}{22.0}$ given that each of the numbers is accurate to 1 d.p.

33.5 lies in the range $33.45 \leqslant x < 33.55$.
22.0 lies in the range $21.95 \leqslant x < 22.05$.

The lower bound of the calculation is obtained by dividing the lower bound of the numerator by the *upper* bound of the denominator. So the minimum value is $33.45 \div 22.05$, i.e. 1.52 (2 d.p.).

The upper bound of the calculation is obtained by dividing the upper bound of the numerator by the *lower* bound of the denominator. So the maximum value is $33.55 \div 21.95$, i.e. 1.53 (2 d.p.).

Exercise 2.2 **1.** Calculate lower and upper bounds for the following calculations, if each of the numbers is given to the nearest whole number.

a) 14×20 b) 135×25 c) 100×50

d) $\dfrac{40}{10}$ e) $\dfrac{33}{11}$ f) $\dfrac{125}{15}$

g) $\dfrac{12 \times 65}{16}$ h) $\dfrac{101 \times 28}{69}$ i) $\dfrac{250 \times 7}{100}$

j) $\dfrac{44}{3^2}$ k) $\dfrac{578}{17 \times 22}$ l) $\dfrac{1000}{4 \times (3 + 8)}$

2. Calculate lower and upper bounds for the following calculations, if each of the numbers is given to 1 d.p.

a) $2.1 + 4.7$ b) 6.3×4.8 c) 10.0×14.9

d) $17.6 - 4.2$ e) $\dfrac{8.5 + 3.6}{6.8}$ f) $\dfrac{7.7 - 6.2}{3.5}$

g) $\dfrac{(16.4)^2}{(3.0 - 0.3)^2}$ h) $\dfrac{100.0}{(50.0 - 40.0)^2}$ i) $(0.1 - 0.2)^2$

3. Calculate lower and upper bounds for the following calculations, if each of the numbers is given to 2 s.f.

a) 64×320 b) 1.7×0.65 c) 4800×240

d) $\dfrac{54\,000}{600}$ e) $\dfrac{4.2}{0.031}$ f) $\dfrac{100}{5.2}$

g) $\dfrac{6.8 \times 42}{120}$ h) $\dfrac{100}{(4.5 \times 6.0)}$ i) $\dfrac{180}{(7.3 - 4.5)}$

Exercise 2.3

6.8 cm

4.2 cm

1. The masses to the nearest 0.5 kg of two parcels are 1.5 kg and 2.5 kg. Calculate the lower and upper bounds of their combined mass.

2. Calculate upper and lower bounds for the perimeter of the rectangle shown (left), if its dimensions are correct to 1 d.p.

3. Calculate upper and lower bounds for the perimeter of the rectangle shown below, whose dimensions are accurate to 2 d.p.

4.86 m

2.00 m

10.0 cm

7.5 cm

4. Calculate upper and lower bounds for the area of the rectangle shown (left), if its dimensions are accurate to 1 d.p.

5. Calculate upper and lower bounds for the area of the rectangle shown below, whose dimensions are correct to 2 s.f.

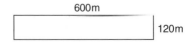

600 m

120 m

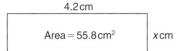

4.2 cm

Area = 55.8 cm² | *x* cm

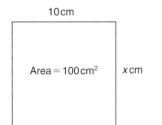

10 cm

Area = 100 cm² | *x* cm

6. Calculate upper and lower bounds for the length marked *x* cm in the rectangle (left). The area and length are both given to 1 d.p.

7. Calculate the upper and lower bounds for the length marked *x* cm in the rectangle (left). The area and length are both accurate to 2 s.f.

8. The radius of the circle shown below is given to 1 d.p. Calculate the upper and lower bounds of:
a) the circumference,
b) the area.

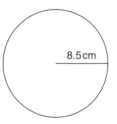

8.5 cm

Area
400 cm²

9. The area of the circle shown (left) is given to 2 s.f. Calculate the upper and lower bounds of:
a) the radius,
b) the circumference.

10. The mass of a cube of side 2 cm is given as 100 g. The side is accurate to the nearest millimetre and the mass accurate to the nearest gram. Calculate the maximum and minimum possible values for the density of the material (density = mass ÷ volume).

11. The distance to the nearest 100 000 km from Earth to the moon is given as 400 000 km. The average speed to the nearest 500 km/h of a rocket to the moon is given as 3500 km/h. Calculate the greatest and least time it could take the rocket to reach the moon.

Extended Section

Student Assessment I

NB: All diagrams are not drawn to scale.

1. Calculate the upper and lower bounds of the following calculations given that each number is written to the nearest whole number.

   a) 20×50
   b) 100×63
   c) $\dfrac{500}{80}$

   d) $\dfrac{14 \times 73}{20}$
   e) $\dfrac{17 - 7}{4 + 6}$
   f) $\dfrac{8 \times (3 + 6)}{10^2}$

2. In the rectangle (left) both dimensions are given to 1 d.p. Calculate the upper and lower bounds for the area.

3. An equilateral triangle has sides of length 4 cm correct to the nearest whole number. Calculate the upper and lower bounds for the perimeter of the triangle.

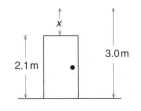

4. The height to 1 d.p. of a room is given as 3.0 m. A door to the room has a height to 1 d.p. of 2.1 m. Write as an inequality the upper and lower bounds for the gap between the top of the door and the ceiling.

5. The mass of 85 oranges is given as 40 kg correct to 2 s.f. Calculate the lower and upper bounds for the average mass of one orange.

Student Assessment 2

NB: All diagrams are not drawn to scale.

1. Calculate the upper and lower bounds of the following calculations given that each number is written to the nearest whole number.

   a) 25×5
   b) 10×100
   c) $\dfrac{400}{60}$

   d) $\dfrac{10 \times 9}{1}$
   e) $\dfrac{7 - 2}{2 + 8}$
   f) $\dfrac{4 \times (8 - 5)}{6^2}$

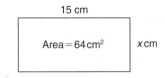

2. In the rectangle (left) both the length and area are given to 2 s.f. Calculate the upper and lower bounds for the width x cm.

3. An equilateral triangle has sides of length 5.2 cm correct to 1 d.p. Calculate the upper and lower bounds for the perimeter of the triangle.

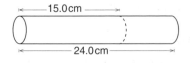

4. A metal rod is 24.0 cm long. A length of 15.0 cm is cut from it. Both measurements are correct to 1 d.p. Give as an inequality the upper and lower bounds of the length l cm of the rod that is left.

5. The mass of 60 potatoes is given as 42 kg correct to 2 s.f. Calculate the lower and upper bounds for the average mass of one potato.

Student Assessment 3

1. Five boys have a mass, given to the nearest 10 kg, of: 40 kg, 50 kg, 50 kg, 60 kg and 80 kg. Calculate the least possible total mass.

2. A water tank measures 30 cm by 50 cm by 20 cm. If each of these measurements is given to the nearest centimetre, calculate the largest possible volume of the tank.

3. The volume of a cube is given as 125 cm^3 to the nearest whole number.
   a) Express as an inequality the upper and lower bounds of the cube's volume.
   b) Express as an inequality the upper and lower bounds of the length of each of the cube's edges.

4. The radius of a circle is given as 4.00 cm to 2 d.p. Express as an inequality the upper and lower bounds for:
   a) the circumference of the circle,
   b) the area of the circle.

5. A cylindrical water tank has a volume of 6000 cm^3 correct to 1 s.f. A full cup of water from the tank has a volume of 300 cm^3 correct to 2 s.f. Calculate the maximum number of full cups of water that can be drawn from the tank.

6. A match measures 5 cm to the nearest centimetre. 100 matches end to end measure 5.43 m correct to 3 s.f.
   a) Calculate the upper and lower limits of the length of one match.
   b) How can the limits of the length of a match be found to 2 d.p.?

Student Assessment 4

1. The masses to 1 d.p. of three books are 0.7 kg, 0.1 kg and 0.2 kg. What are the lower and upper bounds of their combined mass?

2. A box has dimensions 8 cm, 12 cm and 15 cm. Each of these measurements is given to the nearest whole number. Calculate:
   a) the lower and upper bounds of its volume,
   b) the lower and upper bounds of its total surface area.

3. A square has an area of 64 cm^2 given to 2 s.f. Express as an inequality the upper and lower bounds for:
   a) the area of the square,
   b) the perimeter of the square.

4. The mass of a cube of side 15 cm is given as 1 kg. The side is correct to 0.5 cm and the mass accurate to 10 g. Calculate the upper and lower bounds of the density of the material (density = mass ÷ volume).

5. A single ball bearing has a mass of 8 g to the nearest gram. 100 ball bearings have a mass of 807 g to the nearest gram.
    a) State the lower and upper bounds for the value of one ball bearing.
    b) How can the limits of the mass of a ball bearing be found to 2 d.p.?

6. The radius of a circle to 2 d.p. is given as 6.00 cm. Express as an inequality the upper and lower bounds for:
    a) *C* cm, the circumference of the circle,
    b) *A* cm², the area of the circle.

ICT Section

The spreadsheet below shows a circle of radius 6 cm (correct to the nearest whole number). The spreadsheet is arranged in such a way that it automatically calculates the upper and lower bounds of the radius and the upper and lower bounds of the circle's area.

	A	B	C	D	E	F	G
1	Circle	Radius (cm)	Accuracy (decimal places)	Radius (cm)		Area (cm²)	
2				Lower bound	Upper bound	Lower bound	Upper bound
3	1	6	1	5.95	6.05	111.22	114.99

1. Copy the spreadsheet and enter formulae to automatically calculate the upper and lower bounds of the radius and area. Note: These formulae will be dependent on the number of decimal places that the radius is given to.

2. Using your spreadsheet calculate the upper and lower bounds of the radius and area of the following circles:
    a) radius = 10 cm correct to the nearest whole number,
    b) radius = 6.6 cm correct to 1 d.p,
    c) radius = 20.4 cm correct to 1 d.p,
    d) radius = 12.50 cm correct to 2 d.p,
    e) radius = 10.00 cm correct to 2 d.p.

3 RATIO, PROPORTION AND MEASURES OF RATE

Core Section

Demonstrate an understanding of the elementary ideas and notation of ratio, direct and inverse proportion and common measures of rate; divide a quantity in a given ratio.

▨ Direct proportion

Workers in a pottery factory are paid according to how many plates they produce. The wage paid to them is said to be in **direct proportion** to the number of plates made. As the number of plates made increases so does their wage. Other workers are paid for the number of hours worked. For them the wage paid is in **direct proportion** to the number of hours worked. There are two main methods for solving problems involving direct proportion: the ratio method and the unitary method.

Worked example   A bottling machine fills 500 bottles in 15 minutes. How many bottles will it fill in $1\frac{1}{2}$ hours?

Note: The time units must be the same, so for either method the $1\frac{1}{2}$ hours must be changed to 90 minutes.

The ratio method

Let x be the number of bottles filled. Then:

$$\frac{x}{90} = \frac{500}{15}$$

$$\text{so } x = \frac{500 \times 90}{15} = 3000$$

3000 bottles are filled in $1\frac{1}{2}$ hours.

The unitary method

In 15 minutes 500 bottles are filled.
Therefore in 1 minute $\frac{500}{15}$ bottles are filled.
So in 90 minutes $90 \times \frac{500}{15}$ bottles are filled.
In $1\frac{1}{2}$ hours, 3000 bottles are filled.

Exercise 3.1   Use either the ratio method or the unitary method to solve the problems below.

1. A machine prints four books in 10 minutes. How many will it print in 2 hours?

2. A farmer plants five apple trees in 25 minutes. If he continues to work at a constant rate, how long will it take him to plant 200 trees?

3. A television set uses 3 units of electricity in 2 hours. How many units will it use in 7 hours? Give your answer to the nearest unit.

4. A bricklayer lays 1500 bricks in an 8-hour day. Assuming he continues to work at the same rate, calculate:
   a) how many bricks he would expect to lay in a five-day week,
   b) how long to the nearest hour it would take him to lay 10 000 bricks.

5. A machine used to paint white lines on a road uses 250 litres of paint for each 8 km of road marked. Calculate:
   a) how many litres of paint would be needed for 200 km of road,
   b) what length of road could be marked with 4000 litres of paint.

6. An aircraft is cruising at 720 km/h and covers 1000 km. How far would it travel in the same period of time if the speed increased to 800 km/h?

7. A production line travelling at 2 m/s labels 150 tins. In the same period of time how many will it label at:
   a) 6 m/s   b) 1 m/s   c) 1.6 m/s?

If the information is given in the form of a ratio, the method of solution is the same.

Worked example Tin and copper are mixed in the ratio 8 : 3. How much tin is needed to mix with 36 g of copper?

The ratio method

Let x grams be the mass of tin needed.

$$\frac{x}{36} = \frac{8}{3}$$

Therefore $x = \dfrac{8 \times 36}{3}$

$= 96$

So 96 g of tin is needed.

The unitary method

3 g of copper mixes with 8 g of tin.
1 g of copper mixes with $\frac{8}{3}$ g of tin.
So 36 g of copper mixes with $36 \times \frac{8}{3}$ g of tin.
Therefore 36 g of copper mixes with 96 g of tin.

Exercise 3.2 1. Sand and gravel are mixed in the ratio 5 : 3 to form ballast.
   a) How much gravel is mixed with 750 kg of sand?
   b) How much sand is mixed with 750 kg of gravel?

2. A recipe uses 150 g butter, 500 g flour, 50 g sugar and 100 g currants to make 18 small cakes.
    a) How much of each ingredient will be needed to make 6 dozen cakes?
    b) How many whole cakes could be made with 1 kg of butter?

3. A paint mix uses red and white paint in a ratio of 1 : 12.
    a) How much white paint will be needed to mix with 1.4 litres of red paint?
    b) If a total of 15.5 litres of paint is mixed, calculate the amount of white paint and the amount of red paint used. Give your answers to the nearest 0.1 litre.

4. A tulip farmer sells sacks of mixed bulbs to local people. The bulbs develop into two different colours of tulips, red and yellow. The colours are packaged in a ratio of 8 : 5 respectively.
    a) If a sack contains 200 red bulbs, calculate the number of yellow bulbs.
    b) If a sack contains 351 bulbs in total, how many of each colour would you expect to find?
    c) One sack is packaged with a bulb mixture in the ratio 7 : 5 by mistake. If the sack contains 624 bulbs, how many more yellow bulbs would you expect to have compared with a normal sack of 624 bulbs?

5. A pure fruit juice is made by mixing the juices of oranges and mangoes in the ratio of 9 : 2.
    a) If 189 litres of orange juice are used, calculate the number of litres of mango juice needed.
    b) If 605 litres of the juice are made, calculate the number of litres of orange juice and mango juice used.

Divide a quantity in a given ratio

Worked examples a) Divide 20 m in the ratio 3 : 2.

The ratio method

3 : 2 gives 5 parts.
$\frac{3}{5} \times 20$ m $= 12$ m
$\frac{2}{5} \times 20$ m $= 8$ m
20 m divided in the ratio 3 : 2 is 12 m : 8 m.

The unitary method

3 : 2 gives 5 parts.
5 parts is equivalent to 20 m.
1 part is equivalent to $\frac{20}{5}$ m.
Therefore 3 parts is $3 \times \frac{20}{5}$ m; that is 12 m.
Therefore 2 parts is $2 \times \frac{20}{5}$ m; that is 8 m.

b) A factory produces cars in red, blue, white and green in the ratio $7:5:3:1$. Out of a production of 48 000 cars how many are white?

$7 + 5 + 3 + 1$ gives a total of 16 parts.
Therefore the total number of white
cars $= \frac{3}{16} \times 48\,000 = 9000$.

Exercise 3.3

1. Divide 150 in the ratio $2:3$.

2. Divide 72 in the ratio $2:3:4$.

3. Divide 5 kg in the ratio $13:7$.

4. Divide 45 minutes in the ratio $2:3$.

5. Divide 1 hour in the ratio $1:5$.

6. $\frac{7}{8}$ of a can of coke is water, the rest is syrup. What is the ratio of water to syrup?

7. $\frac{5}{9}$ of a litre carton of orange is pure orange juice, the rest is water. How many millilitres of each are in the carton?

8. 55% of students in a school are boys.
    a) What is the ratio of boys to girls?
    b) How many boys and how many girls are there if the school has 800 students?

9. A piece of wood is cut in the ratio $2:3$. What fraction of the length is the longer piece?

10. If the piece of wood in Q.9 is 80 cm long, how long is the shorter piece?

11. A gas pipe is 7 km long. A valve is positioned in such a way that it divides the length of the pipe in the ratio $4:3$. Calculate the distance of the valve from each end of the pipe.

12. The size of the angles of a quadrilateral are in the ratio $1:2:3:3$. Calculate the size of each angle.

13. The angles of a triangle are in the ratio $3:5:4$. Calculate the size of each angle.

14. A millionaire leaves 1.4 million dollars in his will to be shared between his three children in the ratio of their ages. If they are 24, 28 and 32 years old, calculate to the nearest dollar the amount they will each receive.

15. A small company makes a profit of £8000. This is divided between the directors in the ratio of their initial investments. If Alex put £20 000 into the firm, Maria £35 000 and Ahmet £25 000, calculate the amount of the profit they will each receive.

Inverse proportion

Sometimes an increase in one quantity causes a decrease in another quantity. For example, if fruit is to be picked by hand, the more people there are picking the fruit, the less time it will take.

Worked examples
a) If 8 people can pick the apples from the trees in 6 days, how long will it take 12 people?

8 people take 6 days.
1 person will take 6×8 days.
Therefore 12 people will take $\dfrac{6 \times 8}{12}$ days, i.e. 4 days.

b) A cyclist averages a speed of 27 km/h for 4 hours. At what average speed would she need to cycle to cover the same distance in 3 hours?

Completing it in 1 hour would require cycling at 27×4 km/h.
Completing it in 3 hours requires cycling at

$\dfrac{27 \times 4}{3}$ km/h; that is 36 km/h.

Exercise 3.4
1. A teacher shares sweets among 8 students so that they get 6 each. How many sweets would they each have got had there been 12 students?

2. The table below represents the relationship between the speed and the time taken for a train to travel between two stations.

Speed (km/h)	60			120	90	50	10
Time (h)	2	3	4				

Copy and complete the table.

3. A school can buy 150 books costing £12 each. If the price is reduced by 20%, how many more books could be bought?

4. Six people can dig a trench in 8 hours.
a) How long would it take:
   i) 4 people   ii) 12 people   iii) 1 person?
b) How many people would it take to dig the trench in:
   i) 3 hours   ii) 16 hours   iii) 1 hour?

5. Chairs in a hall are arranged in 35 rows of 18.
a) How many rows would there be with 21 chairs to a row?
b) How many chairs would there be in each row if there were 15 rows?

6. A train travelling at 100 km/h takes 4 hours for a journey. How long would it take a train travelling at 60 km/h?

7. A worker in a sugar factory packs 24 cardboard boxes with 15 bags of sugar in each. If he had boxes which held 18 bags of sugar each, how many fewer boxes would be needed?

8. A swimming pool is filled in 30 hours by two identical pumps. How much quicker would it be filled if five similar pumps were used instead?

Core Section

Student Assessment 1

1. A ruler 30 cm long is broken into two parts in the ratio 8 : 7. How long are the two parts?

2. A recipe needs 400 g of flour to make 8 cakes. How much flour would be needed in order to make two dozen cakes?

3. To make 6 jam tarts, 120 g of jam is needed. How much jam is needed to make 10 tarts?

4. The scale of a map is 1 : 25 000.
   a) Two villages are 8 cm apart on the map. How far apart are they in real life? Give your answer in kilometres.
   b) The distance from a village to the edge of a lake is 12 km in real life. How far apart would they be on the map? Give your answer in centimetres.

5. A motorbike uses petrol and oil mixed in the ratio 13 : 2.
   a) How much of each is there in 30 litres of mixture?
   b) How much petrol would be mixed with 500 ml of oil?

6. a) A model car is a $\frac{1}{40}$ scale model. Express this as a ratio.
   b) If the length of the real car is 5.5 m, what is the length of the model car?

7. An aunt gives a brother and sister £2000 to be divided in the ratio of their ages. If the girl is 13 years old and the boy 12 years old, how much will each get?

8. The angles of a triangle are in the ratio 2 : 5 : 8. Find the size of each of the angles.

9. A photocopying machine is capable of making 50 copies each minute.
   a) If four identical copiers are used simultaneously how long would it take to make a total of 50 copies?
   b) How many copiers would be needed to make 6000 copies in 15 minutes?

10. It takes 16 hours for three bricklayers to build a wall. Calculate how long it would take for eight bricklayers to build a similar wall.

Student Assessment 2

1. A piece of wood is cut in the ratio 3 : 7.
   a) What fraction of the whole is the longer piece?
   b) If the wood is 1.5 m long, how long is the shorter piece?

2. A recipe for two people requires $\frac{1}{4}$ kg of rice to 150 g of meat.
   a) How much meat would be needed for five people?
   b) How much rice would there be in 1 kg of the final dish?

3. The scale of a map is 1 : 10 000.
   a) Two rivers are 4.5 cm apart on the map, how far apart are they in real life? Give your answer in metres.
   b) Two towns are 8 km apart in real life. How far apart are they on the map? Give your answer in centimetres.

4. a) A model train is a $\frac{1}{26}$ scale model. Express this as a ratio.
   b) If the length of the model engine is 7 cm, what is the true length of the engine?

5. Divide 3 tonnes in the ratio 2 : 5 : 13.

6. The ratio of the angles of a quadrilateral is 2 : 3 : 3 : 4. Calculate the size of each of the angles.

7. The ratio of the interior angles of a pentagon is 2 : 3 : 4 : 4 : 5. Calculate the size of the largest angle.

8. A large swimming pool takes 36 hours to fill using three identical pumps.
   a) How long would it take to fill using eight identical pumps?
   b) If the pool needs to be filled in 9 hours, how many pumps will be needed?

9. The first triangle is an enlargement of the second. Calculate the size of the missing sides and angles.

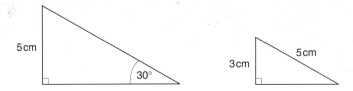

10. A tap issuing water at a rate of 1.2 litres per minute fills a container in 4 minutes.
    a) How long would it take to fill the same container if the rate was decreased to 1 litre per minute? Give your answer in minutes and seconds.
    b) If the container is to be filled in 3 minutes, calculate the rate at which the water should flow.

RATIO, PROPORTION AND MEASURES OF RATE

Extended Section

Express direct and inverse variation in algebraic terms and use this form of expression to find unknown quantities; increase and decrease a quantity by a given ratio.

NB: All diagrams are not drawn to scale.

Direct variation

Consider the tables below:

x	0	1	2	3	5	10
y	0	2	4	6	10	20

$y = 2x$

x	0	1	2	3	5	10
y	0	3	6	9	15	30

$y = 3x$

x	0	1	2	3	5	10
y	0	2.5	5	7.5	12.5	25

$y = 2.5x$

In each case y is directly proportional to x. This is written $y \propto x$. If any of these three tables is shown on a graph, the graph will be a straight line passing through the origin.

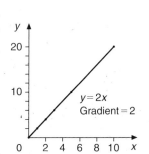

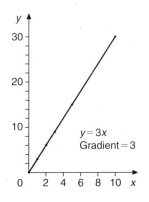

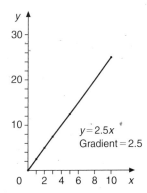

For any statement where $y \propto x$,

$$y = kx$$

where k is a constant equal to the gradient of the graph and is called the **constant of proportionality** or constant of variation.

Inverse variation

If y is inversely proportional to x, then $y \propto \dfrac{1}{x}$ and $y = \dfrac{k}{x}$.

If a graph of y against $\dfrac{1}{x}$ is plotted, this too will be a straight line passing through the origin.

Worked examples   **a)** $y \propto x$. If $y = 7$ when $x = 2$, find y when $x = 5$.

$$y = kx$$
$$7 = k \times 2$$
$$k = 3.5$$
When $x = 5$,
$$y = 3.5 \times 5$$
$$= 17.5$$

b) $y \propto \dfrac{1}{x}$. If $y = 5$ when $x = 3$, find y when $x = 30$.

$$y = \frac{k}{x}$$

$$5 = \frac{k}{3}$$

$$k = 15$$

When $x = 30$,

$$y = \frac{15}{30}$$

$$= 0.5$$

Exercise 3.5   **1.** y is directly proportional to x. If $y = 6$ when $x = 2$, find:
a) the constant of proportionality,
b) the value of y when $x = 7$,
c) the value of y when $x = 9$,
d) the value of x when $y = 9$,
e) the value of x when $y = 30$.

2. y is directly proportional to x^2. If $y = 18$ when $x = 6$, find:
a) the constant of proportionality,
b) the value of y when $x = 4$,
c) the value of y when $x = 7$,
d) the value of x when $y = 32$,
e) the value of x when $y = 128$.

3. y is inversely proportional to x^3. If $y = 3$ when $x = 2$, find:
a) the constant of proportionality,
b) the value of y when $x = 4$,
c) the value of y when $x = 6$,
d) the value of x when $y = 24$.

4. y is inversely proportional to x^2. If $y = 1$ when $x = 0.5$, find:
   a) the constant of proportionality,
   b) the value of y when $x = 0.1$,
   c) the value of y when $x = 0.25$,
   d) the value of x when $y = 64$.

Exercise 3.6

1. Write the following in the form i) $y \propto x$ ii) $y = kx$.
   a) y is directly proportional to x^3.
   b) y is inversely proportional to x^3.
   c) t is directly proportional to P.
   d) s is inversely proportional to t.
   e) A is directly proportional to r^2.
   f) T is inversely proportional to the square root of g.

2. If $y \propto x$ and $y = 6$ when $x = 2$, find y when $x = 3.5$.

3. If $y \propto \dfrac{1}{x}$ and $y = 4$ when $x = 2.5$ find:

   a) y when $x = 20$,
   b) x when $y = 5$.

4. If $p \propto r^2$ and $p = 2$ when $r = 2$, find p when $r = 8$.

5. If $m \propto \dfrac{1}{r^3}$ and $m = 1$ when $r = 2$, find:

   a) m when $r = 4$,
   b) r when $m = 125$.

6. If $y \propto x^2$ and $y = 12$ when $x = 2$, find y when $x = 5$.

Exercise 3.7

1. If a stone is dropped off the edge of a cliff, the height
   (h metres) of the cliff is proportional to the square of
   the time (t seconds) taken for the stone to reach the
   ground.
      A stone takes 5 seconds to reach the ground when
   dropped off a cliff 125 m high.
   a) Write down a relationship between h and t, using k as
      the constant of variation.
   b) Calculate the constant of variation.
   c) Find the height of a cliff if a stone takes 3 seconds to
      reach the ground.
   d) Find the time taken for a stone to fall from a cliff 180 m
      high.

2. The velocity (v metres per second) of a body is known to
   be proportional to the square root of its kinetic energy
   (e joules). When the velocity of a body is 120 m/s its kinetic
   energy is 1600 J.
   a) Write down a relationship between v and e, using k as
      the constant of variation.
   b) Calculate the value of k.
   c) $v = 21$; calculate the kinetic energy of the body in joules.

3. The length (l cm) of an edge of a cube is proportional to the cube root of its mass (m grams). It is known that if $l = 15$, then $m = 125$. Let k be the constant of variation.
   a) Write down the relationship between l, m and k.
   b) Calculate the value of k.
   c) Calculate the value of l when $m = 8$.

4. The power (P) generated in an electrical circuit is proportional to the square of the current (I amps). When the power is 108 watts, the current is 6 amps.
   a) Write down a relationship between P, I and the constant of variation, k.
   b) Calculate the value of I when $P = 75$ watts.

Increase and decrease by a given ratio

Worked examples   a) A photograph is 12 cm wide and 8 cm tall. It is enlarged in the ratio $3 : 2$. What are the dimensions of the enlarged photograph?

$3 : 2$ is an enlargement of $\frac{3}{2}$. Therefore the enlarged width is 12 cm $\times \frac{3}{2}$; that is 18 cm.
The enlarged height is 8 cm $\times \frac{3}{2}$; that is 12 cm.

b) A photographic transparency 5 cm wide and 3 cm tall is projected onto a screen. If the image is 1.5 m wide:
   i) calculate the ratio of the enlargement,
   ii) calculate the height of the image.

   i) 5 cm width is enlarged to become 150 cm.
      So 1 cm width becomes $\frac{150}{5}$ cm; that is 30 cm.
      Therefore the enlargement ratio is $30 : 1$.
   ii) The height of the image = 3 cm $\times$ 30 = 90 cm.

Exercise 3.8   1. Increase 100 by the following ratios:
   a) $8 : 5$ b) $5 : 2$ c) $7 : 4$
   d) $11 : 10$ e) $9 : 4$ f) $32 : 25$

2. Increase 70 by the following ratios:
   a) $4 : 3$ b) $5 : 3$ c) $8 : 7$
   d) $9 : 4$ e) $11 : 5$ f) $17 : 14$

3. Decrease 60 by the following ratios:
   a) $2 : 3$ b) $5 : 6$ c) $7 : 12$
   d) $3 : 5$ e) $1 : 4$ f) $13 : 15$

4. Decrease 30 by the following ratios:
   a) $3 : 4$ b) $2 : 9$ c) $7 : 12$
   d) $3 : 16$ e) $5 : 8$ f) $9 : 20$

5. Increase 40 by a ratio of $5 : 4$.

6. Decrease 40 by a ratio of $4 : 5$.

7. Incrcase 150 by a ratio of $7 : 5$.

8. Decrease 210 by a ratio of $3 : 7$.

Exercise 3.9

1. A photograph measuring 8 cm by 6 cm is enlarged by a ratio of 11 : 4. What are the dimensions of the new print?

2. A photocopier enlarges in the ratio 7 : 4. What would be the new size of a diagram measuring 16 cm by 12 cm?

3. A drawing measuring 10 cm by 16 cm needs to be enlarged. The dimensions of the enlargement need to be 25 cm by 40 cm. Calculate the enlargement needed and express it as a ratio.

4. A banner needs to be enlarged from its original format. The dimensions of the original are 4 cm tall by 25 cm wide. The enlarged banner needs to be at least 8 m wide but no more than 1.4 m tall. Calculate the minimum and maximum ratios of enlargement possible.

5. A rectangle measuring 7 cm by 4 cm is enlarged by a ratio of 2 : 1.
   a) What is the area of:
      i) the original rectangle?
      ii) the enlarged rectangle?
   b) By what ratio has the area been enlarged?

6. A square of side length 3 cm is enlarged by a ratio of 3 : 1.
   a) What is the area of:
      i) the original square?
      ii) the enlarged square?
   b) By what ratio has the area been enlarged?

7. A cuboid measuring 3 cm by 5 cm by 2 cm is enlarged by a ratio of 2 : 1.
   a) What is the volume of:
      i) the original cuboid?
      ii) the enlarged cuboid?
   b) By what ratio has the volume been increased?

8. A cube of side 4 cm is enlarged by a ratio of 3 : 1.
   a) What is the volume of:
      i) the original cube?
      ii) the enlarged cube?
   b) By what ratio has the volume been increased?

9. The triangle (left) is to be reduced by a ratio of 1 : 2.
   a) Calculate the area of the original triangle.
   b) Calculate the area of the reduced triangle.
   c) Calculate the ratio by which the area of the triangle has been reduced.

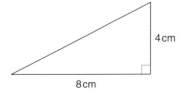

10. From Q.5–9 can you conclude what happens to two- and three-dimensional figures when they are either enlarged or reduced?

Extended Section

Student Assessment 1

1. $y = kx$. When $y = 9$, $x = 3$.
   a) Calculate the value of k.
   b) Calculate y when $x = 4$.
   c) Calculate y when $x = 1$.
   d) Calculate x when $y = 18$.

2. $y = \dfrac{k}{x}$. When $y = 2$, $x = 2$.

   a) Calculate the value of k.
   b) Calculate y when $x = 16$.
   c) Calculate x when $y = 1$.
   d) Calculate x when $y = 0.5$.

3. $p = kq^3$. When $p = 4$, $q = 2$.
   a) Calculate the value of k.
   b) Calculate p when $q = 4$.
   c) Calculate p when $q = 1$.
   d) Calculate q when $p = 108$.

4. $m = \dfrac{k}{\sqrt{n}}$. When $m = \dfrac{5}{12}$, $n = 36$.

   a) Calculate the value of k.
   b) Calculate m when $n = 25$.
   c) Calculate m when $n = 100$.
   d) Calculate n when $m = 10$.

5. $y = \dfrac{k}{x^2}$. When $y = \dfrac{1}{16}$, $x = 2$.

   a) Calculate the value of k.
   b) Calculate y when $x = 1$.
   c) Calculate both values of x when $y = 0.25$.
   d) Calculate both values of x when $y = 0.01$.

Student Assessment 2

1. $y = kx$. When $y = 12$, $x = 8$.
   a) Calculate the value of k.
   b) Calculate y when $x = 10$.
   c) Calculate y when $x = 2$.
   d) Calculate x when $y = 18$.

2. $y = \dfrac{k}{x}$. When $y = 2$, $x = 5$.

   a) Calculate the value of k.
   b) Calculate y when $x = 4$.
   c) Calculate x when $y = 10$.
   d) Calculate x when $y = 0.5$.

3. $p = kq^3$. When $p = 9$, $q = 3$.
   a) Calculate the value of k.
   b) Calculate p when $q = 6$.
   c) Calculate p when $q = 1$.
   d) Calculate q when $p = 576$.

4. $m = \dfrac{k}{\sqrt{n}}$. When $m = 1$, $n = 25$.

   a) Calculate the value of k.
   b) Calculate m when $n = 16$.
   c) Calculate m when $n = 100$.
   d) Calculate n when $m = 5$.

5. $y = \dfrac{k}{x^2}$. When $y = 3$, $x = \dfrac{1}{3}$.

   a) Calculate the value of k.
   b) Calculate y when $x = 0.5$.

   c) Calculate both values of x when $y = \dfrac{1}{12}$.

   d) Calculate both values of x when $y = \dfrac{1}{3}$.

Student Assessment 3

1. Copy and complete the following tables:
   a) $y \propto x$

x	1	2	3	4	5
y			15		

   b) $y \propto \dfrac{1}{x}$

x	1	2	3	4	5
y					6

   c) $y \propto \dfrac{1}{x^2}$

x	1	2	3	4	5
y		5			

2. The braking distance (d metres) of a truck is proportional to the square of its speed (s km/h). If $d = 10$ when $s = 36$, calculate:
   a) d when $s = 60$,
   b) s when $d = 144$.

3. The volume (V cm^3) of a sphere is proportional to the cube of its radius (r cm). When $V = 33.5, r = 2$.
   a) Write down the relationship between V and r, using k as the constant of variation.
   b) Calculate the value of k.
   c) Find the volume of the sphere when $r = 3$.
   d) What is the radius of a sphere of volume 400 cm^3?

4. The volume (V cm^3) of a square-based pyramid is directly proportional to the product of the base area (A cm^2) and the vertical height (h cm).
   a) Write down the relationship between V, A and h, using k as the constant of variation.
   b) A square-based pyramid of base area 25 cm^2 and a vertical height of 9 cm has a volume of 75 cm^3. Find the base area of a square-based pyramid with a volume of 20 cm^3 and a vertical height of 5 cm.

5. A photocopier enlarges by a ratio of 7 : 4. A picture measures 6 cm by 4 cm. How many consecutive enlargements can be made so that the largest possible picture will fit on a sheet measuring 30 cm by 20 cm?

6. A cuboid 3 cm by 3 cm by 5 cm is enlarged by a ratio of 3 : 2.
   a) Express the ratio 'volume of enlarged cuboid : volume of original cuboid' in its simplest form.
   b) Express the ratio 'surface area of enlarged cuboid : surface area of original cuboid' in its simplest form.

Student Assessment 4

1. y is inversely proportional to x.
   a) Copy and complete the table below.

x	1	2	4	8	16	32
y				4		

   b) What is the value of x when $y = 20$?

2. Copy and complete the tables below:
   a) $y \propto x$

x	1	2	4	5	10
y		10			

b) $y \propto \dfrac{1}{x}$

x	1	2	4	5	10
y	20				

c) $y \propto \sqrt{x}$

x	4	16	25	36	64
y	4				

3. The pressure (P) of a given mass of gas is inversely proportional to its volume (V) at a constant temperature. If $P = 4$ when $V = 6$, calculate:
   a) P when $V = 30$,　　　　　b) V when $P = 30$.

4. The gravitational force (F) between two masses is inversely proportional to the square of the distance (d) between them. If $F = 4$ when $d = 5$, calculate:
   a) F when $d = 8$,　　　　　b) d when $F = 25$.

5. A map measuring 60 cm by 25 cm is reduced twice in the ratio 3 : 5. Calculate the final dimensions of the map.

6. A cube of side 1 m is reduced three times in the ratio 1 : 2.
   a) Calculate the edge length of the reduced cube.
   b) What is the area of one face of the reduced cube in relation to that of the original cube? Express your answer as a ratio in its simplest form, 'reduced area : original area'.

ICT section

In this investigation you will use a spreadsheet to help you find the relationship between an enlargement ratio of $n : 1$ and its effect on the area and volume ratios.

1. Set up a spreadsheet similar to the one shown below:

NB: The cells shaded grey indicate where formulae were used to calculate the result. Leave these blank for the moment.

	A	B	C	D	E	F	G	H
1				Original cube				
2	Cube	Side length (cm)	Area of one face (cm²)	Total surface Area (cm²)	Volume (cm³)			
3	1	1	1	6	1			
4								
5				Enarged cube				
6		Side length (cm)	Area of one face (cm²)	Total surface Area (cm²)	Volume (cm³)			
7		2	4	24	8			
8								
9						n	:	1
10				Ratio of enlargement (Length)		2	:	1
11				Ratio of enlargement (Area)		4	:	1
12				Ratio of Enlargement (Volume)		8	:	1

2. Enter a formula into cell C3 to calculate the area of one face of the original cube.

3. Enter formulae into cells D3 and E3 in order to calculate the total surface area and volume respectively of the original cube.

4. In cell F10 enter the number 2. This means we wish to enlarge the side lengths of the original cube in the ratio 2 : 1.

5. Enter formulae in cells B7 to E7 in order to calculate the values for the enlarged cube.

6. Enter a formula in cell F11 to calculate the area ratio of enlargement.

7. Enter a formula in cell F12 to calculate the volume ratio of enlargement.

8. Experiment with different values in cell F10 and see how the scale factor of enlargement affects the area and volume ratios of enlargement.

9. What conclusions can you make from your findings in Q.8 above?

4 PERCENTAGES

Core Section

Calculate a given percentage of a quantity; express one quantity as a percentage of another; calculate percentage increase or decrease; problems involving compound interest.

You should already be familiar with the percentage equivalent of simple fractions and decimals as outlined in the table below.

Fraction	Decimal	Percentage
$\frac{1}{2}$	0.5	50%
$\frac{1}{4}$	0.25	25%
$\frac{3}{4}$	0.75	75%
$\frac{1}{8}$	0.125	12.5%
$\frac{3}{8}$	0.375	37.5%
$\frac{5}{8}$	0.625	62.5%
$\frac{7}{8}$	0.875	87.5%
$\frac{1}{10}$	0.1	10%
$\frac{1}{10}$ or $\frac{1}{5}$	0.2	20%
$\frac{3}{10}$	0.3	30%
$\frac{4}{10}$ or $\frac{2}{5}$	0.4	40%
$\frac{6}{10}$ or $\frac{3}{5}$	0.6	60%
$\frac{7}{10}$	0.7	70%
$\frac{8}{10}$ or $\frac{4}{5}$	0.8	80%
$\frac{9}{10}$	0.9	90%

▦ Simple percentages

Worked examples **a)** Of 100 sheep in a field, 88 are ewes.

i)   What percentage of the sheep are ewes?
88 out of 100 are ewes
$= 88\%$

ii) What percentage are not ewes?
12 out of 100
$= 12\%$

b) A gymnast scored marks out of 10 from five judges.
They were: 8.0, 8.2, 7.9, 8.3, 7.6.
Express these marks as percentages.

$$\frac{8.0}{10} = \frac{80}{100} = 80\% \qquad \frac{8.2}{10} = \frac{82}{100} = 82\% \qquad \frac{7.9}{10} = \frac{79}{100} = 79\%$$

$$\frac{8.3}{10} = \frac{83}{100} = 83\% \qquad \frac{7.6}{10} = \frac{76}{100} = 76\%$$

c) Convert the following percentages into fractions and decimals:

i) 27%   ii) 5%

$$\frac{27}{100} = 0.27 \qquad\qquad \frac{5}{100} = 0.05$$

Exercise 4.1

1. In a survey of 100 cars, 47 were white, 23 were blue and 30 were red. Express each of these numbers as a percentage of the total.

2. $\frac{7}{10}$ of the surface of the Earth is water. Express this as a percentage.

3. There are 200 birds in a flock. 120 of them are female. What percentage of the flock are:
   a) female?   b) male?

4. Write these percentages as fractions of 100:
   a) 73%   b) 28%
   c) 10%   d) 25%

5. Write these fractions as percentages:

   a) $\dfrac{27}{100}$ b) $\dfrac{3}{10}$

   c) $\dfrac{7}{50}$ d) $\dfrac{1}{4}$

6. Convert the following percentages to decimals:
   a) 39%   b) 47%   c) 83%
   d) 7%   e) 2%   f) 20%

7. Convert the following decimals to percentages:
   a) 0.31   b) 0.67   c) 0.09
   d) 0.05   e) 0.2   f) 0.75

▨ Calculating a percentage of a quantity

Worked examples   a) Find 25% of 300 m.

25% can be written as 0.25.
0.25×300 m $= 75$ m.

b) Find 35% of 280 m.

35% can be written as 0.35.
0.35×280 m $= 98$ m.

Exercise 4.2

1. Write the percentage equivalent of the following fractions:
   a) $\frac{1}{4}$ b) $\frac{2}{3}$ c) $\frac{5}{8}$
   d) $1\frac{4}{5}$ e) $4\frac{9}{10}$ f) $3\frac{7}{8}$

2. Write the decimal equivalent of the following:
   a) $\frac{3}{4}$ b) 80%   c) $\frac{1}{5}$
   d) 7%   e) $1\frac{7}{8}$ f) $\frac{1}{6}$

3. Evaluate the following:
a) 25% of 80   b) 80% of 125   c) 62.5% of 80
d) 30% of 120   e) 90% of 5   f) 25% of 30

4. Evaluate the following:
a) 17% of 50   b) 50% of 17   c) 65% of 80
d) 80% of 65   e) 7% of 250   f) 250% of 7

5. In a class of 30 students, 20% have black hair, 10% have blonde hair and 70% have brown hair. Calculate the number of students with
a) black hair,
b) blonde hair,
c) brown hair.

6. A survey conducted among 120 schoolchildren looked at which type of meat they preferred. 55% said they preferred beef, 20% said they preferred chicken, 15% preferred lamb and 10% pork. Calculate the number of children in each category.

7. A survey was carried out in a school to see what nationality its students were. Of the 220 students in the school, 65% were English, 20% were Pakistani, 5% were Greek and 10% belonged to other nationalities. Calculate the number of students of each nationality.

8. A shopkeeper keeps a record of the number of items he sells in one day. Of the 150 items he sold, 46% were newspapers, 24% were pens, 12% were books whilst the remaining 18% were other assorted items. Calculate the number of each item he sold.

Expressing one quantity as a percentage of another

To express one quantity as a percentage of another, first write the first quantity as a fraction of the second and then multiply by 100.

Worked example   In an examination a girl obtains 69 marks out of 75. Express this result as a percentage.

$$\frac{69}{75} \times 100\% = 92\%$$

Exercise 4.3   **1.** Express the first quantity as a percentage of the second.
a) 24 out of 50   b) 46 out of 125
c) 7 out of 20   d) 45 out of 90
e) 9 out of 20   f) 16 out of 40
g) 13 out of 39   h) 20 out of 35

2. A hockey team plays 42 matches. It wins 21, draws 14 and loses the rest. Express each of these results as a percentage of the total number of games played.

3. Four candidates stood in an election:

    A received 24 500 votes
    B received 18 200 votes
    C received 16 300 votes
    D received 12 000 votes

    Express each of these as a percentage of the total votes cast.

4. A car manufacturer produces 155 000 cars a year. The cars
    are available for sale in six different colours. The numbers
    sold of each colour were:

    Red 55 000
    Blue 48 000
    White 27 500
    Silver 10 200
    Green 9300
    Black 5000

    Express each of these as a percentage of the total number
    of cars produced. Give your answers to 1 d.p.

Percentage increases and decreases

Worked examples

a) A garage increases the price of a truck by 12%. If the
    original price was £14 500, calculate its new price.
    Note: the original price represents 100%, therefore the
    increased price can be represented as 112%.

    New price = 112% of £14 500
              = 1.12 × £14 500
              = £16 240

b) A Saudi doctor has a salary of 16 000 Saudi riyals per
    month. If his salary increases by 8%, calculate:
    i)   the amount extra he receives a month,
    ii)   his new monthly salary.

    i)   Increase = 8% of 16 000 riyals
                   = 0.08 × 16 000 riyals = 1280 riyals
    ii)   New salary = old salary + increase
                     = 16 000 + 1280 riyals per month
                     = 17 280 riyals per month

c) A shop is having a sale. It sells a set of tools costing $130 at
    a 15% discount. Calculate the sale price of the tools.
    Note: the old price represents 100%, therefore the new
    price can be represented as (100 − 15)% = 85%.

    85% of $130 = 0.85 × $130
                = $110.50

Exercise 4.4

1. Increase the following by the given percentage:
   a) 150 by 25% b) 230 by 40% c) 7000 by 2%
   d) 70 by 250% e) 80 by 12.5% f) 75 by 62%

2. Decrease the following by the given percentage:
   a) 120 by 25% b) 40 by 5% c) 90 by 90%
   d) 1000 by 10% e) 80 by 37.5% f) 75 by 42%

3. In the following questions the first number is increased to become the second number. Calculate the percentage increase in each case.
   a) $50 \rightarrow 60$ b) $75 \rightarrow 135$ c) $40 \rightarrow 84$
   d) $30 \rightarrow 31.5$ e) $18 \rightarrow 33.3$ f) $4 \rightarrow 13$

4. In the following questions the first number is decreased to become the second number. Calculate the percentage decrease in each case.
   a) $50 \rightarrow 25$ b) $80 \rightarrow 56$ c) $150 \rightarrow 142.5$
   d) $3 \rightarrow 0$ e) $550 \rightarrow 352$ f) $20 \rightarrow 19$

5. A farmer increases the yield on his farm by 15%. If his previous yield was 6500 tonnes, what is his present yield?

6. The cost of a computer in a Brazilian computer store is reduced by 12.5% in a sale. If the computer was priced at 7800 Brazilian real (BRL), what is its price in the sale?

7. A winter coat is priced at £100. In the sale its price is reduced by 25%.
   a) Calculate the sale price of the coat.
   b) After the sale its price is increased by 25% again. Calculate the coat's price after the sale.

8. A farmer takes 250 chickens to be sold at a market. In the first hour he sells 8% of his chickens. In the second hour he sells 10% of those that were left.
   a) How many chickens has he sold in total?
   b) What percentage of the original number did he manage to sell in the two hours?

9. The number of fish on a fish farm increases by approximately 10% each month. If there were originally 350 fish, calculate to the nearest 100 how many fish there would be after 12 months.

Compound interest

Compound interest means that not only is interest paid on the principal amount but interest is paid on the interest. It is compounded (or added to).

This sounds complicated but the example below will make it clear.

A builder is going to build six houses on a plot of land in Spain. He borrows 500 000 euro at 10% and will pay off the loan in full after three years.

At the end of the first year he will owe
€500 000 + 10% of €500 000 i.e. 550 000 euro

At the end of the second year he will owe
€550 000 + 10% of €550 000
i.e. €550 000 + €55 000 = 605 000 euro

At the end of the third year he will owe
€605 000 + 10% of €605 000
i.e. €605 000 + €60 500 = 665 500 euro

His interest will be 665 500 – 500 000 euro i.e. 165 500 euro

The simple interest is 50 000 euro a year
i.e. 150 000 euro in total

The extra 15 500 euro was the compound interest.

The time taken for a debt to grow at compound interest can be calculated as shown in the example below.

How long will it take for a debt to double at a compound interest of 27%?

After 1 year the debt D will be D(1 + 27%) or 1.27D.
After 2 years the debt D will be (1.27 × 1.27)D or 1.61D.
After 3 years the debt D will be (1.27 × 1.27 × 1.27)D or 2.05D.
i.e. the debt will be more than doubled after 3 years.

Exercise 4.5

1. A shipping company borrows £70 million at 5% compound interest to build a new cruise ship. If it repays the debt after three years, how much interest will the company pay?

2. A woman borrows 100 000 euro to improve her house. She borrows the money at 15% interest and repays it in full after three years. What interest will she pay?

3. A man owes £5000 on his credit cards. The annual percentage rate APR is 20%. If he lets the debt grow, how much will he owe in four years?

4. A school increases its intake by 10% each year. If it starts with 1000 students, how many will it have at the beginning of the fourth year of expansion?

5. 8 million tonnes of fish were caught in the North Sea in 2005. If the catch is reduced by 20% each year for four years, what amount can be caught at the end of this time?

6. How many years will it take for a debt to double at 42% compound interest?

7. How many years will it take for a debt to double at 15% compound interest?

8. A car loses value at 27% compound interest. How many years will it take for its value to have halved?

Core Section

Student Assessment I

1. Copy the table below and fill in the missing values:

Fraction	Decimal	Percentage
$\frac{3}{4}$		
	0.8	
$\frac{5}{8}$		
	1.5	

2. Find 40% of 1600 m.

3. A shop increases the price of a television set by 8%. If the present price is £320 what is the new price?

4. A car loses 55% of its value after four years. If it cost $22 500 when new, what is its value after the four years?

5. Express the first quantity as a percentage of the second.
   a) 40 cm, 2 m   b) 25 mins, 1 hour   c) 450 g, 2 kg
   d) 3 m, 3.5 m   e) 70 kg, 1 tonne   f) 75 cl, 2.5 l

6. A house is bought for £75 000 and then resold for £87 000. Calculate the percentage profit.

7. A pair of shoes is priced at £45. During a sale the price is reduced by 20%.
   a) Calculate the sale price of the shoes.
   b) What is the percentage increase in the price if after the sale it is once again restored to £45?

8. The population of a town increases by 5% each year. If in 1997 the population was 86 000, in which year is the population expected to exceed 100 000 for the first time?

9. Find the compound interest on 3 million euro for two years at 8% interest.

10. A house increases in value by 20% a year. How long will it take to double in value?

Student Assessment 2

1. Copy the table below and fill in the missing values:

Fraction	Decimal	Percentage
	0.25	
$\frac{3}{5}$		
		$62\frac{1}{2}\%$
$2\frac{1}{4}$		

2. Find 30% of 2500 m.

3. In a sale a shop reduces its prices by 12.5%. What is the sale price of a desk previously costing 600 euro?

4. In the last six years the value of a house has increased by 35%. If it cost £72 000 six years ago, what is its value now?

5. Express the first quantity as a percentage of the second.
   a) 35 mins, 2 hours   b) 650 g, 3 kg
   c) 5 m, 4 m   d) 15 s, 3 mins
   e) 600 kg, 3 tonnes   f) 35 cl, 3.5 l

6. Shares in a company are bought for $600. After a year the same shares are sold for $550. Calculate the percentage depreciation.

7. In a sale the price of a jacket originally costing 17 000 Japanese yen (¥) is reduced by ¥400. Any item not sold by the last day of the sale is reduced by a further 50%. If the jacket is sold on the last day of the sale:
   a) calculate the price it is finally sold for,
   b) calculate the overall percentage reduction in price.

8. The population of a type of insect increases by approximately 10% each day. How many days will it take for the population to double?

9. Find the compound interest on 5 million euro for three years at 6% interest.

10. A boat loses 15% of its value each year. How long before it has halved in value?

PERCENTAGES

Extended Section

Carry out calculations involving reverse percentages, e.g. finding the cost price given the selling price and the percentage profit.

▨ Reverse percentages

Worked examples   **a)** In a test Ahmed answered 92% of the questions correctly. If he answered 23 questions correctly, how many had he got wrong?

> 92% of the marks is equivalent to 23 questions.
> 1% of the marks therefore is equivalent to $\frac{23}{92}$ questions.

So 100% is equivalent to $\frac{23}{92} \times 100 = 25$ questions.

Ahmed got 2 questions wrong.

b) A boat is sold for £15 360. This represents a profit of 28% to the seller. What did the boat originally cost the seller?

> The selling price is 128% of the original cost to the seller. 128% of the original cost is £15 360.

1% of the original cost is $\dfrac{£15\ 360}{128}$.

100% of the original cost is $\dfrac{£15\ 360}{128} \times 100$, i.e. £12 000.

Exercise 4.6   **1.** Calculate the value of X in each of the following:
a) 40% of X is 240   b) 24% of X is 84
c) 85% of X is 765   d) 4% of X is 10
e) 15% of X is 18.75   f) 7% of X is 0.105

2. Calculate the value of Y in each of the following:
a) 125% of Y is 70   b) 140% of Y is 91
c) 210% of Y is 189   d) 340% of Y is 68
e) 150% of Y is 0.375   f) 144% of Y is -54.72

3. In a geography text book, 35% of the pages are coloured. If there are 98 coloured pages, how many pages are there in the whole book?

4. A town has 3500 families who own a car. If this represents 28% of the families in the town, how many families are there in total?

5. In a test Isabel scored 88%. If she got three questions incorrect, how many did she get correct?

6. Water expands when it freezes. Ice is less dense than water so it floats. If the increase in volume is 4%, what volume of water will make an iceberg of 12 700 000 m³? Give your answer to three significant figures.

Extended Section

Student Assessment 1

1. Calculate the original price in each of the following:

Selling price	Profit
£3780	8%
£14 880	24%
£3.50	250%
£56.56	1%

2. Calculate the original price in each of the following:

Selling price	Loss
£350	30%
£200	20%
£8000	60%
£27 500	80%

3. In a test Ben gained 90% by answering 135 questions correctly. How many questions did he answer incorrectly?

4. A one-year old car is worth £11 250. If its value had depreciated by 25% in that first year, calculate its price when new.

5. This year a farmer's crop yielded 50 000 tonnes. If this represents a 25% increase on last year, what was the yield last year?

6. A company increased its productivity by 10% each year for the last two years. If it produced 56 265 units this year, how many units did it produce two years ago?

Student Assessment 2

1. Calculate the original price in each of the following:

Selling price	Profit
$224	12%
$62.50	150%
$660.24	26%
$38.50	285%

2. Calculate the original price in each of the following:

Selling price	Loss
$392.70	15%
$2480	38%
$3937.50	12.5%
$4675	15%

3. In an examination Sarah obtained 87.5% by gaining 105 marks. How many marks did she lose?

4. At the end of a year a factory has produced 38 500 television sets. If this represents a 10% increase in productivity on last year, calculate the number of sets that were made last year.

5. A computer manufacturer is expected to have produced 24 000 units by the end of this year. If this represents a 4% decrease on last year's output, calculate the number of units produced last year.

6. A farmer increased his yield by 5% each year over the last five years. If he produced 600 tonnes this year, calculate to the nearest tonne his yield five years ago.

ICT Section

In this activity you will be using a spreadsheet to track the price of a company's shares over a period of time.

1. a) Using the Internet or a newspaper as a resource, find the value of a particular company's shares.
   b) Over a period of a month (or week), record the value of the company's shares. This should be carried out on a daily basis.

2. When you have collected all the results, enter them into a spreadsheet similar to the one shown below:

	A	B	C
1		Company Name	
2	Day	Share Price	Percentage Value
3	1	3.26	100
4	2	3.29	
5	3	4.11	
6	4		
7	5		
8			
9			
10			
11	etc	etc	

3. In column C enter formulae that will calculate the value of the shares as a percentage of their value on day 1.

4. When the spreadsheet is complete produce a graph showing how the percentage value of the share price changed over time.

5. Write a short report explaining the performance of the company's shares during that time.

ALGEBRA

5 GRAPHS IN PRACTICAL SITUATIONS

Core Section

Interpret and use graphs in practical situations including conversion graphs, distance–time graphs and travel graphs.

▨ Conversion graphs

A straight line graph can be used to convert one set of units to another. Examples include converting from one currency to another, converting distance in miles to kilometres and converting temperature from degrees Celsius to degrees Fahrenheit.

Worked example The graph below converts Mexican pesos into euro based on an exchange rate of €1 = 8.80 pesos.

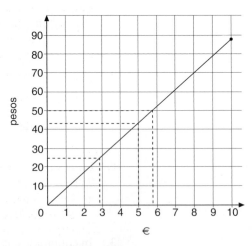

i) Using the graph estimate the number of pesos equivalent to €5.

A line is drawn up from €5 until it reaches the plotted line, then across to the *y*-axis.
From the graph it can be seen that €5 ≈ 44 pesos.
(≈ is the symbol for 'is approximately equal to')

ii) Using the graph, what would be the cost in euro of a drink costing 25 pesos?

A line is drawn across from 25 pesos until it reaches the plotted line, then down to the *x*-axis.
From the graph it can be seen that the cost of the drink ≈ €2.80.

iii) If a meal costs 200 pesos, use the graph to estimate its cost in euro.

The graph does not go up to 200 pesos, therefore a factor of 200 needs to be used e.g. 50 pesos.
From the graph 50 pesos ≈ €5.70, therefore it can be deduced that 200 pesos ≈ €22.80 (i.e. 4 × €5.70).

Exercise 5.1

1. Given that 80 km = 50 miles, draw a conversion graph up to 100 km. Using your graph estimate:
   a) how many miles is 50 km,
   b) how many kilometres is 80 miles,
   c) the speed in miles per hour (mph) equivalent to 100 km/h,
   d) the speed in km/h equivalent to 40 mph.

2. You can roughly convert temperature in degrees Celsius to degrees Fahrenheit by doubling the degrees Celsius and adding 30.
   Draw a conversion graph up to 50 °C. Use your graph to estimate the following:
   a) the temperature in °F equivalent to 25 °C,
   b) the temperature in °C equivalent to 100 °F,
   c) the temperature in °F equivalent to 0 °C,
   d) the temperature in °C equivalent to 200 °F.

3. Given that 0 °C = 32 °F and 50 °C = 122 °F, on the same graph as in Q.2, draw a true conversion graph.
   i) Use the true graph to calculate the conversions in Q.2.
   ii) Where would you say the rough conversion is most useful?

4. Long-distance calls from New York to Harare are priced at 85 cents/min off peak and $1.20/min at peak times.
   a) Draw, on the same axes, conversion graphs for the two different rates.
   b) From your graph estimate the cost of an 8 minute call made off peak.
   c) Estimate the cost of the same call made at peak rate.
   d) A call is to be made from a telephone box. If the caller has only $4 to spend, estimate how much more time he can talk for if he rings at off peak instead of at peak times.

5. A maths exam is marked out of 120. Draw a conversion graph to change the following marks to percentages.
   a) 80   b) 110   c) 54   d) 72

Speed, distance and time

Students need to be aware of the following formulae:

$$\text{distance} = \text{speed} \times \text{time}$$

Rearranging the formula gives:

$$\text{speed} = \frac{\text{distance}}{\text{time}}$$

Where the speed is not constant:

$$\text{average speed} = \frac{\text{total distance}}{\text{total time}}$$

Exercise 5.2

1. Find the average speed of an object moving:
   a) 30 m in 5 s   b) 48 m in 12 s
   c) 78 km in 2 h   d) 50 km in 2.5 h
   e) 400 km in 2 h 30 min   f) 110 km in 2 h 12 min

2. How far will an object travel during:
   a) 10 s at 40 m/s   b) 7 s at 26 m/s
   c) 3 hours at 70 km/h   d) 4 h 15 min at 60 km/h
   e) 10 min at 60 km/h   f) 1 h 6 min at 20 m/s?

3. How long will it take to travel:
   a) 50 m at 10 m/s   b) 1 km at 20 m/s
   c) 2 km at 30 km/h   d) 5 km at 70 m/s
   e) 200 cm at 0.4 m/s   f) 1 km at 15 km/h?

The graph of an object travelling at a constant speed is a straight line as shown (left).

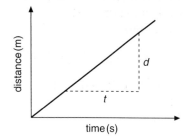

$$\text{Gradient} = \frac{d}{t}$$

The units of the gradient are m/s, hence the gradient of a distance–time graph represents the speed at which the object is travelling.

Worked example

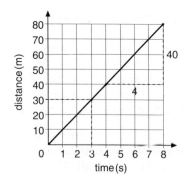

The graph (left) represents an object travelling at constant speed.

i) From the graph calculate how long it took to cover a distance of 30 m.
   The time taken to travel 30 m is 3 seconds.

ii) Calculate the gradient of the graph.
   Taking two points on the line, gradient $= \dfrac{40}{4} = 10$.

iii) Calculate the speed at which the object was travelling.
   Gradient of a distance–time graph = speed.
   Therefore the speed is 10 m/s.

Exercise 5.3

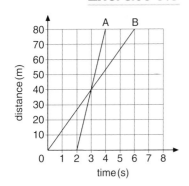

1. Draw a distance–time graph for the first 10 seconds of an object travelling at 6 m/s.

2. Draw a distance–time graph for the first 10 seconds of an object travelling at 5 m/s. Use your graph to estimate:
   a) the time taken to travel 25 m,
   b) how far the object travels in 3.5 seconds.

3. Two objects A and B set off from the same point and move in the same straight line. B sets off first, whilst A sets off 2 seconds later. Using the distance–time graph (left) estimate:
   a) the speed of each of the objects,
   b) how far apart the objects would be 20 seconds after the start.

4. Three objects A, B and C move in the same straight line away from a point X. Both A and C change their speed during the journey, whilst B travels at the same constant speed throughout. From the distance–time graph (left) estimate:
   a) the speed of object B,
   b) the two speeds of object A,
   c) the average speed of object C,
   d) how far object C is from X 3 seconds from the start,
   e) how far apart objects A and C are 4 seconds from the start.

Travel graphs

The graphs of two or more journeys can be shown on the same axes. The shape of the graph gives a clear picture of the movement of each of the objects.

Worked example   Car X and Car Y both reach point B 100 km from A at 11 a.m.

i) Calculate the speed of Car X between 7 a.m. and 8 a.m.

$$\text{speed} = \frac{\text{distance}}{\text{time}}$$

$$= \frac{60}{1} \text{ km/h}$$

$$= 60 \text{ km/h}$$

ii) Calculate the speed of Car Y between 9 a.m. and 11 a.m.

$$\text{speed} = \frac{100}{2} \text{ km/h}$$

$$= 50 \text{ km/h}$$

iii) Explain what is happening to Car X between 8 a.m. and 9 a.m.

No distance has been travelled, therefore Car X is stationary.

Exercise 5.4

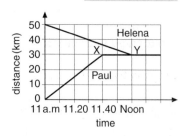

1. Two friends Paul and Helena arrange to meet for lunch at noon. They live 50 km apart and the restaurant is 30 km from Paul's home. The travel graph (left) illustrates their journeys.

   a) What is Paul's average speed between 11 a.m. and 11.40 a.m.?
   b) What is Helena's average speed between 11 a.m. and noon?
   c) What does the line XY represent?

2. A car travels at a speed of 60 km/h for 1 hour. It then stops for 30 minutes and then continues at a constant speed of 80 km/h for a further 1.5 hours. Draw a distance–time graph for this journey.

3. A girl cycles for 1.5 hours at 10 km/h. She then stops for an hour and then travels for a further 15 km in 1 hour. Draw a distance–time graph of the girl's journey.

4. Two friends leave their houses at 4 p.m. The houses are 4 km apart and the friends travel towards each other on the same road. Fyodor walks at 7 km/h and Yin at 5 km/h.
   a) On the same axes, draw a distance–time graph of their journeys.
   b) From your graph estimate the time at which they meet.
   c) Estimate the distance from Fyodor's house to the point where they meet.

5. A train leaves a station P at 6 p.m. and travels to station Q 150 km away. It travels at a steady speed of 75 km/h. At 6.10 p.m. another train leaves Q for P at a steady speed of 100 km/h.
   a) On the same axes draw a distance–time graph to show both journeys.
   b) From the graph estimate the time at which both trains pass each other.
   c) At what distance from station Q do both trains pass each other?
   d) Which train arrives at its destination first?

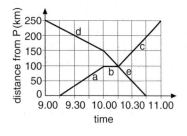

6. A train sets off from town P at 9.15 a.m. and heads towards town Q 250 km away. Its journey is split into the three stages a, b and c. At 9.00 a.m. a second train left town Q heading for town P. Its journey was split into the two stages d and e. Using the graph (left) calculate the following:
   a) the speed of the first train during stages a, b and c,
   b) the speed of the second train during stages d and e.

Core Section

Student Assessment I

1. 1 euro had an exchange rate of 8 Chinese yuan and 200 Hungarian forint.
   a) Draw a conversion graph for yuan to forint up to 80 yuan.
   b) Estimate from your graph how many forint you would get for 50 yuan.
   c) Estimate from your graph how many yuan you would get for 1600 forint.

2. A South African taxi driver has a fixed charge of 20 rand and then charges 6 rand per km.
   a) Draw a conversion graph to enable you to estimate the cost of the following taxi rides:
      i) 5 km
      ii) 8.5 km
   b) If a trip cost 80 rand, estimate from your graph the distance travelled.

3. An electricity account can work in two ways:
   ■ account A which involves a fixed charge of $5 and then a rate of 7c per unit,
   ■ account B which involves no fixed charge but a rate of 9.5c per unit.
   a) On the same axes draw a graph up to 400 units for each type of account, converting units used to cost.
   b) Use your graph to advise a customer on which account to use.

4. A car travels at 60 km/h for 1 hour. The driver then takes a 30 minute break. After her break, she continues at 80 km/h for 90 minutes.
   a) Draw a distance–time graph for her journey.
   b) Calculate the total distance travelled.

5. Two trains depart at the same time from cities M and N, which are 200 km apart. One train travels from M to N, the other from N to M. The train departing from M travels a distance of 60 km in the first hour, 120 km in the next 1.5 hours and then the rest of the journey at 40 km/h. The train departing from N travels the whole distance at a speed of 100 km/h. Assuming all speeds are constant:
   a) draw a travel graph to show both journeys,
   b) estimate how far from city M the trains are when they pass each other,
   c) estimate how long after the start of the journey it is when the trains pass each other.

Student Assessment 2

1. Absolute zero, 0 degrees Kelvin (K), is equivalent to −273 °C. 0 °C is equivalent to 273 K. Draw a conversion graph which will convert K into °C. Use your graph to estimate the following:
   a) the temperature in K equivalent to −40 °C,
   b) the temperature in °C equivalent to 100 K.

2. A Canadian plumber has a call-out charge of 70 Canadian dollars and then charges a rate of $50 per hour.
   a) Draw a conversion graph and estimate the cost of the following:
      i) a job lasting $4\frac{1}{2}$ hours,
      ii) a job lasting $6\frac{3}{4}$ hours.
   b) If a job cost $245, estimate from your graph how long it took to complete.

3. A boy lives 3.5 km from his school. He walks home at a constant speed of 9 km/h for the first 10 minutes. He then stops and talks to his friends for 5 minutes. He finally runs the rest of his journey home at a constant speed of 12 km/h.
   a) Illustrate this information on a distance–time graph.
   b) Use your graph to estimate the total time it took the boy to get home that day.

4. Below are four distance–time graphs A, B, C and D. Two of them are not possible.
   a) Which two graphs are impossible?
   b) Explain why the two you have chosen are not possible.

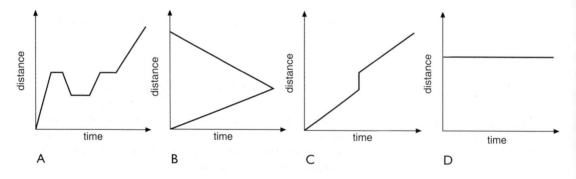

A   B   C   D

GRAPHS IN PRACTICAL SITUATIONS

Extended Section

Apply the idea of rate of change to easy kinematics involving speed–time graphs, acceleration and deceleration; calculate distance travelled as area under a linear speed–time graph.

NB: All diagrams are not drawn to scale.

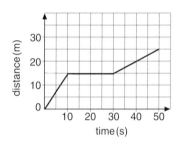

▨ Speed–time graphs, acceleration and deceleration

So far the graphs that have been dealt with have been similar to the one shown (left) i.e. distance–time graphs.

If the graph were of a girl walking it would indicate that initially she was walking at a constant speed of 1.5 m/s for 10 seconds, then she stopped for 20 seconds and finally she walked at a constant speed of 0.5 m/s for 20 seconds.

The student should be aware at this stage that for a distance–time graph the following is true:

■ a straight line represents constant speed,
■ a horizontal line indicates no movement,
■ the gradient of a line gives the speed.

This section also deals with the interpretation of travel graphs, but where the *y*-axis represents the object's speed.

Worked example

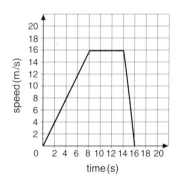

The graph (left) shows the speed of a car over a period of 16 seconds.

i) Explain the shape of the graph.

For the first 8 seconds the speed of the car is increasing uniformly with time. This means it is **accelerating** at a constant rate. Between 8 and 14 seconds the car is travelling at a constant speed of 16 m/s. Between 14 and 16 seconds the speed of the car decreases uniformly. This means that it is **decelerating** at a constant rate.

ii) Calculate the rate of acceleration during the first 8 seconds.

From a speed–time graph, the acceleration is found by calculating the gradient of the line. Therefore:

$$\text{acceleration} = \frac{16}{8} = 2 \text{ m/s}^2$$

c) Calculate the rate of deceleration between 14 and 16 seconds:

$$\text{deceleration} = \frac{16}{2} = 8 \text{ m/s}^2$$

Exercise 5.5 Using the graphs below, calculate the acceleration/deceleration in each case.

1.
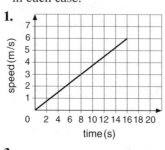

2.
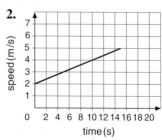

3.
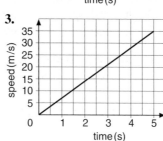

4.
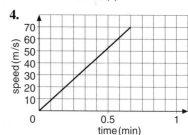

5.
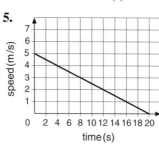

6.
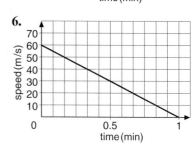

7. Sketch a graph to show an aeroplane accelerating from rest at a constant rate of 5 m/s² for 10 seconds.

8. A train travelling at 30 m/s starts to decelerate at a constant rate of 3 m/s². Sketch a speed–time graph showing the train's motion until it stops.

Exercise 5.6

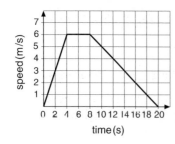

1. The graph (left) shows the speed–time graph of a boy running for 20 seconds. Calculate:
a) the acceleration during the first four seconds,
b) the acceleration during the second period of four seconds,
c) the deceleration during the final twelve seconds.

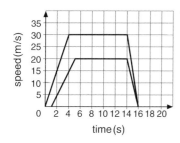

2. The speed–time graph (left) represents a cheetah chasing a gazelle.

    a) Does the top graph represent the cheetah or the gazelle?

    b) Calculate the cheetah's acceleration in the initial stages of the chase.

    c) Calculate the gazelle's acceleration in the initial stages of the chase.

    d) Calculate the cheetah's deceleration at the end.

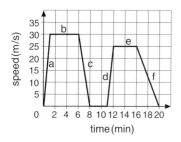

3. The speed–time graph (left) represents a train travelling from one station to another.

    a) Calculate the acceleration during stage a.

    b) Calculate the deceleration during stage c.

    c) Calculate the deceleration during stage f.

    d) Describe the train's motion during stage b.

    e) Describe the train's motion 10 minutes from the start.

Area under a speed–time graph

The area under a speed–time graph gives the distance travelled.

Worked example   The table below shows the speed of a train over a 30 second period.

Time (s)	0	5	10	15	20	25	30
Speed (m/s)	20	20	20	22.5	25	27.5	30

i) Plot a speed–time graph for the first 30 seconds.

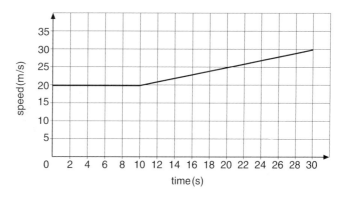

ii) Calculate the train's acceleration after the first 10 seconds.

$$\text{Acceleration} = \frac{10}{20} = \frac{1}{2}\ \text{m/s}^2$$

iii) Calculate the distance travelled during the 30 seconds.

    This is calculated by working out the area under the graph. The graph can be split into two regions as shown below.

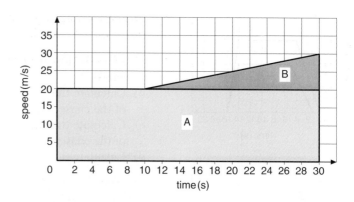

Distance represented by region A = (20 × 30) m
                                 = 600 m
Distance represented by region B = ($\frac{1}{2}$ × 20 × 10) m
                                 = 100 m
Total distance travelled = (600 + 100) m
                         = 700 m

Exercise 5.7

1. The table below gives the speed of a boat over a 10 second period.

Time (s)	0	2	4	6	8	10
Speed (m/s)	5	6	7	8	9	10

a) Plot a speed–time graph for the 10 second period.
b) Calculate the acceleration of the boat.
c) Calculate the total distance travelled during the 10 seconds.

2. A cyclist travelling at 6 m/s applies the brakes and decelerates at a constant rate of 2 m/s².
a) Copy and complete the table below.

Time (s)	0	0.5	1	1.5	2	2.5	3
Speed (m/s)	6						0

b) Plot a speed–time graph for the 3 seconds shown in the table above.
c) Calculate the distance travelled during the 3 seconds of deceleration.

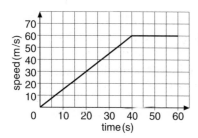

3. A car accelerates as shown in the graph (left).

a) Calculate the rate of acceleration in the first 40 seconds.
b) Calculate the distance travelled over the 60 seconds shown.
c) After what time had the motorist travelled half the distance?

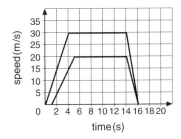

time (s)

4. The graph (left) represents the cheetah and gazelle chase from Q.2 in Exercise 5.6.

   a) Calculate the distance run by the cheetah during the chase.
   b) Calculate the distance run by the gazelle during the chase.

5. The graph (right) represents the train journey from Q.3 in Exercise 5.6. Calculate, in km, the distance travelled during the 20 minutes shown.

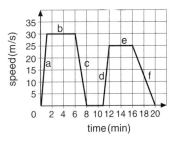

6. An aircraft accelerates uniformly from rest at a rate of 10 m/s² for 12 seconds before it takes off. Calculate the distance it travels along the runway.

7. The speed–time graph below depicts the motion of two motorbikes A and B over a 15 second period.

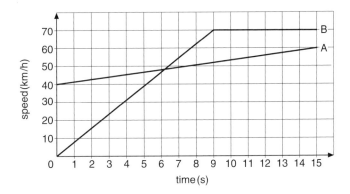

time (s)

At the start of the graph motorbike A overtakes a stationary motorbike B. Assume they then travel in the same direction.

   a) Calculate motorbike A's acceleration over the 15 seconds in m/s².
   b) Calculate motorbike B's acceleration over the first 9 seconds in m/s².
   c) Calculate the distance travelled by A during the 15 seconds (give your answer to the nearest metre).
   d) Calculate the distance travelled by B during the 15 seconds (give your answer to the nearest metre).
   e) How far apart were the two motorbikes at the end of the 15 second period?

Extended Section

Student Assessment 1

1. The graph below is a speed–time graph for a car accelerating from rest.

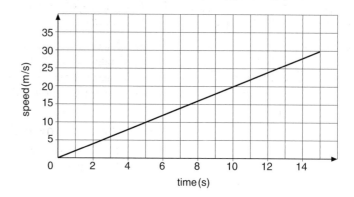

a) Calculate the car's acceleration in m/s^2.
b) Calculate, in metres, the distance the car travels in 15 seconds.
c) How long did it take the car to travel half the distance?

2. The speed–time graph below represents a 100 m sprinter during a race.

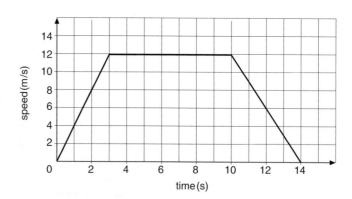

a) Calculate the sprinter's acceleration during the first two seconds of the race.
b) Calculate the sprinter's deceleration at the end of the race.
c) Calculate the distance the sprinter ran in the first 10 seconds.
d) Calculate the sprinter's time for the 100 m race. Give your answer to 2 d.p.

3. A motorcyclist accelerates uniformly from rest to 50 km/h in 8 seconds. He then accelerates to 110 km/h in a further 6 seconds.
   a) Draw a speed–time graph for the first 14 seconds.
   b) Use your graph to find the total distance the motorcyclist travels. Give your answer in metres.

4. The graph (left) shows the speed of a car over a period of 50 seconds.

   a) Calculate the car's acceleration in the first 15 seconds.
   b) Calculate the distance travelled whilst the car moved at constant speed.
   c) Calculate the total distance travelled.

Student Assessment 2

1. The graph below is a speed–time graph for a car decelerating to rest.

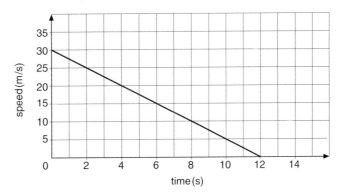

   a) Calculate the car's deceleration in m/s^2.
   b) Calculate, in metres, the distance the car travels in 12 seconds.
   c) How long did it take the car to travel half the distance?

2. The graph below shows the speeds of two cars A and B over a 15 second period.

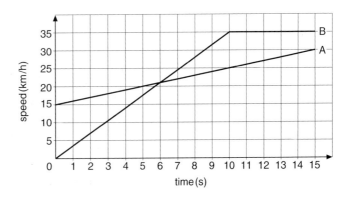

a) Calculate the acceleration of car A in m/s².
b) Calculate the distance travelled in metres during the 15 seconds by car A.
c) Calculate the distance travelled in metres during the 15 seconds by car B.

3. A motor cycle accelerates uniformly from rest to 30 km/h in 3 seconds. It then accelerates to 150 km/h in a further 6 seconds.
   a) Draw a speed–time graph for the first 9 seconds.
   b) Use your graph to find the total distance the motor cycle travels. Give your answer in metres.

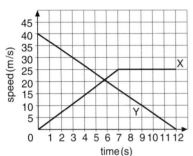

4. Two cars X and Y are travelling in the same direction. The speed–time graph (left) shows their speeds over 12 seconds.

   a) Calculate the deceleration of Y during the 12 seconds.
   b) Calculate the distance travelled by Y in the 12 seconds.
   c) Calculate the total distance travelled by X in the 12 seconds.

ICT Section

The following activity refers to the graphing package Autograph; however, a similar package may be used.

The velocity of a pupil at different parts of a 100 m sprint will be analysed.

A racecourse is set out as shown below:

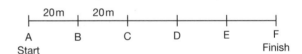

1. Choose a pupil to start at A. He/she is to run the 100 m and is timed as he/she runs past each of the points B – F by other pupils with stop watches at those points.

2. In Autograph, plot a distance–time graph of the results by entering the data as pairs of coordinates. i.e. (time, distance)

3. Ensuring that all the points are selected, draw a curve of best fit through them.

4. Select the curve and plot a coordinate of your choice on it. This point can now be moved along the curve using the cursor keys on the keyboard.

5. Draw a tangent to the curve through the point.

6. What does the gradient of the tangent represent?

7. At what point of the race was the pupil running fastest? How did you reach this answer?

8. Collect similar data for other pupils. Compare their graphs and running speeds.

9. Analyse carefully one of the graphs and write a brief report to the runner identifying, giving reasons, the parts of the race he/she needs to improve on.

6 ALGEBRAIC REPRESENTATION AND MANIPULATION

Core Section

Use letters to express generalised numbers and express basic arithmetic processes algebraically, substitute numbers for words and letters in formulae; transform simple formulae; manipulate directed numbers; use brackets and extract common factors.

▓ Expanding brackets

When removing brackets, every term inside the bracket must be multiplied by whatever is outside the bracket.

Worked examples

a) $3(x + 4)$
$= 3x + 12$

b) $5x(2y + 3)$
$= 10xy + 15x$

c) $2a(3a + 2b - 3c)$
$= 6a^2 + 4ab - 6ac$

d) $-4p(2p - q + r^2)$
$= -8p^2 + 4pq - 4pr^2$

e) $-2x^2\left(x + 3y - \dfrac{1}{x}\right)$

$= -2x^3 - 6x^2y + 2x$

f) $\dfrac{-2}{x}\left(-x + 4y + \dfrac{1}{x}\right)$

$= 2 - \dfrac{8y}{x} - \dfrac{2}{x^2}$

Exercise 6.1 Expand the following:

1. a) $4(x - 3)$
    c) $-6(7x - 4y)$
    e) $-7(2m - 3n)$
    b) $5(2p - 4)$
    d) $3(2a - 3b - 4c)$
    f) $-2(8x - 3y)$

2. a) $3x(x - 3y)$
    c) $4m(2m - n)$
    e) $-4x(-x + y)$
    b) $a(a + b + c)$
    d) $-5a(3a - 4b)$
    f) $-8p(-3p + q)$

3. a) $-(2x^2 - 3y^2)$
    c) $-(-7p + 2q)$
    e) $\frac{3}{4}(4x - 2y)$
    b) $-(-a + b)$
    d) $\frac{1}{2}(6x - 8y + 4z)$
    f) $\frac{1}{5}x(10x - 15y)$

4. a) $3r(4r^2 - 5s + 2t)$.
    c) $3a^2(2a - 3b)$
    e) $m^2(m - n + nm)$
    b) $a^2(a + b + c)$
    d) $pq(p + q - pq)$
    f) $a^3(a^3 + a^2b)$

Exercise 6.2 Expand and simplify the following:

1. a) $3a - 2(2a + 4)$
    c) $3(p - 4) - 4$
    e) $6x - 3(2x - 1)$
    b) $8x - 4(x + 5)$
    d) $7(3m - 2n) + 8n$
    f) $5p - 3p(p + 2)$

2. a) $7m(m + 4) + m^2 + 2$
    c) $6(p + 3) - 4(p - 1)$
    e) $3a(a + 2) - 2(a^2 - 1)$
    b) $3(x - 4) + 2(4 - x)$
    d) $5(m - 8) - 4(m - 7)$
    f) $7a(b - 2c) - c(2a - 3)$

3. a) $\frac{1}{2}(6x + 4) + \frac{1}{3}(3x + 6)$
  b) $\frac{1}{4}(2x + 6y) + \frac{3}{4}(6x - 4y)$
  c) $\frac{1}{8}(6x - 12y) + \frac{1}{2}(3x - 2y)$
  d) $\frac{1}{5}(15x + 10y) + \frac{3}{10}(5x - 5y)$
  e) $\frac{2}{3}(6x - 9y) + \frac{1}{3}(9x + 6y)$
  f) $\frac{x}{7}(14x - 21y) - \frac{x}{2}(4x - 6y)$

Simple factorising

When factorising, the largest possible factor is removed from each of the terms and placed outside the brackets.

Worked examples Factorise the following expressions:

a) $10x + 15$

   $= 5(2x + 3)$

b) $8p - 6q + 10r$

   $= 2(4p - 3q + 5r)$

c) $-2q - 6p + 12$

   $= 2(-q - 3p + 6)$

d) $2a^2 + 3ab - 5ac$

   $= a(2a + 3b - 5c)$

e) $6ax - 12ay - 18a^2$
   $= 6a(x - 2y - 3a)$

f) $3b + 9ba - 6bd$
   $= 3b(1 + 3a - 2d)$

Exercise 6.3

Factorise the following:

1. a) $4x - 6$
   c) $6y - 3$
   e) $3p - 3q$
   b) $18 - 12p$
   d) $4a + 6b$
   f) $8m + 12n + 16r$

2. a) $3ab + 4ac - 5ad$
   c) $a^2 - ab$
   e) $abc + abd + fab$
   b) $8pq + 6pr - 4ps$
   d) $4x^2 - 6xy$
   f) $3m^2 + 9m$

3. a) $3pqr - 9pqs$
   c) $8x^2y - 4xy^2$
   e) $12p - 36$
   b) $5m^2 - 10mn$
   d) $2a^2b^2 - 3b^2c^2$
   f) $42x - 54$

4. a) $18 + 12y$
   c) $11x + 11xy$
   e) $5pq - 10qr + 15qs$
   b) $14a - 21b$
   d) $4s - 16t + 20r$
   f) $4xy + 8y^2$

5. a) $m^2 + mn$
   c) $pqr + qrs$
   e) $3p^3 - 4p^4$
   b) $3p^2 - 6pq$
   d) $ab + a^2b + ab^2$
   f) $7b^3c + b^2c^2$

6. a) $m^3 - m^2n + mn^2$
   c) $56x^2y - 28xy^2$
   b) $4r^3 - 6r^2 + 8r^2s$
   d) $72m^2n + 36mn^2 - 18m^2n^2$

Substitution

Worked examples Evaluate the expressions below if $a = 3, b = 4, c = -5$:

a) $2a + 3b - c$
   $= 6 + 12 + 5$
   $= 23$

b) $3a - 4b + 2c$
   $= 9 - 16 - 10$
   $= -17$

c) $\quad -2a + b - 3c$
$\quad = -6 + 8 + 15$
$\quad = 17$

d) $\quad a^2 + b^2 + c^2$
$\quad = 9 + 16 + 25$
$\quad = 50$

e) $\quad 3a(2b - 3c)$
$\quad = 9(8 + 15)$
$\quad = 9 \times 23$
$\quad = 207$

f) $\quad -2c(-a + 2b)$
$\quad = 10(-3 + 8)$
$\quad = 10 \times 5$
$\quad = 50$

Exercise 6.4

Evaluate the following expressions if $p = 4$, $q = -2$, $r = 3$ and $s = -5$:

1. a) $2p + 4q$
   c) $3q - 4s$
   e) $3r - 3p + 5q$
   b) $5r - 3s$
   d) $6p - 8q + 4s$
   f) $-p - q + r + s$

2. a) $2p - 3q - 4r + s$
   c) $p^2 + q^2$
   e) $p(q - r + s)$
   b) $3s - 4p + r + q$
   d) $r^2 - s^2$
   f) $r(2p - 3q)$

3. a) $2s(3p - 2q)$
   c) $2pr - 3rq$
   e) $s^3 - p^3$
   b) $pq + rs$
   d) $q^3 - r^2$
   f) $r^4 - q^5$

4. a) $-2pqr$
   c) $-2rq + r$
   e) $(p + s)(r - q)$
   b) $-2p(q + r)$
   d) $(p + q)(r - s)$
   f) $(r + q)(p - s)$

5. a) $(2p + 3q)(p - q)$
   c) $q^2 - r^2$
   e) $(p + r)(p - r)$
   b) $(q + r)(q - r)$
   d) $p^2 - r^2$
   f) $(-s + p)q^2$

Transformation of formulae

In the formula $a = 2b + c$, 'a' is the subject. In order to make either b or c the subject, the formula has to be rearranged.

Worked examples

Rearrange the following formulae to make the **bold** letter the subject:

a) $\quad a = 2b + \mathbf{c}$

$\quad a - 2b = \mathbf{c}$

c) $\quad ab = \mathbf{c}d$

$\quad \dfrac{ab}{d} = \mathbf{c}$

b) $\quad 2r + \mathbf{p} = q$

$\quad \mathbf{p} = q - 2r$

d) $\quad \dfrac{a}{b} = \dfrac{c}{\mathbf{d}}$

$\quad a\mathbf{d} = cb$

$\quad \mathbf{d} = \dfrac{cb}{a}$

Exercise 6.5 In the following questions, make the letter in **bold** the subject of the formula:

1. a) $m + n = r$ b) $m + n = p$ c) $2m + n = 3p$
    d) $3x = 2p + q$ e) $ab = cd$ f) $ab = cd$

2. a) $3xy = 4m$ b) $7pq = 5r$ c) $3x = c$
    d) $3x + 7 = y$ e) $5y - 9 = 3r$ f) $5y - 9 = 3x$

3. a) $6b = 2a - 5$ b) $6b = 2a - 5$ c) $3x - 7y = 4z$
    d) $3x - 7y = 4z$ e) $3x - 7y = 4z$ f) $2pr - q = 8$

4. a) $\dfrac{p}{4} = r$ b) $\dfrac{4}{p} = 3r$ c) $\dfrac{1}{5}n = 2p$

    d) $\dfrac{1}{5}n = 2p$ e) $p(q + r) = 2t$ f) $p(q + r) = 2t$

5. a) $3m - n = rt(p + q)$ b) $3m - n = rt(p + q)$

    c) $3m - n = rt(p + q)$ d) $3m - n = rt(p + q)$

    e) $3m - n = rt(p + q)$ f) $3m - n = rt(p + q)$

6. a) $\dfrac{ab}{c} = de$ b) $\dfrac{ab}{c} = de$ c) $\dfrac{ab}{c} = de$

    d) $\dfrac{a + b}{c} = d$ e) $\dfrac{a}{c} + b = d$ f) $\dfrac{a}{c} + b = d$

Core Section

Student Assessment 1

1. Expand the following and simplify where possible:
   a) $5(2a - 6b + 3c)$ b) $3x(5x - 9)$
   c) $-5y(3xy + y^2)$ d) $3x^2(5xy + 3y^2 - x^3)$
   e) $5p - 3(2p - 4)$
   f) $4m(2m - 3) + 2(3m^2 - m)$
   g) $\frac{1}{3}(6x - 9) + \frac{1}{4}(8x + 24)$
   h) $\dfrac{m}{4}(6m - 8) + \dfrac{m}{2}(10m - 2)$

2. Factorise the following:
   a) $12a - 4b$ b) $x^2 - 4xy$
   c) $8p^3 - 4p^2q$ d) $24xy - 16x^2y + 8xy^2$

3. If $x = 2$, $y = -3$ and $z = 4$, evaluate the following:
   a) $2x + 3y - 4z$ b) $10x + 2y^2 - 3z$
   c) $z^2 - y^3$ d) $(x + y)(y - z)$
   e) $z^2 - x^2$ f) $(z + x)(z - x)$

4. Rearrange the following formulae to make the **bold** letter the subject:
   a) $x = 3p + \boldsymbol{q}$ b) $3m - 5\boldsymbol{n} = 8r$ c) $2m = \dfrac{3\boldsymbol{y}}{t}$
   d) $x(\boldsymbol{w} + y) = 2y$ e) $\dfrac{xy}{2\boldsymbol{p}} = \dfrac{rs}{t}$ f) $\dfrac{\boldsymbol{x} + y}{w} = m + n$

Student Assessment 2

1. Expand the following and simplify where possible:
   a) $3(2x - 3y + 5z)$ b) $4p(2m - 7)$
   c) $-4m(2mn - n^2)$ d) $4p^2(5pq - 2q^2 - 2p)$
   e) $4x - 2(3x + 1)$ f) $4x(3x - 2) + 2(5x^2 - 3x)$
   g) $\frac{1}{5}(15x - 10) - \frac{1}{3}(9x - 12)$ h) $\dfrac{x}{2}(4x - 6) + \dfrac{x}{4}(2x + 8)$

2. Factorise the following:
   a) $16p - 8q$ b) $p^2 - 6pq$
   c) $5p^2q - 10pq^2$ d) $9pq - 6p^2q + 12q^2p$

3. If $a = 4$, $b = 3$ and $c = -2$, evaluate the following:
   a) $3a - 2b + 3c$ b) $5a - 3b^2$
   c) $a^2 + b^2 + c^2$ d) $(a + b)(a - b)$
   e) $a^2 - b^2$ f) $b^3 - c^3$

4. Rearrange the following formulae to make the **bold** letter the subject:
   a) $p = 4m + \boldsymbol{n}$ b) $4x - 3\boldsymbol{y} = 5z$
   c) $2x = \dfrac{3\boldsymbol{y}}{5p}$ d) $m(x + \boldsymbol{y}) = 3w$
   e) $\dfrac{pq}{4\boldsymbol{r}} = \dfrac{mn}{t}$ f) $\dfrac{p + \boldsymbol{q}}{r} = m - n$

ALGEBRAIC REPRESENTATION AND MANIPULATION

Extended Section

Expand products of algebraic expressions; factorise and simplify expressions; manipulate algebraic fractions; transform more complicated formulae.

Further expansion

When multiplying together expressions in brackets, it is necessary to multiply all the terms in one bracket by all the terms in the other bracket.

Worked examples Expand the following:

a) $(x + 3)(x + 5)$ 　　　　　　　**b)** $(x + 2)(x + 1)$

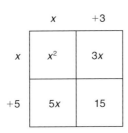

$$= x^2 + 5x + 3x + 15$$
$$= x^2 + 8x + 15$$

$$= x^2 + x + 2x + 2$$
$$= x^2 + 3x + 2$$

Exercise 6.6 Expand the following and simplify your answer:

1. a) $(x + 2)(x + 3)$ 　　　　b) $(x + 3)(x + 4)$
    c) $(x + 5)(x + 2)$ 　　　　d) $(x + 6)(x + 1)$
    e) $(x - 2)(x + 3)$ 　　　　f) $(x + 8)(x - 3)$

2. a) $(x - 4)(x + 6)$ 　　　　b) $(x - 7)(x + 4)$
    c) $(x + 5)(x - 7)$ 　　　　d) $(x + 3)(x - 5)$
    e) $(x + 1)(x - 3)$ 　　　　f) $(x - 7)(x + 9)$

3. a) $(x - 2)(x - 3)$ 　　　　b) $(x - 5)(x - 2)$
    c) $(x - 4)(x - 8)$ 　　　　d) $(x + 3)(x + 3)$
    e) $(x - 3)(x - 3)$ 　　　　f) $(x - 7)(x - 5)$

4. a) $(x + 3)(x - 3)$ 　　　　b) $(x + 7)(x - 7)$
    c) $(x - 8)(x + 8)$ 　　　　d) $(x + y)(x - y)$
    e) $(a + b)(a - b)$ 　　　　f) $(p - q)(p + q)$

5. a) $(y + 2)(2y + 3)$ 　　　b) $(y + 7)(3y + 4)$
    c) $(2y + 1)(y + 8)$ 　　　d) $(2y + 1)(2y + 2)$
    e) $(3y + 4)(2y + 5)$ 　　　f) $(6y + 3)(3y + 1)$

6. a) $(2p - 3)(p + 8)$ b) $(4p - 5)(p + 7)$
     c) $(3p - 4)(2p + 3)$ d) $(4p - 5)(3p + 7)$
     e) $(6p + 2)(3p - 1)$ f) $(7p - 3)(4p + 8)$

7. a) $(2x - 1)(2x - 1)$ b) $(3x + 1)^2$
     c) $(4x - 2)^2$ d) $(5x - 4)^2$
     e) $(2x + 6)^2$ f) $(2x + 3)(2x - 3)$

8. a) $(3 + 2x)(3 - 2x)$ b) $(4x - 3)(4x + 3)$
     c) $(3 + 4x)(3 - 4x)$ d) $(7 - 5y)(7 + 5y)$
     e) $(3 + 2y)(4y - 6)$ f) $(7 - 5y)^2$

Further factorisation

Factorisation by grouping

Worked examples   Factorise the following expressions:

a) $6x + 3 + 2xy + y$
$= 3(2x + 1) + y(2x + 1)$
$= (3 + y)(2x + 1)$
Note that $(2x + 1)$ was a common factor of both terms.

b) $ax + ay - bx - by$
$= a(x + y) - b(x + y)$
$= (a - b)(x + y)$

c) $2x^2 - 3x + 2xy - 3y$
$= x(2x - 3) + y(2x - 3)$
$= (x + y)(2x - 3)$

Exercise 6.7   Factorise the following by grouping:

1. a) $ax + bx + ay + by$ b) $ax + bx - ay - by$
     c) $3m + 3n + mx + nx$ d) $4m + mx + 4n + nx$
     e) $3m + mx - 3n - nx$ f) $6x + xy + 6z + zy$

2. a) $pr - ps + qr - qs$ b) $pq - 4p + 3q - 12$
     c) $pq + 3q - 4p - 12$ d) $rs + rt + 2ts + 2t^2$
     e) $rs - 2ts + rt - 2t^2$ f) $ab - 4cb + ac - 4c^2$

3. a) $xy + 4y + x^2 + 4x$ b) $x^2 - xy - 2x + 2y$
     c) $ab + 3a - 7b - 21$ d) $ab - b - a + 1$
     e) $pq - 4p - 4q + 16$ f) $mn - 5m - 5n + 25$

4. a) $mn - 2m - 3n + 6$ b) $mn - 2mr - 3rn - 6r^2$
     c) $pr - 4p - 4qr + 16q$ d) $ab - a - bc + c$
     e) $x^2 - 2xz - 2xy + 4yz$ f) $2a^2 + 2ab + b^2 + ab$

Difference of two squares

On expanding $(x + y)(x - y)$
$$= x^2 - xy + xy - y^2$$
$$= x^2 - y^2$$
The reverse is that $x^2 - y^2$ factorises to $(x + y)(x - y)$. x^2 and y^2 are both square and therefore $x^2 - y^2$ is known as the **difference of two squares**.

Worked examples **a)** $p^2 - q^2$
$= (p + q)(p - q)$

b) $4a^2 - 9b^2$
$= (2a)^2 - (3b)^2$
$= (2a + 3b)(2a - 3b)$

c) $(mn)^2 - 25k^2$
$= (mn)^2 - (5k)^2$
$= (mn + 5k)(mn - 5k)$

d) $4x^2 - (9y)^2$
$= (2x)^2 - (9y)^2$
$= (2x + 9y)(2x - 9y)$

Exercise 6.8

Factorise the following:

1. a) $a^2 - b^2$ b) $m^2 - n^2$ c) $x^2 - 25$
    d) $m^2 - 49$ e) $81 - x^2$ f) $100 - y^2$

2. a) $144 - y^2$ b) $q^2 - 169$ c) $m^2 - 1$
    d) $1 - t^2$ e) $4x^2 - y^2$ f) $25p^2 - 64q^2$

3. a) $9x^2 - 4y^2$ b) $16p^2 - 36q^2$ c) $64x^2 - y^2$
    d) $x^2 - 100y^2$ e) $(pq)^2 - 4p^2$ f) $(ab)^2 - (cd)^2$

4. a) $m^2n^2 - 9y^2$ b) $\frac{1}{4}x^2 - \frac{1}{9}y^2$ c) $p^4 - q^4$
    d) $4m^4 - 36y^4$ e) $16x^4 - 81y^4$ f) $(2x)^2 - (3y)^4$

Evaluation

Once factorised, numerical expressions can be evaluated.

Worked examples Evaluate the following expressions:

a) $13^2 - 7^2$
$= (13 + 7)(13 - 7)$
$= 20 \times 6$
$= 120$

b) $6.25^2 - 3.75^2$
$= (6.25 + 3.75)(6.25 - 3.75)$
$= 10 \times 2.5$
$= 25$

Exercise 6.9

By factorising, evaluate the following:

1. a) $8^2 - 2^2$ b) $16^2 - 4^2$ c) $49^2 - 1$
    d) $17^2 - 3^2$ e) $88^2 - 12^2$ f) $96^2 - 4^2$

2. a) $45^2 - 25$ b) $99^2 - 1$ c) $27^2 - 23^2$
    d) $66^2 - 34^2$ e) $999^2 - 1$ f) $225 - 8^2$

3. a) $8.4^2 - 1.6^2$ b) $9.3^2 - 0.7^2$ c) $42.8^2 - 7.2^2$
    d) $(8\frac{1}{2})^2 - (1\frac{1}{2})^2$ e) $(7\frac{3}{4})^2 - (2\frac{1}{4})^2$ f) $5.25^2 - 4.75^2$

4. a) $8.62^2 - 1.38^2$ b) $0.9^2 - 0.1^2$ c) $3^4 - 2^4$
    d) $2^4 - 1$ e) $1111^2 - 111^2$ f) $2^8 - 25$

Factorising quadratic expressions

$x^2 + 5x + 6$ is known as a quadratic expression as the highest power of any of its terms is squared – in this case x^2.
    It can be factorised by writing it as a product of two brackets.

Worked examples **a)** Factorise $x^2 + 5x + 6$.

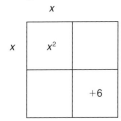

On setting up a 2×2 grid, some of the information can immediately be entered.

As there is only one term in x^2, this can be entered, as can the constant $+6$. The only two values which multiply to give x^2 are x and x. These too can be entered.

We now need to find two values which multiply to give $+6$ and which add to give $+5x$.

The only two values which satisfy both these conditions are $+3$ and $+2$.

Therefore $x^2 + 5x + 6 = (x + 3)(x + 2)$

b) Factorise $x^2 + 2x - 24$.

Therefore $x^2 + 2x - 24 = (x + 6)(x - 4)$

c) Factorise $2x^2 + 11x + 12$.

Therefore $2x^2 + 11x + 12 = (2x + 3)(x + 4)$

d) Factorise $3x^2 + 7x - 6$.

Therefore $3x^2 + 7x - 6 = (3x - 2)(x + 3)$

Exercise 6.10   Factorise the following quadratic expressions:

1. a) $x^2 + 7x + 12$ b) $x^2 + 8x + 12$ c) $x^2 + 13x + 12$
    d) $x^2 - 7x + 12$ e) $x^2 - 8x + 12$ f) $x^2 - 13x + 12$

2. a) $x^2 + 6x + 5$ b) $x^2 + 6x + 8$ c) $x^2 + 6x + 9$
    d) $x^2 + 10x + 25$ e) $x^2 + 22x + 121$ f) $x^2 - 13x + 42$

3. a) $x^2 + 14x + 24$ b) $x^2 + 11x + 24$ c) $x^2 - 10x + 24$
    d) $x^2 + 15x + 36$ e) $x^2 + 20x + 36$ f) $x^2 - 12x + 36$

4. a) $x^2 + 2x - 15$ b) $x^2 - 2x - 15$ c) $x^2 + x - 12$
    d) $x^2 - x - 12$ e) $x^2 + 4x - 12$ f) $x^2 - 15x + 36$

5. a) $x^2 - 2x - 8$ b) $x^2 - x - 20$ c) $x^2 + x - 30$
    d) $x^2 - x - 42$ e) $x^2 - 2x - 63$ f) $x^2 + 3x - 54$

6. a) $2x^2 + 4x + 2$ b) $2x^2 + 7x + 6$ c) $2x^2 + x - 6$
    d) $2x^2 - 7x + 6$ e) $3x^2 + 8x + 4$ f) $3x^2 + 11x - 4$
    g) $4x^2 + 12x + 9$ h) $9x^2 - 6x + 1$ i) $6x^2 - x - 1$

▨ Transformation of complex formulae

Worked examples   Make the letters in **bold** the subject of each formula:

a) $C = 2\pi \boldsymbol{r}$

$\dfrac{C}{2\pi} = \boldsymbol{r}$

b) $A = \pi \boldsymbol{r}^2$

$\dfrac{A}{\pi} = \boldsymbol{r}^2$

$\sqrt{\dfrac{A}{\pi}} = \boldsymbol{r}$

c) $R\boldsymbol{x}^2 = p$

$\boldsymbol{x}^2 = \dfrac{p}{R}$

$\boldsymbol{x} = \sqrt{\dfrac{p}{R}}$

d) $x^2 + \boldsymbol{y}^2 = h^2$

$\boldsymbol{y}^2 = h^2 - x^2$

$\boldsymbol{y} = \sqrt{h^2 - x^2}$

> Note:
> not $y = h - x$

e) $\sqrt{\boldsymbol{x}} = tv$

$\boldsymbol{x} = t^2v^2$

or $\boldsymbol{x} = (tv)^2$

f) $f = \sqrt{\dfrac{\boldsymbol{x}}{k}}$

$f^2 = \dfrac{\boldsymbol{x}}{k}$

$f^2k = \boldsymbol{x}$

g) $\quad m = 3a\sqrt{\dfrac{p}{x}}$ $\qquad$ **h)** $\qquad A = \dfrac{y + x}{p + q^2}$

Square both sides

$m^2 = \dfrac{9a^2p}{x}$ $\qquad A(p + q^2) = y + x$

$m^2x = 9a^2p$ $\qquad p + q^2 = \dfrac{y + x}{A}$

$x = \dfrac{9a^2p}{m^2}$ $\qquad q^2 = \dfrac{y + x}{A} - p$

$\qquad q = \sqrt{\dfrac{y + x}{A} - p}$

Exercise 6.11 In the formulae below, make x the subject:

1. a) $P = 2mx$ $\qquad$ b) $T = 3x^2$
   c) $mx^2 = y^2$ $\qquad$ d) $x^2 + y^2 = p^2 - q^2$
   e) $m^2 + x^2 = y^2 - n^2$ $\qquad$ f) $p^2 - q^2 = 4x^2 - y^2$

2. a) $\dfrac{P}{Q} = rx$ $\qquad$ b) $\dfrac{P}{Q} = rx^2$

   c) $\dfrac{P}{Q} = \dfrac{x^2}{r}$ $\qquad$ d) $\dfrac{m}{n} = \dfrac{1}{x^2}$

   e) $\dfrac{r}{st} = \dfrac{w}{x^2}$ $\qquad$ f) $\dfrac{p + q}{r} = \dfrac{w}{x^2}$

3. a) $\sqrt{x} = rp$ $\qquad$ b) $\dfrac{mn}{p} = \sqrt{x}$

   c) $g = \sqrt{\dfrac{k}{x}}$ $\qquad$ d) $r = 2\pi\sqrt{\dfrac{x}{g}}$

   e) $p^2 = \dfrac{4m^2r}{x}$ $\qquad$ f) $p = 2m\sqrt{\dfrac{r}{x}}$

Exercise 6.12 In the following questions, make the letter in **bold** the subject of the formula:

1. a) $v = u + a\boldsymbol{t}$ $\qquad$ b) $v^2 = \boldsymbol{u}^2 + 2as$ $\quad$ c) $v^2 = u^2 + 2a\boldsymbol{s}$
   d) $s = u\boldsymbol{t} + \frac{1}{2}at^2$ $\qquad$ e) $s = ut + \frac{1}{2}\boldsymbol{a}t^2$ $\quad$ f) $s = ut + \frac{1}{2}at^2$

2. a) $A = \pi r\sqrt{\boldsymbol{s}^2 + t^2}$ $\qquad$ b) $A = \pi r\sqrt{\boldsymbol{h}^2 + r^2}$

   c) $\dfrac{1}{f} = \dfrac{1}{\boldsymbol{u}} + \dfrac{1}{v}$ $\qquad$ d) $\dfrac{1}{f} = \dfrac{1}{u} + \dfrac{1}{\boldsymbol{v}}$

   e) $t = 2\pi\sqrt{\dfrac{\boldsymbol{l}}{g}}$ $\qquad$ f) $t = 2\pi\sqrt{\dfrac{l}{\boldsymbol{g}}}$

▨ **Algebraic fractions**
Simplifying algebraic fractions

The rules for fractions involving algebraic terms are the same as those for numeric fractions. However the actual calculations are often easier when using algebra.

Worked examples

a) $\dfrac{3}{4} \times \dfrac{5}{7} = \dfrac{15}{28}$

b) $\dfrac{a}{c} \times \dfrac{b}{d} = \dfrac{ab}{cd}$

c) $\dfrac{\cancel{3}}{4} \times \dfrac{5}{\cancel{6}_2} = \dfrac{5}{8}$

d) $\dfrac{\cancel{a}}{c} \times \dfrac{b}{2\cancel{a}} = \dfrac{b}{2c}$

e) $\dfrac{\cancel{a}b}{e\cancel{c}} \times \dfrac{\cancel{c}d}{f\cancel{a}} = \dfrac{bd}{ef}$

f) $\dfrac{m^2}{m} = \dfrac{m \times \cancel{m}}{\cancel{m}} = m$

g) $\dfrac{x^5}{x_3} = \dfrac{\cancel{x} \times \cancel{x} \times \cancel{x} \times x \times x}{\cancel{x} \times \cancel{x} \times \cancel{x}} = x^2$

Exercise 6.13

Simplify the following algebraic fractions:

1. a) $\dfrac{x}{y} \times \dfrac{p}{q}$　　**b)** $\dfrac{x}{y} \times \dfrac{q}{x}$　　**c)** $\dfrac{p}{q} \times \dfrac{q}{r}$

d) $\dfrac{ab}{c} \times \dfrac{d}{ab}$　　**e)** $\dfrac{ab}{c} \times \dfrac{d}{ac}$　　**f)** $\dfrac{p^2}{q^2} \times \dfrac{q^2}{p}$

2. a) $\dfrac{m^3}{m}$　　**b)** $\dfrac{r^7}{r^2}$　　**c)** $\dfrac{x^9}{x^3}$

d) $\dfrac{x^2y^4}{xy^2}$　　**e)** $\dfrac{a^2b^3c^4}{ab^2c}$　　**f)** $\dfrac{pq^2r^4}{p^2q^3r}$

3. a) $\dfrac{4ax}{2ay}$　　**b)** $\dfrac{12pq^2}{3p}$　　**c)** $\dfrac{15mn^2}{3mn}$

d) $\dfrac{24x^5y^3}{8x^2y^2}$　　**e)** $\dfrac{36p^2qr}{12pqr}$　　**f)** $\dfrac{16m^2n}{24m^3n^2}$

4. a) $\dfrac{2}{b} \times \dfrac{a}{3}$　　**b)** $\dfrac{4}{x} \times \dfrac{y}{2}$　　**c)** $\dfrac{8}{x} \times \dfrac{x}{4}$

d) $\dfrac{9y}{2} \times \dfrac{2x}{3}$　　**e)** $\dfrac{12x}{7} \times \dfrac{7}{4x}$　　**f)** $\dfrac{4x^3}{3y} \times \dfrac{9y^2}{2x^2}$

5. a) $\dfrac{2ax}{3bx} \times \dfrac{4by}{a}$　　　　**b)** $\dfrac{3p^2}{2q} \times \dfrac{5q}{3p}$

c) $\dfrac{p^2q}{rs} \times \dfrac{pr}{q}$　　　　**d)** $\dfrac{a^2b}{fc^2} \times \dfrac{cd}{bd} \times \dfrac{ef^2}{ca^2}$

e) $\dfrac{2pq^2}{3rs} \times \dfrac{5m}{4q} \times \dfrac{8rs}{15p^2}$　　**f)** $\dfrac{x^4}{wy^2} \times \dfrac{yz^2}{x^2} \times \dfrac{wx}{z^3}$

Addition and subtraction of fractions

In arithmetic it is easy to add or subtract fractions with the same denominator. It is the same process when dealing with algebraic fractions.

Worked examples

a) $\dfrac{4}{11} + \dfrac{3}{11}$

$= \dfrac{7}{11}$

b) $\dfrac{a}{11} + \dfrac{b}{11}$

$= \dfrac{a+b}{11}$

c) $\dfrac{4}{x} + \dfrac{3}{x}$

$= \dfrac{7}{x}$

If the denominators are different, the fractions need to be changed to form fractions with the same denominator.

d) $\dfrac{2}{9} + \dfrac{1}{3}$

$= \dfrac{2}{9} + \dfrac{3}{9}$

$= \dfrac{5}{9}$

e) $\dfrac{a}{9} + \dfrac{b}{3}$

$= \dfrac{a}{9} + \dfrac{3b}{9}$

$= \dfrac{a+3b}{9}$

f) $\dfrac{4}{5a} + \dfrac{7}{10a}$

$= \dfrac{8}{10a} + \dfrac{7}{10a}$

$= \dfrac{15}{10a}$

$= \dfrac{3}{2a}$

Similarly, with subtraction, the denominators need to be the same.

g) $\dfrac{7}{a} - \dfrac{1}{2a}$

$= \dfrac{14}{2a} - \dfrac{1}{2a}$

$= \dfrac{13}{2a}$

h) $\dfrac{p}{3} - \dfrac{q}{15}$

$= \dfrac{5p}{15} - \dfrac{q}{15}$

$= \dfrac{5p-q}{15}$

i) $\dfrac{5}{3b} - \dfrac{8}{9b}$

$= \dfrac{15}{9b} - \dfrac{8}{9b}$

$= \dfrac{7}{9b}$

Exercise 6.14 Simplify the following fractions:

1.
  a) $\dfrac{1}{7} + \dfrac{3}{7}$
  b) $\dfrac{a}{7} + \dfrac{b}{7}$
  c) $\dfrac{5}{13} + \dfrac{6}{13}$

  d) $\dfrac{c}{13} + \dfrac{d}{13}$
  e) $\dfrac{x}{3} + \dfrac{y}{3} + \dfrac{z}{3}$
  f) $\dfrac{p^2}{5} + \dfrac{q^2}{5}$

2.
  a) $\dfrac{5}{11} - \dfrac{2}{11}$
  b) $\dfrac{c}{11} - \dfrac{d}{11}$
  c) $\dfrac{6}{a} - \dfrac{2}{a}$

  d) $\dfrac{2a}{3} - \dfrac{5b}{3}$
  e) $\dfrac{2x}{7} - \dfrac{3y}{7}$
  f) $\dfrac{3}{4x} - \dfrac{5}{4x}$

3.
  a) $\dfrac{5}{6} - \dfrac{1}{3}$
  b) $\dfrac{5}{2a} - \dfrac{1}{a}$
  c) $\dfrac{2}{3c} + \dfrac{1}{c}$

  d) $\dfrac{2}{x} + \dfrac{3}{2x}$
  e) $\dfrac{5}{2p} - \dfrac{1}{p}$
  f) $\dfrac{1}{w} - \dfrac{3}{2w}$

4. a) $\dfrac{p}{4} - \dfrac{q}{12}$ b) $\dfrac{x}{4} - \dfrac{y}{2}$ c) $\dfrac{m}{3} - \dfrac{n}{9}$

    d) $\dfrac{x}{12} - \dfrac{y}{6}$ e) $\dfrac{r}{2} + \dfrac{m}{10}$ f) $\dfrac{s}{3} - \dfrac{t}{15}$

5. a) $\dfrac{3x}{4} - \dfrac{2x}{12}$ b) $\dfrac{3x}{5} - \dfrac{2y}{15}$ c) $\dfrac{3m}{7} + \dfrac{m}{14}$

    d) $\dfrac{4m}{3p} - \dfrac{3m}{9p}$ e) $\dfrac{4x}{3y} - \dfrac{5x}{6y}$ f) $\dfrac{3r}{7s} + \dfrac{2r}{14s}$

Often one denominator is not a multiple of the other. In these cases the **lowest common multiple** of both denominators has to be found.

Worked examples

a) $\dfrac{1}{4} + \dfrac{1}{3}$

$= \dfrac{3}{12} + \dfrac{4}{12}$

$= \dfrac{7}{12}$

b) $\dfrac{1}{5} + \dfrac{2}{3}$

$= \dfrac{3}{15} + \dfrac{10}{15}$

$= \dfrac{13}{15}$

c) $\dfrac{a}{3} + \dfrac{b}{4}$

$= \dfrac{4a}{12} + \dfrac{3b}{12}$

$= \dfrac{4a + 3b}{12}$

d) $\dfrac{2a}{3} + \dfrac{3b}{5}$

$= \dfrac{10a}{15} + \dfrac{9b}{15}$

$= \dfrac{10a + 9b}{15}$

Exercise 6.15   Simplify the following fractions:

1. a) $\dfrac{1}{2} + \dfrac{1}{3}$ b) $\dfrac{1}{3} + \dfrac{1}{5}$ c) $\dfrac{1}{4} + \dfrac{1}{7}$

    d) $\dfrac{2}{5} + \dfrac{1}{3}$ e) $\dfrac{1}{4} + \dfrac{5}{9}$ f) $\dfrac{2}{7} + \dfrac{2}{5}$

2. a) $\dfrac{a}{2} + \dfrac{b}{3}$ b) $\dfrac{a}{3} + \dfrac{b}{5}$ c) $\dfrac{p}{4} + \dfrac{q}{7}$

    d) $\dfrac{2a}{5} + \dfrac{b}{3}$ e) $\dfrac{x}{4} + \dfrac{5y}{9}$ f) $\dfrac{2x}{7} + \dfrac{2y}{5}$

3. a) $\dfrac{a}{2} - \dfrac{a}{3}$ b) $\dfrac{a}{3} - \dfrac{a}{5}$ c) $\dfrac{p}{4} + \dfrac{p}{7}$

    d) $\dfrac{2a}{5} + \dfrac{a}{3}$ e) $\dfrac{x}{4} + \dfrac{5x}{9}$ f) $\dfrac{2x}{7} + \dfrac{2x}{5}$

4. a) $\dfrac{3m}{5} - \dfrac{m}{2}$ b) $\dfrac{3r}{5} - \dfrac{r}{2}$ c) $\dfrac{5x}{4} - \dfrac{3x}{2}$

d) $\dfrac{2x}{7} + \dfrac{3x}{4}$ e) $\dfrac{11x}{2} - \dfrac{5x}{3}$ f) $\dfrac{2p}{3} - \dfrac{p}{2}$

5. a) $p - \dfrac{p}{2}$ b) $c - \dfrac{c}{3}$ c) $x - \dfrac{x}{5}$

d) $m - \dfrac{2m}{3}$ e) $q - \dfrac{4q}{5}$ f) $w - \dfrac{3w}{4}$

6. a) $2m - \dfrac{m}{2}$ b) $3m - \dfrac{2m}{3}$ c) $2m - \dfrac{5m}{2}$

d) $4m - \dfrac{3m}{2}$ e) $2p - \dfrac{5p}{3}$ f) $6q - \dfrac{6q}{7}$

7. a) $p - \dfrac{p}{r}$ b) $\dfrac{x}{y} + x$ c) $m + \dfrac{m}{n}$

d) $\dfrac{a}{b} + a$ e) $2x - \dfrac{x}{y}$ f) $2p - \dfrac{3p}{q}$

With more complex algebraic fractions, the method of getting a common denominator is still required.

Worked examples

a)
$$\frac{2}{x+1} + \frac{3}{x+2}$$

$$= \frac{2(x+2)}{(x+1)(x+2)} + \frac{3(x+1)}{(x+1)(x+2)}$$

$$= \frac{2(x+2) + 3(x+1)}{(x+1)(x+2)}$$

$$= \frac{2x+4+3x+3}{(x+1)(x+2)}$$

$$= \frac{5x+7}{(x+1)(x+2)}$$

b)
$$\frac{5}{p+3} - \frac{3}{p-5}$$

$$= \frac{5(p-5)}{(p+3)(p-5)} - \frac{3(p+3)}{(p+3)(p-5)}$$

$$= \frac{5(p-5) - 3(p+3)}{(p+3)(p-5)}$$

$$= \frac{5p-25-3p-9}{(p+3)(p-5)}$$

$$= \frac{2p-34}{(p+3)(p-5)}$$

c) $\dfrac{x^2 - 2x}{x^2 + x - 6}$ **d)** $\dfrac{x^2 - 3x}{x^2 + 2x - 15}$

$= \dfrac{x(x - 2)}{(x + 3)(x - 2)}$ $= \dfrac{x(x - 3)}{(x - 3)(x + 5)}$

$= \dfrac{x}{x + 3}$ $= \dfrac{x}{x + 5}$

Exercise 6.16   Simplify the following algebraic fractions:

1. **a)** $\dfrac{1}{x + 1} + \dfrac{2}{x + 2}$ **b)** $\dfrac{3}{m + 2} - \dfrac{2}{m - 1}$

   c) $\dfrac{2}{p - 3} + \dfrac{1}{p - 2}$ **d)** $\dfrac{3}{w - 1} - \dfrac{2}{w + 3}$

   e) $\dfrac{4}{y + 4} - \dfrac{1}{y + 1}$ **f)** $\dfrac{2}{m - 2} - \dfrac{3}{m + 3}$

2. **a)** $\dfrac{x(x - 4)}{(x - 4)(x + 2)}$ **b)** $\dfrac{y(y - 3)}{(y + 3)(y - 3)}$

   c) $\dfrac{(m + 2)(m - 2)}{(m - 2)(m - 3)}$ **d)** $\dfrac{p(p + 5)}{(p - 5)(p + 5)}$

   e) $\dfrac{m(2m + 3)}{(m + 4)(2m + 3)}$ **f)** $\dfrac{(m + 1)(m - 1)}{(m + 2)(m - 1)}$

3. **a)** $\dfrac{x^2 - 5x}{(x + 3)(x - 5)}$ **b)** $\dfrac{x^2 - 3x}{(x + 4)(x - 3)}$

   c) $\dfrac{y^2 - 7y}{(y - 7)(y - 1)}$ **d)** $\dfrac{x(x - 1)}{x^2 + 2x - 3}$

   e) $\dfrac{x(x + 2)}{x^2 + 4x + 4}$ **f)** $\dfrac{x(x + 4)}{x^2 + 5x + 4}$

4. **a)** $\dfrac{x^2 - x}{x^2 - 1}$ **b)** $\dfrac{x^2 + 2x}{x^2 + 5x + 6}$

   c) $\dfrac{x^2 + 4x}{x^2 + x - 12}$ **d)** $\dfrac{x^2 - 5x}{x^2 - 3x - 10}$

   e) $\dfrac{x^2 + 3x}{x^2 - 9}$ **f)** $\dfrac{x^2 - 7x}{x^2 - 49}$

Extended Section

Student Assessment 1

1. Factorise the following fully:
   a) $mx - 5m - 5nx + 25n$ b) $4x^2 - 81y^2$
   c) $88^2 - 12^2$ d) $x^4 - y^4$

2. Expand the following and simplify where possible:
   a) $(x + 3)(x + 5)$ b) $(x - 7)(x - 7)$
   c) $(x + 5)^2$ d) $(x - 7)(x + 2)$
   e) $(2x - 1)(3x + 8)$ f) $(7 - 5y)^2$

3. Factorise the following:
   a) $x^2 - 18x + 32$ b) $x^2 - 2x - 24$
   c) $x^2 - 9x + 18$ d) $x^2 - 2x + 1$
   e) $2x^2 + 5x - 3$ f) $9x^2 - 12x + 4$

4. Make the letter in **bold** the subject of the formula:
   a) $v^2 = u^2 + 2as$ b) $r^2 + h^2 = p^2$

   c) $\dfrac{m}{n} = \dfrac{r}{s^2}$ d) $t = 2\pi\sqrt{\dfrac{l}{g}}$

5. Simplify the following algebraic fractions:

   a) $\dfrac{ab}{c} \times \dfrac{bc}{a}$ b) $\dfrac{(x^3)^2}{x^2}$

   c) $\dfrac{12mn^2}{3m^2}$ d) $\dfrac{p^2q^4}{r^4} \times \dfrac{pr^5}{qr^3}$

6. Simplify the following algebraic fractions:

   a) $\dfrac{p}{5} + \dfrac{6p}{5} + \dfrac{4p}{5}$ b) $\dfrac{3m}{4} + \dfrac{5n}{16}$

   c) $\dfrac{5m}{4y} - \dfrac{3m}{6y}$ d) $\dfrac{2r}{3x} + \dfrac{5r}{4x} - \dfrac{3r}{2x}$

7. Simplify the following:

   a) $\dfrac{x}{3} + \dfrac{x}{7}$ b) $\dfrac{2x}{3} - \dfrac{3y}{4}$ c) $r - \dfrac{r}{7}$

8. Simplify the following:

   a) $\dfrac{3}{m + 2} + \dfrac{2}{m + 3}$ b) $\dfrac{(y + 3)(y - 3)}{(y - 3)^2}$

   c) $\dfrac{x^2 - 3x}{x^2 + 4x - 21}$

Student Assessment 2

1. Factorise the following fully:
   a) $pq - 3rq + pr - 3r^2$ b) $1 - t^4$
   c) $875^2 - 125^2$ d) $7.5^2 - 2.5^2$

2. Expand the following and simplify where possible:
   a) $(x - 4)(x + 2)$
   b) $(x - 8)^2$
   c) $(x + y)^2$
   d) $(x - 11)(x + 11)$
   e) $(3x - 2)(2x - 3)$
   f) $(5 - 3x)^2$

3. Factorise the following:
   a) $x^2 - 4x - 77$
   b) $x^2 - 6x + 9$
   c) $x^2 - 144$
   d) $3x^2 + 3x - 18$
   e) $2x^2 + 5x - 12$
   f) $4x^2 - 20x + 25$

4. Make the letter in **bold** the subject of the formula:
   a) $m\mathbf{f}^2 = p$
   b) $m = 5\mathbf{t}^2$

   c) $A = \pi r \sqrt{\mathbf{p} + q}$
   d) $\dfrac{1}{\mathbf{x}} + \dfrac{1}{y} = \dfrac{1}{t}$

5. Simplify the following algebraic fractions:
   a) $\dfrac{x^7}{x^3}$
   b) $\dfrac{mn}{p} \times \dfrac{pq}{m}$
   c) $\dfrac{(y^3)^3}{(y^2)^3}$
   d) $\dfrac{28pq^2}{7pq^3}$

6. Simplify the following algebraic fractions:
   a) $\dfrac{m}{11} + \dfrac{3m}{11} - \dfrac{2m}{11}$
   b) $\dfrac{3p}{8} - \dfrac{9p}{16}$
   c) $\dfrac{4x}{3y} - \dfrac{7x}{12y}$
   d) $\dfrac{3m}{15p} + \dfrac{4n}{5p} - \dfrac{11n}{30p}$

7. Simplify the following:
   a) $\dfrac{p}{5} + \dfrac{p}{4}$
   b) $\dfrac{3m}{5} - \dfrac{2m}{4}$
   c) $\dfrac{2p}{3} - \dfrac{3p}{4}$

8. Simplify the following:
   a) $\dfrac{4}{(x - 5)} + \dfrac{3}{(x - 2)}$

   b) $\dfrac{a^2 - b^2}{(a + b)^2}$

   c) $\dfrac{x - 2}{x^2 + x - 6}$

Student Assessment 3

1. The volume V of a cylinder is given by the formula $V = \pi r^2 h$, where h is the height of the cylinder and r is the radius.
   a) Find the volume of a cylindrical post 6.5 m long and with a diameter of 20 cm.
   b) Make r the subject of the formula.
   c) A cylinder of height 60 cm has a volume of 5500 cm³: find its radius correct to 3 s.f.

2. The formula for the surface area of a closed cylinder is
   $A = 2\pi r(r + h)$, where r is the radius of the cylinder and h
   is its height.
   a) Find the surface area of a cylinder of radius 12 cm and
      height 20 cm, giving your answer to 2 s.f.
   b) Rearrange the formula to make h the subject.
   c) What is the height of a cylinder of surface area 500 cm²
      and radius 5 cm. Give your answer to 3 s.f.?

3. Mario and Louisa plan a holiday to the USA. Mario
   changes 250 euro into US dollars at a rate of $0.88 to €1. A
   charge of 2% is deducted.
   a) How many dollars does Mario receive?
   Louisa changes her €250 into dollars at a rate of $0.8 to €1,
   but is not charged by the bank.
   b) Who has more money to spend?
   c) They bring back a total of 50 dollars which they change
      back to euro at a rate of $0.87 to €1. How much do they
      receive to the nearest pound?
   d) The exchange rate for euro to Indian rupees is
      €1 = 40 rupees. Assuming no bank charges and an
      exchange rate of $0.87 to €1, how many rupees would
      you receive for $250?

4. The formula for finding the length d of the body diagonal
   of a cuboid whose dimensions are x, y and z is:

   $$d = \sqrt{x^2 + y^2 + z^2}$$

   a) Find d when $x = 2$, $y = 3$ and $z = 4$.
   b) How long is the body diagonal of a block of concrete in
      the shape of a rectangular prism of dimensions 2 m, 3 m
      and 75 cm?
   c) Rearrange the formula to make x the subject.
   d) Find x when $d = 0.86$, $y = 0.25$ and $z = 0.41$.

5. A pendulum of length l metres takes T seconds to complete
   one full oscillation. The formula for T is:

   $$T = 2\pi\sqrt{\frac{l}{g}}$$

   where g m/s² is the acceleration due to gravity.
   a) Find T if $l = 5$ and $g = 10$.
   b) Rearrange the formula to make l the subject of the
      formula.
   c) How long is a pendulum which takes 3 seconds for one
      oscillation, if $g = 10$?

Student Assessment 4

1. The volume of a cylinder is given by the formula $V = \pi r^2 h$, where h is the height of the cylinder and r is the radius.
   a) Find the volume of a cylindrical post of length 7.5 m and a diameter of 30 cm.
   b) Make r the subject of the formula.
   c) A cylinder of height 75 cm has a volume of 6000 cm³, find its radius correct to 3 s.f.

2. The formula $C = \frac{5}{9}(F - 32)$ can be used to convert temperatures in degrees Fahrenheit (°F) into degrees Celsius (°C).
   a) What temperature in °C is equivalent to 150 °F?
   b) What temperature in °C is equivalent to 12 °F?
   c) Make F the subject of the formula.
   d) Use your rearranged formula to find what temperature in °F is equivalent to 160 °C.

3. The height of Mount Kilamanjaro is given as 5900 m. The formula for the time taken, T hours, to climb to a height H metres is:

$$T = \frac{H}{1200} + k$$

   where k is a constant.
   a) Calculate the time taken, to the nearest hour, to climb to the top of the mountain if $k = 9.8$.
   b) Make H the subject of the formula.
   c) How far up the mountain, to the nearest 100 m, could you expect to be after 14 hours?

4. The formula for the volume V of a sphere is given as $V = \frac{4}{3}\pi r^3$.
   a) Find V if $r = 5$.
   b) Make r the subject of the formula.
   c) Find the radius of a sphere of volume 2500 m³.

5. The cost £x of printing n newspapers is given by the formula $x = 1.50 + 0.05n$.
   a) Calculate the cost of printing 5000 newspapers.
   b) Make n the subject of the formula.
   c) How many newspapers can be printed for £25?

ICT Section

Produce a PowerPoint presentation that explains how to factorise algebraic expressions.

7 EQUATIONS AND INEQUALITIES

Core Section

Solve simple linear equations in one unknown; solve simultaneous linear equations in two unknowns; construct simple expressions and set up simple equations.

An equation is formed when the value of an unknown quantity is needed.

Simple linear equations

Worked examples   Solve the following linear equations:

a)
$$3x + 8 = 14$$
$$3x = 6$$
$$x = 2$$

b)
$$12 = 20 + 2x$$
$$-8 = 2x$$
$$-4 = x$$

c)
$$3(p + 4) = 21$$
$$3p + 12 = 21$$
$$3p = 9$$
$$p = 3$$

d)
$$4(x - 5) = 7(2x - 5)$$
$$4x - 20 = 14x - 35$$
$$4x + 15 = 14x$$
$$15 = 10x$$
$$1.5 = x$$

Exercise 7.1   Solve the following linear equations:

1.
a) $3x = 2x - 4$
b) $5y = 3y + 10$
c) $2y - 5 = 3y$
d) $p - 8 = 3p$
e) $3y - 8 = 2y$
f) $7x + 11 = 5x$

2.
a) $3x - 9 = 4$
b) $4 = 3x - 11$
c) $6x - 15 = 3x + 3$
d) $4y + 5 = 3y - 3$
e) $8y - 31 = 13 - 3y$
f) $4m + 2 = 5m - 8$

3.
a) $7m - 1 = 5m + 1$
b) $5p - 3 = 3 + 3p$
c) $12 - 2k = 16 + 2k$
d) $6x + 9 = 3x - 54$
e) $8 - 3x = 18 - 8x$
f) $2 - y = y - 4$

4.
a) $\dfrac{x}{2} = 3$
b) $\frac{1}{2}y = 7$

c) $\dfrac{x}{4} = 1$
d) $\frac{1}{4}m = 3$

e) $7 = \dfrac{x}{5}$
f) $4 = \frac{1}{5}p$

5.
a) $\dfrac{x}{3} - 1 = 4$
b) $\dfrac{x}{5} + 2 = 1$

c) $\frac{2}{3}x = 5$
d) $\frac{3}{4}x = 6$

e) $\frac{1}{5}x = \frac{1}{2}$
f) $\dfrac{2x}{5} = 4$

6. a) $\dfrac{x + 1}{2} = 3$ b) $4 = \dfrac{x - 2}{3}$

c) $\dfrac{x - 10}{3} = 4$ d) $8 = \dfrac{5x - 1}{3}$

e) $\dfrac{2(x - 5)}{3} = 2$ f) $\dfrac{3(x - 2)}{4} = 4x - 8$

7. a) $6 = \dfrac{2(y - 1)}{3}$ b) $2(x + 1) = 3(x - 5)$

c) $5(x - 4) = 3(x + 2)$ d) $\dfrac{3 + y}{2} = \dfrac{y + 1}{4}$

e) $\dfrac{7 + 2x}{3} = \dfrac{9x - 1}{7}$ f) $\dfrac{2x + 3}{4} = \dfrac{4x - 2}{6}$

NB: All diagrams are not drawn to scale.

Constructing equations

In many cases, when dealing with the practical applications of mathematics, equations need to be constructed first before they can be solved. Often the information is either given within the context of a problem or in a diagram.

Worked examples

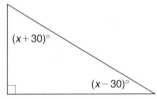

a) Find the size of each of the angles in the triangle (left) by constructing an equation and solving it to find the value of x.

The sum of the angles of a triangle is 180°.

$$(x + 30) + (x - 30) + 90 = 180$$
$$2x + 90 = 180$$
$$2x = 90$$
$$x = 45$$

The three angles are therefore: $90°$, $x + 30 = 75°$, $x - 30 = 15°$.
Check: $90° + 75° + 15° = 180°$.

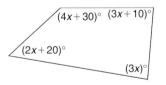

b) Find the size of each of the angles in the quadrilateral (left) by constructing an equation and solving it to find the value of x.

The sum of the angles of a quadrilateral is 360°.

$$4x + 30 + 3x + 10 + 3x + 2x + 20 = 360$$
$$12x + 60 = 360$$
$$12x = 300$$
$$x = 25$$

The angles are:

$$4x + 30 = (4 \times 25) + 30 = 130°$$
$$3x + 10 = (3 \times 25) + 10 = 85°$$
$$3x = 3 \times 25 = 75°$$
$$2x + 20 = (2 \times 25) + 20 = \underline{70°}$$
$$\text{Total} = 360°$$

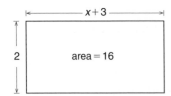

c) Construct an equation and solve it to find the value of *x* in the diagram (left).

Area of rectangle = base × height

$$2(x + 3) = 16$$
$$2x + 6 = 16$$
$$2x = 10$$
$$x = 5$$

Exercise 7.2   In questions 1–3:

i)   construct an equation in terms of *x*,
ii) solve the equation,
iii) calculate the value of each of the angles,
iv) check your answers.

1. a)
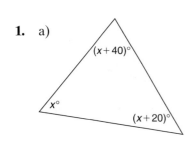

b)
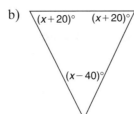

c)
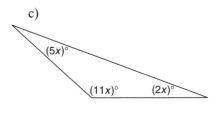

d)
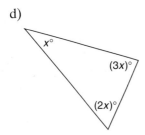

e)
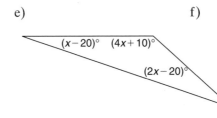

f)
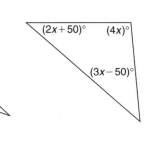

2. a)
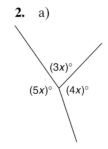

b)
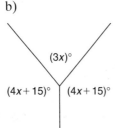

c)
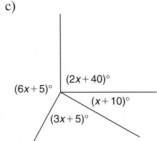

d)
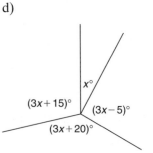

3. a)

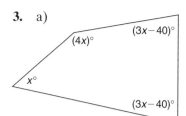

b)

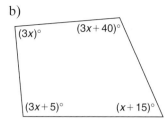

c)

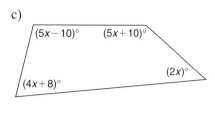

d)

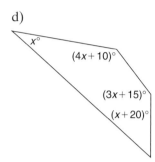

e)

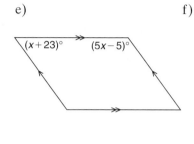

f)

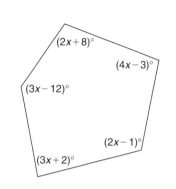

4. By constructing an equation and solving it, find the value of x in each of these isosceles triangles:

a)

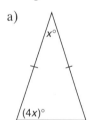

b)

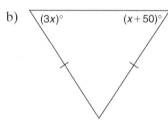

c)

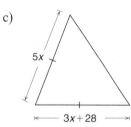

d)

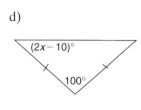

e)

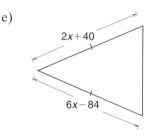

f)

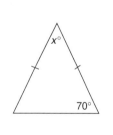

5. Using angle properties, calculate the value of x in each of these questions:

a)   b)   c)   d)

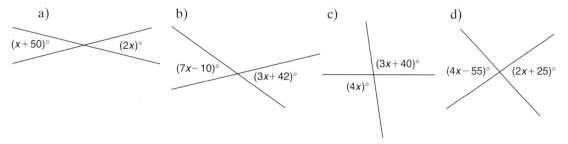

6. Calculate the value of *x*:

a)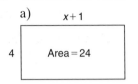

b)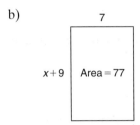

c)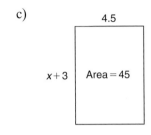

d)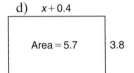

e)

f)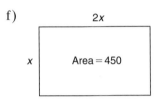

▪ Simultaneous equations

When the values of two unknowns are needed, two equations need to be formed and solved. The process of solving two equations and finding a common solution is known as solving equations simultaneously.

The two most common ways of solving simultaneous equations algebraically are by **elimination** and by **substitution**.

By elimination

The aim of this method is to eliminate one of the unknowns by either adding or subtracting the two equations.

Worked examples Solve the following simultaneous equations by finding the values of *x* and *y* which satisfy both equations.

a) $3x + y = 9$ $\qquad$ (1)
$5x - y = 7$ $\qquad$ (2)

By adding equations (1) + (2) we eliminate the variable *y*:

$$8x = 16$$
$$x = 2$$

To find the value of *y* we substitute $x = 2$ into either equation (1) or (2).
Substituting $x = 2$ into equation (1):

$$3x + y = 9$$
$$6 + y = 9$$
$$y = 3$$

To check that the solution is correct, the values of *x* and *y* are substituted into equation (2). If it is correct then the left-hand side of the equation will equal the right-hand side.

$$5x - y = 7$$
$$10 - 3 = 7$$
$$7 \quad = 7$$

b) $4x + y = 23$ (1)
     $x + y = 8$ (2)

By subtracting the equations i.e. (1) − (2), we eliminate the variable y:

$$3x = 15$$
$$x = 5$$

By substituting $x = 5$ into equation (2), y can be calculated:

$$x + y = 8$$
$$5 + y = 8$$
$$y = 3$$

Check by substituting both values into equation (1):

$$4x + y = 23$$
$$20 + 3 = 23$$
$$23 \quad = 23$$

By substitution

The same equations can also be solved by the method known as **substitution**.

Worked examples   **a)** $3x + y = 9$ (1)
                   $5x - y = 7$ (2)

Equation (2) can be rearranged to give: $y = 5x - 7$
This can now be substituted into equation (1):

$$3x + (5x - 7) = 9$$
$$3x + 5x - 7 = 9$$
$$8x - 7 = 9$$
$$8x = 16$$
$$x = 2$$

To find the value of y, $x = 2$ is substituted into either equation (1) or (2) as before giving **$y = 3$**.

b) $4x + y = 23$ (1)
     $x + y = 8$ (2)

Equation (2) can be rearranged to give $y = 8 - x$.
This can be substituted into equation (1):

$$4x + (8 - x) = 23$$
$$4x + 8 - x = 23$$
$$3x + 8 = 23$$
$$3x = 15$$
$$x = 5$$

y can be found as before, giving a result of **$y = 3$**.

Exercise 7.3 Solve the following simultaneous equations either by elimination or by substitution:

1. a) $x + y = 6$ b) $x + y = 11$ c) $x + y = 5$
          $x - y = 2$ $x - y - 1 = 0$ $x - y = 7$
      d) $2x + y = 12$ e) $3x + y = 17$ f) $5x + y = 29$
          $2x - y = 8$ $3x - y = 13$ $5x - y = 11$

2. a) $3x + 2y = 13$ b) $6x + 5y = 62$ c) $x + 2y = 3$
          $4x = 2y + 8$ $4x - 5y = 8$ $8x - 2y = 6$
      d) $9x + 3y = 24$ e) $7x - y = -3$ f) $3x = 5y + 14$
          $x - 3y = -14$ $4x + y = 14$ $6x + 5y = 58$

3. a) $2x + y = 14$ b) $5x + 3y = 29$ c) $4x + 2y = 50$
          $x + y = 9$ $x + 3y = 13$ $x + 2y = 20$
      d) $x + y = 10$ e) $2x + 5y = 28$ f) $x + 6y = -2$
          $3x = -y + 22$ $4x + 5y = 36$ $3x + 6y = 18$

4. a) $x - y = 1$ b) $3x - 2y = 8$ c) $7x - 3y = 26$
          $2x - y = 6$ $2x - 2y = 4$ $2x - 3y = 1$
      d) $x = y + 7$ e) $8x - 2y = -2$ f) $4x - y = -9$
          $3x - y = 17$ $3x - 2y = -7$ $7x - y = -18$

5. a) $x + y = -7$ b) $2x + 3y = -18$
          $x - y = -3$ $2x = 3y + 6$
      c) $5x - 3y = 9$ d) $7x + 4y = 42$
          $2x + 3y = 19$ $9x - 4y = -10$
      e) $4x - 4y = 0$ f) $x - 3y = -25$
          $8x + 4y = 12$ $5x - 3y = -17$

6. a) $2x + 3y = 13$ b) $2x + 4y = 50$
          $2x - 4y + 8 = 0$ $2x + y = 20$
      c) $x + y = 10$ d) $5x + 2y = 28$
          $3y = 22 - x$ $5x + 4y = 36$
      e) $2x - 8y = 2$ f) $x - 4y = 9$
          $2x - 3y = 7$ $x - 7y = 18$

7. a) $-4x = 4y$ b) $3x = 19 + 2y$
          $4x - 8y = 12$ $-3x + 5y = 5$
      c) $3x + 2y = 12$ d) $3x + 5y = 29$
          $-3x + 9y = -12$ $3x + y = 13$
      e) $-5x + 3y = 14$ f) $-2x + 8y = 6$
          $5x + 6y = 58$ $2x = 3 - y$

If neither x nor y can be eliminated by simply adding or subtracting the two equations then it is necessary to multiply one or both of the equations. The equations are multiplied by a number in order to make the coefficients of x (or y) numerically equal.

Worked examples   **a)** $3x + 2y = 22$ (1)
$\qquad x + y = 9$ (2)

To eliminate y, equation (2) is multiplied by 2:

$\qquad 3x + 2y = 22$ (1)
$\qquad 2x + 2y = 18$ (3)

By subtracting (3) from (1), the variable y is eliminated:

$\qquad x = 4$

Substituting $x = 4$ into equation (2), we have:

$\qquad x + y = 9$
$\qquad 4 + y = 9$
$\qquad\quad y = 5$

Check by substituting both values into equation (1):

$\qquad 3x + 2y = 22$
$\qquad 12 + 10 = 22$
$\qquad\quad 22 = 22$

b) $5x - 3y = 1$ (1)
$\quad 3x + 4y = 18$ (2)

To eliminate the variable y, equation (1) is multiplied by 4, and equation (2) is multiplied by 3.

$\qquad 20x - 12y = 4$ (3)
$\qquad\; 9x + 12y = 54$ (4)

By adding equations (3) and (4) the variable y is eliminated:

$\qquad 29x = 58$
$\qquad\quad x = 2$

Substituting $x = 2$ into equation (2) gives:

$\qquad 3x + 4y = 18$
$\qquad 6 + 4y = 18$
$\qquad\quad 4y = 12$
$\qquad\quad\; y = 3$

Check by substituting both values into equation (1):

$\qquad 5x - 3y = 1$
$\qquad 10 - 9 = 1$
$\qquad\quad 1 = 1$

Exercise 7.4   Solve the following:

1. a) $2x + y = 7$ b) $5x + 4y = 21$ c) $x + y = 7$
$\quad\;\; 3x + 2y = 12$ $\quad\;\, x + 2y = 9$ $\quad\; 3x + 4y = 23$
$\;$ d) $2x - 3y = -3$ e) $4x = 4y + 8$ f) $x + 5y = 11$
$\quad\;\; 3x + 2y = 15$ $\quad\;\; x + 3y = 10$ $\quad\; 2x - 2y = 10$

2. a) $x + y = 5$ b) $2x - 2y = 6$ c) $2x + 3y = 15$
      $3x - 2y + 5 = 0$ $x - 5y = -5$ $2y = 15 - 3x$
    d) $x - 6y = 0$ e) $2x - 5y = -11$ f) $x + y = 5$
      $3x - 3y = 15$ $3x + 4y = 18$ $2x - 2y = -2$

3. a) $3y = 9 + 2x$ b) $x + 4y = 13$ c) $2x = 3y - 19$
      $3x + 2y = 6$ $3x - 3y = 9$ $3x + 2y = 17$
    d) $2x - 5y = -8$ e) $5x - 2y = 0$ f) $8y = 3 - x$
      $-3x - 2y = -26$ $2x + 5y = 29$ $3x - 2y = 9$

4. a) $4x + 2y = 5$ b) $4x + y = 14$ c) $10x - y = -2$
      $3x + 6y = 6$ $6x - 3y = 3$ $-15x + 3y = 9$
    d) $-2y = 0.5 - 2x$ e) $x + 3y = 6$ f) $5x - 3y = -0.5$
      $6x + 3y = 6$ $2x - 9y = 7$ $3x + 2y = 3.5$

Exercise 7.5

1. The sum of two numbers is 17 and their difference is 3. Find the two numbers by forming two equations and solving them simultaneously.

2. The difference between two numbers is 7. If their sum is 25, find the two numbers by forming two equations and solving them simultaneously.

3. Find the values of x and y.

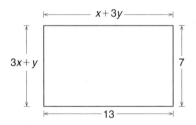

4. Find the values of x and y.

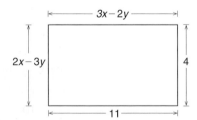

5. A man's age is three times his son's age. Ten years ago he was five times his son's age. By forming two equations and solving them simultaneously, find both of their ages.

6. A grandfather is ten times older than his granddaughter. He is also 54 years older than her. How old is each of them?

Core Section

Student Assessment 1

1. Solve the following equations:

    a) $x + 7 = 16$ b) $2x - 9 = 13$

    c) $8 - 4x = 24$ d) $5 - 3x = -13$

2. a) $7 - m = 4 + m$ b) $5m - 3 = 3m + 11$

    c) $6m - 1 = 9m - 13$ d) $18 - 3p = 6 + p$

3. a) $\dfrac{x}{-5} = 2$ b) $4 = \frac{1}{3}x$

    c) $\dfrac{x + 2}{3} = 4$ d) $\dfrac{2x - 5}{7} = \dfrac{5}{2}$

4. a) $\frac{2}{3}(x - 4) = 8$ b) $4(x - 3) = 7(x + 2)$

    c) $4 = \frac{2}{7}(3x + 8)$ d) $\frac{3}{4}(x - 1) = \frac{5}{8}(2x - 4)$

5. Solve the following simultaneous equations:

    a) $2x + 3y = 16$ b) $4x + 2y = 22$

       $2x - 3y = 4$ $-2x + 2y = 2$

    c) $x + y = 9$ d) $2x - 3y = 7$

       $2x + 4y = 26$ $-3x + 4y = -11$

Student Assessment 2

1. Solve the following equations:

    a) $y + 9 = 3$ b) $3x - 5 = 13$

    c) $12 - 5p = -8$ d) $2.5y + 1.5 = 7.5$

2. a) $5 - p = 4 + p$ b) $8m - 9 = 5m + 3$

    c) $11p - 4 = 9p + 15$ d) $27 - 5r = r - 3$

3. a) $\dfrac{p}{-2} = -3$ b) $6 = \frac{2}{5}x$

    c) $\dfrac{m - 7}{5} = 3$ d) $\dfrac{4t - 3}{3} = 7$

4. a) $\frac{2}{5}(t - 1) = 3$ b) $5(3 - m) = 4(m - 6)$

    c) $5 = \frac{2}{3}(x - 1)$ d) $\frac{4}{5}(t - 2) = \frac{1}{4}(2t + 8)$

5. Solve the following simultaneous equations:

    a) $x + y = 11$ b) $5p - 3q = -1$

       $x - y = 3$ $-2p - 3q = -8$

    c) $3x + 5y = 26$ d) $2m - 3n = -9$

       $x - y = 6$ $3m + 2n = 19$

EQUATIONS AND INEQUALITIES

Extended Section

> Construct and transform more complicated formulae and equations; solve quadratic equations by factorisation and either by use of the formula or by completing the square; solve simple linear inequalities.

▨ Constructing equations

In the core section of this chapter we looked at some simple examples of constructing and solving equations when we were given geometrical diagrams. This section extends this work with more complicated formulae and equations.

Construct and solve the equations below.

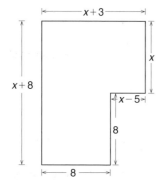

a) Using the shape (left), construct an equation for the perimeter in terms of x. Find the value of x by solving the equation.

$$x + 3 + x + x - 5 + 8 + 8 + x + 8 = 54$$
$$4x + 22 = 54$$
$$4x = 32$$
$$x = 8$$

b) A number is doubled, 5 is subtracted from it, and the total is 17. Find the number.

Let x be the unknown number.

$$2x - 5 = 17$$
$$2x = 22$$
$$x = 11$$

c) 3 is added to a number. The result is multiplied by 8. If the answer is 64 calculate the value of the original number.

Let x be the unknown number.

$$8(x + 3) = 64$$
$$8x + 24 = 64$$
$$8x = 40$$
$$x = 5$$

or
$$8(x + 3) = 64$$
$$x + 3 = 8$$
$$x = 5$$

The original number $= 5$

Exercise 7.6 **1.** Calculate the value of *x*:

a)

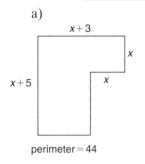

perimeter = 44

b)

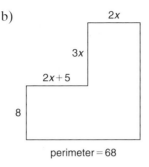

perimeter = 68

c)

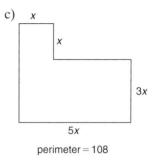

perimeter = 108

d)

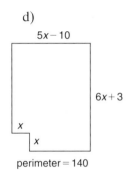

perimeter = 140

e)

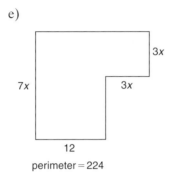

perimeter = 224

f)

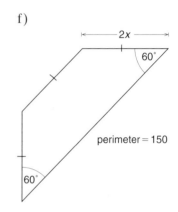

perimeter = 150

2. a) A number is trebled and then 7 is added to it. If the total is 28, find the number.

   b) Multiply a number by 4 and then add 5 to it. If the total is 29, find the number.

   c) If 31 is the result of adding 1 to 5 times a number, find the number.

   d) Double a number and then subtract 9. If the answer is 11, what is the number?

   e) If 9 is the result of subtracting 12 from 7 times a number, find the number.

3. a) Add 3 to a number and then double the result. If the total is 22, find the number.

   b) 27 is the answer when you add 4 to a number and then treble it. What is the number?

   c) Subtract 1 from a number and multiply the result by 5. If the answer is 35, what is the number?

   d) Add 3 to a number. If the result of multiplying this total by 7 is 63, find the number.

   e) Add 3 to a number. Quadruple the result. If the answer is 36, what is the number?

4. a) Gabriella is x years old. Her brother is 8 years older and her sister is 12 years younger than she is. If their total age is 50 years, how old are they?

b) A series of mathematics textbooks consists of four volumes. The first volume has x pages, the second 54 pages more. The third and fourth volume each have 32 pages more than the second. If the total number of pages in all four volumes is 866, calculate the number of pages in each of the volumes.

c) The five interior angles (in °) of a pentagon are x, $x + 30, 2x, 2x + 40$ and $3x + 20$. The sum of the interior angles of a pentagon is 540°. Calculate the size of each of the angles.

d) A hexagon consists of three interior angles of equal size and a further three which are double the size. The sum of all six angles is 720°. Calculate the size of each of the angles.

e) Four of the exterior angles of an octagon are the same size. The other four are twice as big. If the sum of the exterior angles is 360°, calculate the size of the interior angles.

Solving quadratic equations by factorising

Students will need to be familiar with the work covered in Chapter 6 on the factorising of quadratics.

$x^2 - 3x - 10 = 0$ is a quadratic equation, which when factorised can be written as $(x - 5)(x + 2) = 0$.

Therefore either $x - 5 = 0$ or $x + 2 = 0$ since, if two things multiply to make zero, then one of them must be zero.

$$\begin{array}{lll} x - 5 = 0 & \text{or} & x + 2 = 0 \\ x \quad = 5 & \text{or} & x \quad = -2 \end{array}$$

Worked examples Solve the following equations to give two solutions for x:

a)
$$x^2 - x - 12 = 0$$
$$(x - 4)(x + 3) = 0$$
so either
$$\begin{array}{lll} x - 4 = 0 & \text{or} & x + 3 = 0 \\ x \quad = 4 & \text{or} & x \quad = -3 \end{array}$$

b)
$$x^2 + 2x = 24$$
This becomes
$$x^2 + 2x - 24 = 0$$
$$(x + 6)(x - 4) = 0$$
so either
$$\begin{array}{lll} x + 6 = 0 & \text{or} & x - 4 = 0 \\ x \quad = -6 & \text{or} & x \quad = 4 \end{array}$$

c)
$$x^2 - 6x = 0$$
$$x(x - 6) = 0$$
so either
$$\begin{array}{lll} x = 0 & \text{or} & x - 6 = 0 \\ & \text{or} & x \quad = 6 \end{array}$$

d)
$$x^2 - 4 = 0$$
$$(x - 2)(x + 2) = 0$$
so either $\quad x - 2 = 0 \qquad\qquad$ or $\quad x + 2 = 0$
$$\qquad\qquad\quad x \quad\ = 2 \qquad\qquad$$ or $\quad x \quad\ = -2$

Exercise 7.7

Solve the following quadratic equations by factorising:

1.
a) $x^2 + 7x + 12 = 0$
b) $x^2 + 8x + 12 = 0$
c) $x^2 + 13x + 12 = 0$
d) $x^2 - 7x + 10 = 0$
e) $x^2 - 5x + 6 = 0$
f) $x^2 - 6x + 8 = 0$

2.
a) $x^2 + 3x - 10 = 0$
b) $x^2 - 3x - 10 = 0$
c) $x^2 + 5x - 14 = 0$
d) $x^2 - 5x - 14 = 0$
e) $x^2 + 2x - 15 = 0$
f) $x^2 - 2x - 15 = 0$

3.
a) $x^2 + 5x = -6$
b) $x^2 + 6x = -9$
c) $x^2 + 11x = -24$
d) $x^2 - 10x = -24$
e) $x^2 + x = 12$
f) $x^2 - 4x = 12$

4.
a) $x^2 - 2x = 8$
b) $x^2 - x = 20$
c) $x^2 + x = 30$
d) $x^2 - x = 42$
e) $x^2 - 2x = 63$
f) $x^2 + 3x = 54$

Exercise 7.8

Solve the following quadratic equations:

1.
a) $x^2 - 9 = 0$
b) $x^2 - 16 = 0$
c) $x^2 = 25$
d) $x^2 = 121$
e) $x^2 - 144 = 0$
f) $x^2 - 220 = 5$

2.
a) $4x^2 - 25 = 0$
b) $9x^2 - 36 = 0$
c) $25x^2 = 64$
d) $x^2 = \frac{1}{4}$
e) $x^2 - \frac{1}{9} = 0$
f) $16x^2 - \frac{1}{25} = 0$

3.
a) $x^2 + 5x + 4 = 0$
b) $x^2 + 7x + 10 = 0$
c) $x^2 + 6x + 8 = 0$
d) $x^2 - 6x + 8 = 0$
e) $x^2 - 7x + 10 = 0$
f) $x^2 + 2x - 8 = 0$

4.
a) $x^2 - 3x - 10 = 0$
b) $x^2 + 3x - 10 = 0$
c) $x^2 - 3x - 18 = 0$
d) $x^2 + 3x - 18 = 0$
e) $x^2 - 2x - 24 = 0$
f) $x^2 - 2x - 48 = 0$

5.
a) $x^2 + x = 12$
b) $x^2 + 8x = -12$
c) $x^2 + 5x = 36$
d) $x^2 + 2x = -1$
e) $x^2 + 4x = -4$
f) $x^2 + 17x = -72$

6.
a) $x^2 - 8x = 0$
b) $x^2 - 7x = 0$
c) $x^2 + 3x = 0$
d) $x^2 + 4x = 0$
e) $x^2 - 9x = 0$
f) $4x^2 - 16x = 0$

7.
a) $2x^2 + 5x + 3 = 0$
b) $2x^2 - 3x - 5 = 0$
c) $3x^2 + 2x - 1 = 0$
d) $2x^2 + 11x + 5 = 0$
e) $2x^2 - 13x + 15 = 0$
f) $12x^2 + 10x - 8 = 0$

8.
a) $x^2 + 12x = 0$
b) $x^2 + 12x + 27 = 0$
c) $x^2 + 4x = 32$
d) $x^2 + 5x = 14$
e) $2x^2 = 72$
f) $3x^2 - 12 = 288$

Exercise 7.9 In the following questions construct equations from the information given and then solve to find the unknown.

1. When a number x is added to its square, the total is 12. Find two possible values for x.

2. A number x is equal to its own square minus 42. Find two possible values for x.

3. If the area of the rectangle below is 10 cm², calculate the only possible value for x.

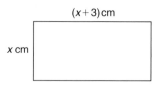

$(x+3)$cm

x cm

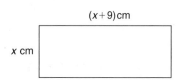

$(x+9)$cm

x cm

4. If the area of the rectangle (left) is 52 cm², calculate the only possible value for x.

5. A triangle has a base length of $2x$ cm and a height of $(x - 3)$ cm. If its area is 18 cm², calculate its height and base length.

6. A triangle has a base length of $(x - 8)$ cm and a height of $2x$ cm. If its area is 20 cm² calculate its height and base length.

7. A right-angled triangle has a base length of x cm and a height of $(x - 1)$ cm. If its area is 15 cm² calculate the base length and height.

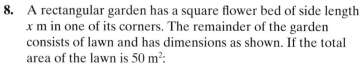

7m

x m

x m

2m

8. A rectangular garden has a square flower bed of side length x m in one of its corners. The remainder of the garden consists of lawn and has dimensions as shown. If the total area of the lawn is 50 m²:
    a) form an equation in terms of x,
    b) solve the equation,
    c) calculate the length and width of the whole garden.

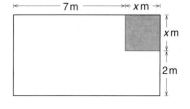

▮ The quadratic formula

In general a quadratic equation takes the form $ax^2 + bx + c = 0$ where a, b and c are integers. Quadratic equations can be solved by the use of the quadratic formula which states that:

$$x = \frac{-b \pm \sqrt{b^2 - 4ac}}{2a}$$

Worked examples   **a)** Solve the quadratic equation $x^2 + 7x + 3 = 0$.

$a = 1, b = 7$ and $c = 3$.
Substituting these values into the quadratic formula gives:

$$x = \frac{-7 \pm \sqrt{7^2 - 4 \times 1 \times 3}}{2 \times 1}$$

$$x = \frac{-7 \pm \sqrt{49 - 12}}{2}$$

$$x = \frac{-7 \pm \sqrt{37}}{2}$$

Therefore $\quad x = \dfrac{-7 + 6.08}{2} \quad$ or $\quad x = \dfrac{-7 - 6.08}{2}$

$\qquad\qquad x = -0.46$ (2 d.p.)   or $x = -6.54$ (2 d.p.)

b) Solve the quadratic equation $x^2 - 4x - 2 = 0$.

$a = 1, b = -4$ and $c = -2$.
Substituting these values into the quadratic formula gives:

$$x = \frac{-(-4) \pm \sqrt{(-4)^2 - (4 \times 1 \times -2)}}{2 \times 1}$$

$$x = \frac{4 \pm \sqrt{16 + 8}}{2}$$

$$x = \frac{4 \pm \sqrt{24}}{2}$$

Therefore $\quad x = \dfrac{4 + 4.90}{2} \quad$ or $\quad x = \dfrac{4 - 4.90}{2}$

$\qquad\qquad x = 4.45$ (2 d.p.)   or $x = -0.45$ (2 d.p.)

Completing the square

Quadratics can also be solved by expressing them in terms of a
perfect square. We look once again at the quadratic
$x^2 - 4x - 2 = 0$.

The perfect square $(x - 2)^2$ can be expanded to give
$x^2 - 4x + 4$. Notice that the x^2 and x terms are the same as
those in the original quadratic.

Therefore $(x - 2)^2 - 6 = x^2 - 4x - 2$ and can be used to
solve the quadratic.

$$
\begin{aligned}
(x - 2)^2 - 6 &= 0 \\
(x - 2)^2 &= 6 \\
x - 2 &= \pm \sqrt{6} \\
x &= 2 \pm \sqrt{6} \\
x = 4.45 \quad \text{or} \quad & x = -0.45
\end{aligned}
$$

Exercise 7.10 Solve the following quadratic equations using either the quadratic formula or by completing the square. Give your answers to 2 d.p.

1. a) $x^2 - x - 13 = 0$ b) $x^2 + 4x - 11 = 0$
    c) $x^2 + 5x - 7 = 0$ d) $x^2 + 6x + 6 = 0$
    e) $x^2 + 5x - 13 = 0$ f) $x^2 - 9x + 19 = 0$

2. a) $x^2 + 7x + 9 = 0$ b) $x^2 - 35 = 0$
    c) $x^2 + 3x - 3 = 0$ d) $x^2 - 5x - 7 = 0$
    e) $x^2 + x - 18 = 0$ f) $x^2 - 8 = 0$

3. a) $x^2 - 2x - 2 = 0$ b) $x^2 - 4x - 11 = 0$
    c) $x^2 - x - 5 = 0$ d) $x^2 + 2x - 7 = 0$
    e) $x^2 - 3x + 1 = 0$ f) $x^2 - 8x + 3 = 0$

4. a) $2x^2 - 3x - 4 = 0$ b) $4x^2 + 2x - 5 = 0$
    c) $5x^2 - 8x + 1 = 0$ d) $-2x^2 - 5x - 2 = 0$
    e) $3x^2 - 4x - 2 = 0$ f) $-7x^2 - x + 15 = 0$

Linear inequalities

The statement

   6 is less than 8

can be written as:

   $6 < 8$

This inequality can be manipulated in the following ways:

adding 2 to each side:	$8 < 10$	this inequality is still true
subtracting 2 from each side:	$4 < 6$	this inequality is still true
multiplying both sides by 2:	$12 < 16$	this inequality is still true
dividing both sides by 2:	$3 < 4$	this inequality is still true
multiplying both sides by -2:	$-12 < -16$	this inequality is not true
dividing both sides by -2:	$-3 < -4$	this inequality is not true

As can be seen, when both sides of an inequality are either multiplied or divided by a negative number, the inequality is no longer true. For it to be true the inequality sign needs to be changed around.

   i.e. $-12 > -16$ and $-3 > -4$

When solving linear inequalities, the procedure is very similar to that of solving linear equations.

Worked examples Remember:

implies that the number is not included in the solution. It is associated with > and <.

implies that the number is included in the solution. It is associated with ⩾ and ⩽.

Solve the following inequalities and represent the solution on a number line:

a) $15 + 3x < 6$
$$3x < -9$$
$$x < -3$$

b) $17 ⩽ 7x + 3$
$$14 ⩽ 7x$$
$$2 ⩽ x \qquad \text{that is} \qquad x ⩾ 2$$

c) $9 - 4x ⩾ 17$
$$-4x ⩾ 8$$
$$x ⩽ -2 \qquad \text{Note the inequality sign has changed direction.}$$

Exercise 7.11   Solve the following inequalities and illustrate your solution on a number line:

1. a) $x + 3 < 7$
    c) $4 + 2x ⩽ 10$
    e) $5 > 3 + x$

   b) $5 + x > 6$
   d) $8 ⩽ x + 1$
   f) $7 < 3 + 2x$

2. a) $x - 3 < 4$
    c) $8 + 3x > -1$
    e) $12 > -x - 12$

   b) $x - 6 ⩾ -8$
   d) $5 ⩾ -x - 7$
   f) $4 ⩽ 2x + 10$

3. a) $\dfrac{x}{2} < 1$
    c) $1 ⩽ \dfrac{x}{2}$

   b) $4 ⩾ \dfrac{x}{3}$
   d) $9x ⩾ -18$

    e) $-4x + 1 < 3$

   f) $1 ⩾ -3x + 7$

Worked example   Find the range of values for which $7 < 3x + 1 \leqslant 13$ and illustrate the solutions on a number line.

This is in fact two inequalities which can therefore be solved separately.

$$7 < 3x + 1 \qquad \text{and} \qquad 3x + 1 \leqslant 13$$
$$(-1) \rightarrow \quad 6 < 3x \qquad\qquad\qquad (-1) \rightarrow \quad 3x \quad\ \leqslant 12$$
$$(\div 3) \rightarrow \quad 2 < x \ \text{ that is } \ x > 2 \qquad (\div 3) \rightarrow \quad x \quad\ \leqslant 4$$

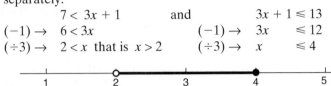

Exercise 7.12   Find the range of values for which the following inequalities are satisfied. Illustrate each solution on a number line:

1.   a) $4 < 2x \leqslant 8$ b) $3 \leqslant 3x < 15$
      c) $7 \leqslant 2x < 10$ d) $10 \leqslant 5x < 21$

2.   a) $5 < 3x + 2 \leqslant 17$ b) $3 \leqslant 2x + 5 < 7$
      c) $12 < 8x - 4 < 20$ d) $15 \leqslant 3(x - 2) < 9$

Extended Section

Student Assessment I

1. The angles of a triangle are x, $2x$ and $(x + 40)$ degrees.
   a) Construct an equation in terms of x.
   b) Solve the equation.
   c) Calculate the size of each of the three angles.

2. Seven is added to three times a number. The result is then doubled. If the answer is 68, calculate the number.

3. A decagon has six equal exterior angles. Each of the remaining four is fifteen degrees larger than these six angles. Construct an equation and then solve it to find the sizes of the angles.

4. Solve the following quadratic equation by factorisation:
   $$x^2 + 6x = -5$$

5. Solve the following quadratic equation by using the quadratic formula:
   $$x^2 + 6 = 8x$$

6. Solve the inequality below and illustrate your answer on a number line:
   $$12 \leq 3(x - 2) < 15$$

7. For what values of p is the following inequality true?
   $$\frac{1}{p^2} \leq -1$$

Student Assessment 2

1. The angles of a quadrilateral are x, $3x$, $(2x - 40)$ and $(3x - 50)$ degrees.
   a) Construct an equation in terms of x.
   b) Solve the equation.
   c) Calculate the size of the four angles.

2. Three is subtracted from seven times a number. The result is multiplied by 5. If the answer is 55, calculate the value of the number by constructing an equation and solving it.

3. The interior angles of a pentagon are $9x$, $5x + 10$, $6x + 5$, $8x - 25$ and $10x - 20$ degrees. If the sum of the interior angles of a pentagon is $540°$, find the size of each of the angles.

4. Solve the following quadratic equation by factorisation:
   $$x^2 - x = 20$$

5. Solve the following quadratic equation by using the quadratic formula:

$$2x^2 - 7 = 3x$$

6. Solve the inequality below and illustrate your answer on a number line

$$6 < 2x \leqslant 10$$

7. For what values of m is the following inequality true?

$$\frac{1}{m^2} > 0$$

Student Assessment 3

1. A rectangle is x cm long. The length is 3 cm more than the width. The perimeter of the rectangle is 54 cm.
   a) Draw a diagram to illustrate the above information.
   b) Construct an equation in terms of x.
   c) Solve the equation and hence calculate the length and width of the rectangle.

2. At the end of a football season the leading goal scorer in a league has scored eight more goals than the second leading goal scorer. The second has scored fifteen more than the third. The total number of goals scored by all three players is 134.
   a) Write an expression for each of the three scores.
   b) Form an equation and then solve it to find the number of goals scored by each player.

3. a) Show that $x = 1 + \dfrac{7}{x - 5}$ can be written as

$$x^2 - 6x - 2 = 0.$$

   b) Use the quadratic formula to solve the equation

$$x = 1 + \frac{7}{x - 5}.$$

4. The angles of a quadrilateral are $x°$, $y°$, $70°$ and $40°$. The difference between the two unknowns is $18°$.
   a) Write two equations from the information given above.
   b) Solve the equations to find x and y.

5. A right-angled triangle has sides of length x, $x - 1$ and $x - 8$.
   a) Illustrate this information on a diagram.
   b) Show from the information given that
      $$x^2 - 18x + 65 = 0.$$
   c) Solve the quadratic equation and calculate the length of each of the three sides.

Student Assessment 4

1. The angles of a triangle are $x°$, $y°$ and $40°$. The difference between the two unknown angles is $30°$.
    a) Write down two equations from the information given above.
    b) What is the size of the two unknown angles?

2. The interior angles of a pentagon increase by $10°$ as you progress clockwise.
    a) Illustrate this information in a diagram.
    b) Write an expression for the sum of the interior angles.
    c) The sum of the interior angles of a pentagon is $540°$. Use this to calculate the largest **exterior** angle of the pentagon.
    d) Illustrate on your diagram the size of each of the five exterior angles.
    e) Show that the sum of the exterior angles is $360°$.

3. A flat sheet of card measures 12 cm by 10 cm. It is made into an open box by cutting a square of side x cm from each corner and then folding up the sides.
    a) Illustrate the box and its dimensions on a simple 3D sketch.
    b) Write an expression for the surface area of the outside of the box.
    c) If the surface area is 56 cm², form and solve a quadratic equation to find the value of x.

4. a) Show that $x - 2 = \dfrac{4}{x - 3}$ can be written as

     $x^2 - 5x + 2 = 0$.

    b) Use the quadratic formula to solve $x - 2 = \dfrac{4}{x - 3}$.

5. A right-angled triangle ABC has side lengths as follows: $AB = x$ cm, AC is 2 cm shorter than AB, and BC is 2 cm shorter than AC.
    a) Illustrate this information on a diagram.
    b) Using this information show that $x^2 - 12x + 20 = 0$.
    c) Solve the above quadratic and hence find the length of each of the three sides of the triangle.

ICT Section

In this activity you will be using a graphing package to analyse the relationship between the graph of a quadratic equation and the equation in its factorised form.

The graph below shows the equation $y = x^2 - 5x + 6$

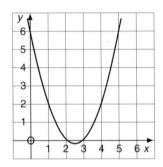

1. a) Factorise the expression $x^2 - 5x + 6$
    b) Compare the factorised form of the expression with the graph above. Write down any similarities that you notice.

2. For each of the following equations:
    i) use a graphing package to plot the graph,
    ii) re-write them in factorised form.
    a) $y = x^2 + 3x - 4$
    b) $y = x^2 - 7x + 10$
    c) $y = x^2 - 2x - 3$
    d) $y = x^2 - 10x + 24$
    e) $y = x^2 - 4x + 4$
    f) $y = -x^2 + 1$

3. By comparing the graphs with the factorised form of each of your equations in Q.2 above, describe how they relate to each other.

8 STRAIGHT LINE GRAPHS

> Calculate the gradient of a straight line from the coordinates of two points on it; calculate the length of a straight line segment from the coordinates of its end points; calculate the length and coordinates of the midpoint of a straight line segment from the coordinates of its end points; interpret and obtain the equation of a straight line graph in the form $y = mx + c$; graphical solution of simultaneous equations.

Lines are made of an infinite number of points. This chapter deals with those points which form a straight line. Each point on a straight line, if plotted on a pair of axes, will have particular coordinates. The relationship between the coordinates of the points on a straight line indicates the characteristics of that straight line.

Gradient

The **gradient** of a straight line refers to its 'steepness' or 'slope'. The gradient of a straight line is **constant**, i.e. it does not change. The gradient can be calculated by considering the coordinates of any two points (x_1, y_1), (x_2, y_2) on the line.

It is calculated using the following formula:

$$\text{gradient} = \frac{\text{vertical distance between the two points}}{\text{horizontal distance between the two points}}$$

By considering the x and y coordinates of the two points this can be rewritten as:

$$\text{gradient} = \frac{y_2 - y_1}{x_2 - x_1}$$

Worked examples

a) The coordinates of two points on a straight line are $(1, 3)$ and $(5, 7)$. Plot the two points on a pair of axes and calculate the gradient of the line joining them.

$$\text{gradient} = \frac{7 - 3}{5 - 1} = \frac{4}{4}$$
$$= 1$$

Note: It does not matter which point we choose to be (x_1, y_1) or (x_2, y_2) as the gradient will be the same. In the example above, choosing the other way round:

$$\text{gradient} = \frac{3 - 7}{1 - 5} = \frac{-4}{-4}$$
$$= 1$$

b) The coordinates of two points on a straight line are $(2, 6)$ and $(4, 2)$. Plot the two points on a pair of axes and calculate the gradient of the line joining them.

$$\text{gradient} = \frac{6 - 2}{2 - 4} = \frac{4}{-2} = -2$$

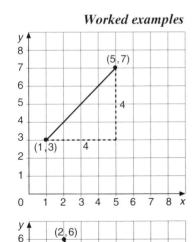

To check whether or not the sign of the gradient is correct the following guideline is useful:

A line sloping this way will have a positive gradient

A line sloping this way will have a negative gradient

Exercise 8.1

1. With the aid of axes if necessary, calculate the gradient of the line joining the following pairs of points:
   a) $(5, 6)$ $(1, 2)$ b) $(6, 4)$ $(3, 1)$
   c) $(1, 4)$ $(5, 8)$ d) $(0, 0)$ $(4, 8)$
   e) $(2, 1)$ $(4, 7)$ f) $(0, 7)$ $(-3, 1)$
   g) $(-3, -3)$ $(-1, 5)$ h) $(4, 2)$ $(-4, -2)$
   i) $(-3, 5)$ $(4, 5)$ j) $(2, 0)$ $(2, 6)$
   k) $(-4, 3)$ $(4, 5)$ l) $(3, 6)$ $(-3, -3)$

2. With the aid of axes if necessary, calculate the gradient of the line joining the following pairs of points:
   a) $(1, 4)$ $(4, 1)$ b) $(3, 6)$ $(7, 2)$
   c) $(2, 6)$ $(6, -2)$ d) $(1, 2)$ $(9, -2)$
   e) $(0, 3)$ $(-3, 6)$ f) $(-3, -5)$ $(-5, -1)$
   g) $(-2, 6)$ $(2, 0)$ h) $(2, -3)$ $(8, 1)$
   i) $(6, 1)$ $(-6, 4)$ j) $(-2, 2)$ $(4, -4)$
   k) $(-5, -3)$ $(6, -3)$ l) $(3, 6)$ $(5, -2)$

▮ Length of a straight line segment

To calculate the length of a line segment, the coordinates of its end points need to be given. Once these are known, Pythagoras' theorem can be used to calculate its length.

Worked example The coordinates of two points are $(1, 3)$ and $(5, 6)$. Plot them on a pair of axes and calculate the distance between them.

By dropping a perpendicular from the point $(5, 6)$ and drawing a line across from $(1, 3)$ a right-angled triangle is formed. The length of the hypotenuse of the triangle is the length we wish to find.

Using Pythagoras' theorem, we have

$$a^2 = 3^2 + 4^2$$
$$a^2 = 25$$
$$a = \sqrt{25}$$
$$a = 5$$

Length is 5 units

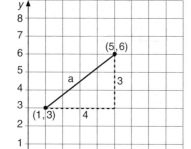

To find the coordinates of the midpoint add the two x coordinates and divide by two to get the x coordinate of the midpoint. Repeat with the y coordinates to find the y coordinate of the midpoint.

Worked examples   **a)** Find the coordinates of the midpoint of the line segment on p.140 with coordinates $(1, 3)$ and $(5, 6)$.

The x coordinate at the midpoint will be $\dfrac{1 + 5}{2} = 3$

The y coordinate at the midpoint will be $\dfrac{3 + 6}{2} = 4.5$

So the coordinates are $(3, 4.5)$.

b) Find the coordinates of the midpoint of a line segment of coordinates $(-2, -5)$ and $(4, 7)$.

The x coordinate at the midpoint will be $\dfrac{-2 + 4}{2} = 1$

The y coordinate at the midpoint will be $\dfrac{-5 + 7}{2} = 1$

So the coordinates are $(1, 1)$.

c) Find the coordinates of the midpoint of a line segment of coordinates $(-2, -5)$ and $(-6, -7)$.

The x coordinate at the midpoint will be $\dfrac{-2 + (-6)}{2} = -4$

The y coordinate at the midpoint will be $\dfrac{-5 + (-7)}{2} = -6$

So the coordinates are $(-4, -6)$.

Exercise 8.2   **1.** With the aid of axes if necessary,
i) calculate the length of the line segment between each of the following pairs of points,
ii) find the coordinates of the midpoint of the line segment.

a) $(5, 6)$	$(1, 2)$	b) $(6, 4)$	$(3, 1)$
c) $(1, 4)$	$(5, 8)$	d) $(0, 0)$	$(4, 8)$
e) $(2, 1)$	$(4, 7)$	f) $(0, 7)$	$(-3, 1)$
g) $(-3, -3)$	$(-1, 5)$	h) $(4, 2)$	$(-4, -2)$
i) $(-3, 5)$	$(4, 5)$	j) $(2, 0)$	$(2, 6)$
k) $(-4, 3)$	$(4, 5)$	l) $(3, 6)$	$(-3, -3)$

2. With the aid of axes if necessary,
   i) calculate the length of the line segment between each of the following pairs of points,
   ii) find the coordinates of the midpoint of the line segment.
   a) $(1, 4)$ $(4, 1)$ b) $(3, 6)$ $(7, 2)$
   c) $(2, 6)$ $(6, -2)$ d) $(1, 2)$ $(9, -2)$
   e) $(0, 3)$ $(-3, 6)$ f) $(-3, -5)$ $(-5, -1)$
   g) $(-2, 6)$ $(2, 0)$ h) $(2, -3)$ $(8, 1)$
   i) $(6, 1)$ $(-6, 4)$ j) $(-2, 2)$ $(4, -4)$
   k) $(-5, -3)$ $(6, -3)$ l) $(3, 6)$ $(5, -2)$

Equations of a straight line

The coordinates of every point on a straight line all have a common relationship. This relationship when expressed algebraically as an equation in terms of x and/or y is known as the equation of the straight line.

Worked examples **a)** By looking at the coordinates of some of the points on the line below, establish the equation of the straight line.

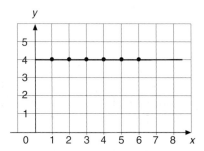

x	y
1	4
2	4
3	4
4	4
5	4
6	4

Some of the points on the line have been identified and their coordinates entered in a table above. By looking at the table it can be seen that the only rule all the points have in common is that $y = 4$.

Hence the equation of the straight line is $y = 4$.

b) By looking at the coordinates of some of the points on the line (left), establish the equation of the straight line.

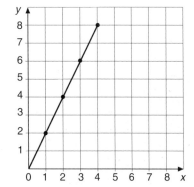

x	y
1	2
2	4
3	6
4	8

Once again, by looking at the table it can be seen that the relationship between the x and y coordinates is that each y coordinate is twice the corresponding x coordinate.

Hence the equation of the straight line is $y = 2x$.

Exercise 8.3   **1.**   In each of the following identify the coordinates of some of the points on the line and use these to find the equation of the straight line.

a)

b)

c)

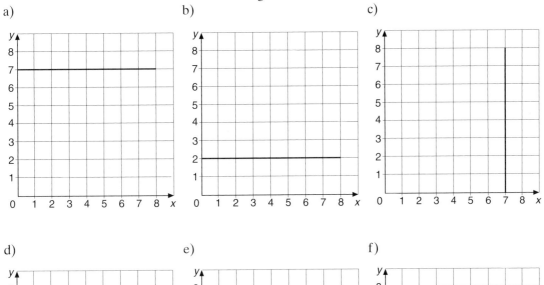

d)

e)

f)

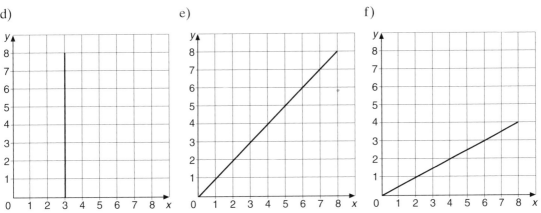

g)

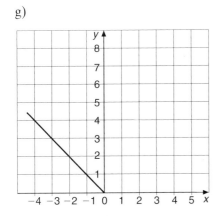

h)

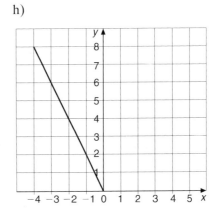

Exercise 8.4   **1.**   In each of the following identify the coordinates of some of the points on the line and use these to find the equation of the straight line.

a)

b)
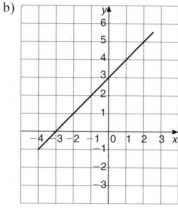

c)
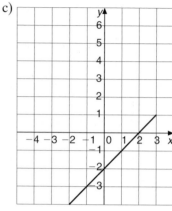

d)
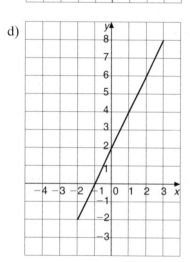

e)
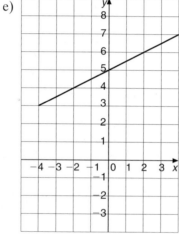

f)
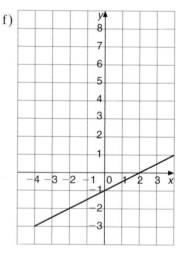

2.   In each of the following identify the coordinates of some of the points on the line and use these to find the equation of the straight line.

a)
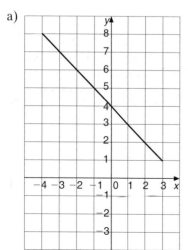

b)
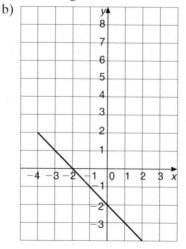

c)
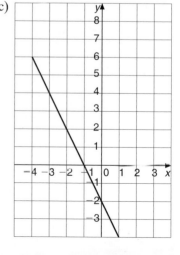

d)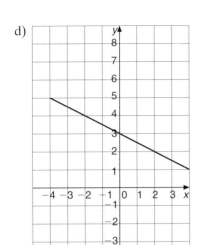

e)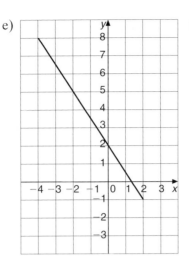

f)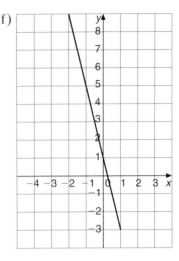

3. a) For each of the graphs in Q.1 and 2 calculate the gradient of the straight line.
   b) What do you notice about the gradient of each line and its equation?
   c) What do you notice about the equation of the straight line and where the line intersects the y-axis?

4. Copy the diagrams in Q.1. Draw two lines on the diagram parallel to the given line.
   a) Write the equation of these new lines in the form $y = mx + c$.
   b) What do you notice about the equations of these new parallel lines?

5. In Q.2 you have an equation for these lines in the form $y = mx + c$. Change the value of the intercept c and then draw the new line.
   What do you notice about this new line and the first line?

In general the equation of any straight line can be written in the form:

$$y = mx + c$$

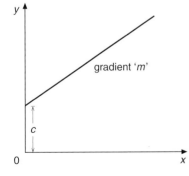

where 'm' represents the gradient of the straight line and 'c' the intercept with the y-axis. This is shown in the diagram (left).

By looking at the equation of a straight line written in the form $y = mx + c$, it is therefore possible to deduce the line's gradient and intercept with the y-axis without having to draw it.

Worked examples **a)** Calculate the gradient and *y*-intercept of the following
straight lines:

    i) $y = 3x - 2$ gradient $= 3$

                               y-intercept $= -2$

    ii) $y = -2x + 6$ gradient $= -2$

                               y-intercept $= 6$

b) Calculate the gradient and *y*-intercept of the following
straight lines:

    i) $2y = 4x + 2$

This needs to be rearranged into **gradient-intercept** form
(i.e. $y = mx + c$).

$$y = 2x + 1 \qquad\qquad \text{gradient} = 2$$
$$\text{*y*-intercept} = 1$$

    ii) $y - 2x = -4$

Rearranging into gradient-intercept form, we have:

$$y = 2x - 4 \qquad\qquad \text{gradient} = 2$$
$$\text{*y*-intercept} = -4$$

    iii) $-4y + 2x = 4$

Rearranging into gradient-intercept form, we have:

$$y = \tfrac{1}{2}x - 1 \qquad\qquad \text{gradient} = \tfrac{1}{2}$$
$$\text{*y*-intercept} = -1$$

    iv) $\dfrac{y + 3}{4} = -x + 2$

Rearranging into gradient-intercept form, we have:

$$y + 3 = -4x + 8$$
$$y = -4x + 5 \qquad\qquad \text{gradient} = -4$$
$$\text{*y*-intercept} = 5$$

Exercise 8.5   For the following linear equations, calculate both the gradient
and *y*-intercept in each case.

1.   a) $y = 2x + 1$ b) $y = 3x + 5$ c) $y = x - 2$
      d) $y = \tfrac{1}{2}x + 4$ e) $y = -3x + 6$ f) $y = -\tfrac{2}{3}x + 1$
      g) $y = -x$ h) $y = -x - 2$ i) $y = -(2x - 2)$

2.   a) $y - 3x = 1$ b) $y + \tfrac{1}{2}x - 2 = 0$
      c) $y + 3 = -2x$ d) $y + 2x + 4 = 0$
      e) $y - \tfrac{1}{4}x - 6 = 0$ f) $-3x + y = 2$
      g) $2 + y = x$ h) $8x - 6 + y = 0$
      i) $-(3x + 1) + y = 0$

3.   a) $2y = 4x - 6$ b) $2y = x + 8$
      c) $\tfrac{1}{2}y = x - 2$ d) $\tfrac{1}{4}y = -2x + 3$
      e) $3y - 6x = 0$ f) $\tfrac{1}{3}y + x = 1$
      g) $6y - 6 = 12x$ h) $4y - 8 + 2x = 0$
      i) $2y - (4x - 1) = 0$

4. a) $2x - y = 4$ b) $x - y + 6 = 0$
  c) $-2y = 6x + 2$ d) $12 - 3y = 3x$
  e) $5x - \frac{1}{2}y = 1$ f) $-\frac{2}{3}y + 1 = 2x$
  g) $9x - 2 = -y$ h) $-3x + 7 = -\frac{1}{2}y$
  i) $-(4x - 3) = -2y$

5. a) $\dfrac{y + 2}{4} = \dfrac{1}{2}x$ b) $\dfrac{y - 3}{x} = 2$ c) $\dfrac{y - x}{8} = 0$

  d) $\dfrac{2y - 3x}{2} = 6$ e) $\dfrac{3y - 2}{x} = -3$ f) $\dfrac{\frac{1}{2}y - 1}{x} = -2$

  g) $\dfrac{3x - y}{2} = 6$ h) $\dfrac{6 - 2y}{3} = 2$ i) $\dfrac{-(x + 2y)}{5x} = 1$

6. a) $\dfrac{3x - y}{y} = 2$ b) $\dfrac{-x + 2y}{4} = y + 1$

  c) $\dfrac{y - x}{x + y} = 2$ d) $\dfrac{1}{y} = \dfrac{1}{x}$

  e) $\dfrac{-(6x + y)}{2} = y + 1$ f) $\dfrac{2x - 3y + 4}{4} = 4$

  g) $\dfrac{y + 1}{x} + \dfrac{3y - 2}{2x} = -1$ h) $\dfrac{x}{y + 1} + \dfrac{1}{2y + 2} = -3$

  i) $\dfrac{-(-y + 3x)}{-(6x - 2y)} = 1$

  j) $\dfrac{-(x - 2y) - (-x - 2y)}{4 + x - y} = -2$

▨ The equation of a line through two points

The equation of a straight line can be deduced once the coordinates of two points on the line are known.

Worked example   Calculate the equation of the straight line passing through the points $(-3, 3)$ and $(5, 5)$.

Plotting the two points gives:

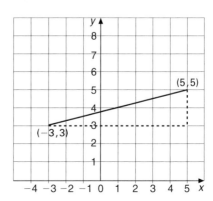

The equation of any straight line can be written in the general form $y = mx + c$. Here we have:

$$\text{gradient} = \frac{5 - 3}{5 - (-3)} = \frac{2}{8}$$

$$\text{gradient} = \frac{1}{4}$$

The equation of the line now takes the form $y = \frac{1}{4}x + c$.

Since the line passes through the two given points, their coordinates must satisfy the equation. So to calculate the value of 'c' the x and y coordinates of one of the points are substituted into the equation. Substituting $(5, 5)$ into the equation gives:

$$5 = \tfrac{1}{4} \times 5 + c$$
$$5 = \tfrac{5}{6} + c$$

Therefore $c = 5 - 1\frac{1}{4} = 3\frac{3}{4}$

The equation of the straight line passing through $(-3, 3)$ and $(5, 5)$ is:

$$y = \tfrac{1}{4}x + 3\tfrac{3}{4}$$

Exercise 8.6 Find the equation of the straight line which passes through each of the following pairs of points:

1. a) $(1, 1)$ $(4, 7)$ b) $(1, 4)$ $(3, 10)$
    c) $(1, 5)$ $(2, 7)$ d) $(0, -4)$ $(3, -1)$
    e) $(1, 6)$ $(2, 10)$ f) $(0, 4)$ $(1, 3)$
    g) $(3, -4)$ $(10, -18)$ h) $(0, -1)$ $(1, -4)$
    i) $(0, 0)$ $(10, 5)$

2. a) $(-5, 3)$ $(2, 4)$ b) $(-3, -2)$ $(4, 4)$
    c) $(-7, -3)$ $(-1, 6)$ d) $(2, 5)$ $(1, -4)$
    e) $(-3, 4)$ $(5, 0)$ f) $(6, 4)$ $(-7, 7)$
    g) $(-5, 2)$ $(6, 2)$ h) $(1, -3)$ $(-2, 6)$
    i) $(6, -4)$ $(6, 6)$

Drawing straight line graphs

To draw a straight line graph only two points need to be known. Once these have been plotted the line can be drawn between them and extended if necessary at both ends.

Worked examples **a)** Plot the line $y = x + 3$.

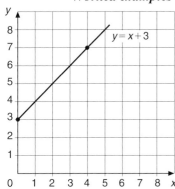

To identify two points simply choose two values of x. Substitute these into the equation and calculate their corresponding y values.

> When $x = 0$, $y = 3$.
> When $x = 4$, $y = 7$.

Therefore two of the points on the line are $(0, 3)$ and $(4, 7)$. The straight line $y = x + 3$ is plotted on the left.

b) Plot the line $y = -2x + 4$.

> When $x = 2$, $y = 0$.
> When $x = -1$, $y = 6$.

The coordinates of two points on the line are $(2, 0)$ and $(-1, 6)$

Note that, in questions of this sort, it is often easier to rearrange the equation into gradient-intercept form first.

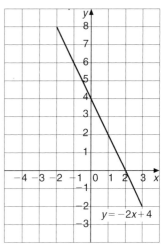

Exercise 8.7

1. Plot the following straight lines.
   a) $y = 2x + 3$ b) $y = x - 4$ c) $y = 3x - 2$
   d) $y = -2x$ e) $y = -x - 1$ f) $-y = x + 1$
   g) $-y = 3x - 3$ h) $2y = 4x - 2$ i) $y - 4 = 3x$

2. Plot the following straight lines:
   a) $-2x + y = 4$ b) $-4x + 2y = 12$
   c) $3y = 6x - 3$ d) $2x = x + 1$
   e) $3y - 6x = 9$ f) $2y + x = 8$
   g) $x + y + 2 = 0$ h) $3x + 2y - 4 = 0$
   i) $4 = 4y - 2x$

3. Plot the following straight lines:
   a) $\dfrac{x + y}{2} = 1$ b) $x + \dfrac{y}{2} = 1$

   c) $\dfrac{x}{3} + \dfrac{y}{2} = 1$ d) $y + \dfrac{x}{2} = 3$

   e) $\dfrac{y}{5} + \dfrac{x}{3} = 0$ f) $\dfrac{-(2x + y)}{4} = 1$

   g) $\dfrac{y - (x - y)}{3x} = -1$ h) $\dfrac{y}{2x + 3} - \dfrac{1}{2} = 0$

   i) $-2(x + y) + 4 = -y$

Graphical solution of simultaneous equations

When solving two equations simultaneously the aim is to find a solution which works for both equations. In Chapter 7 it was shown how to arrive at the solution algebraically. It is, however, possible to arrive at the same solution graphically.

Worked example

i) By plotting both of the following equations on the same axes, find a common solution.

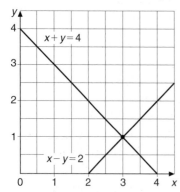

$$x + y = 4$$
$$x - y = 2$$

When both lines are plotted, the point at which they cross gives the common solution as it is the only point which lies on both lines.

Therefore the common solution is the point $(3, 1)$.

ii) Check the result obtained above by solving the equations algebraically.

$$x + y = 4 \quad\quad (1)$$
$$x - y = 2 \quad\quad (2)$$

$$\text{eq.(1)} + \text{eq.(2)} \rightarrow 2x = 6$$
$$x = 3$$

Substituting $x = 3$ into equation (1) we have:

$$3 + y = 4$$
$$y = 1$$

Therefore the common solution occurs at $(3, 1)$.

Exercise 8.8

Solve the simultaneous equations below:
i) by graphical means,
ii) by algebraic means.

1. a) $x + y = 5$ b) $x + y = 7$
    $\quad x - y = 1$ $x - y = 3$
    c) $2x + y = 5$ d) $2x + 2y = 6$
    $\quad x - y = 1$ $2x - y = 3$
    e) $x + 3y = -1$ f) $x - y = 6$
    $\quad x - 2y = -6$ $x + y = 2$

2. a) $3x - 2y = 13$ b) $4x - 5y = 1$
    $\quad 2x + y = 4$ $2x + y = -3$
    c) $x + 5 = y$ d) $x = y$
    $\quad 2x + 3y - 5 = 0$ $x + y + 6 = 0$
    e) $2x + y = 4$ f) $y - 3x = 1$
    $\quad 4x + 2y = 8$ $y = 3x - 3$

Student Assessment 1

1. Sketch the following graphs on the same pair of axes labelling each clearly.

   a) $x = -2$ b) $y = 3$

   c) $y = 2x$ d) $y = -\dfrac{x}{2}$

2. For each of the following linear equations:
   i) calculate the gradient and y-intercept,
   ii) plot the graph.

        a) $y = x + 1$ b) $y = 3 - 3x$

        c) $2x - y = -4$ d) $2y - 5x = 8$

3. Find the equation of the straight line which passes through each of the following pairs of points:

   a) $(1, -1)$ $(4, 8)$ b) $(0, 7)$ $(3, 1)$

4. The coordinates of the end points of two line segments are given below. Calculate the length of each line segment.

   a) $(-6, -1)$ $(6, 4)$ b) $(1, 2)$ $(7, 10)$

5. Solve the following pairs of simultaneous equations graphically:

        a) $x + y = 4$ b) $3x + y = 2$
           $x - y = 0$ $x - y = 2$

        c) $y + 4x + 4 = 0$ d) $x - y = -2$
           $x + y = 2$ $3x + 2y + 6 = 0$

Student Assessment 2

1. Sketch the following graphs on the same pair of axes, labelling each clearly.

   a) $x = 3$ b) $y = -2$

   c) $y = -3x$ d) $y = \dfrac{x}{4} + 4$

2. For each of the following linear equations:
   i) calculate the gradient and y-intercept,
   ii) plot the graph.

        a) $y = 2x + 3$ b) $y = 4 - x$

        c) $2x - y = 3$ d) $-3x + 2y = 5$

3. Find the equation of the straight line which passes through each of the following pairs of points:

   a) $(-2, -9)$ $(5, 5)$ b) $(1, -1)$ $(-1, 7)$

4. The coordinates of the end points of two line segments are given below. Calculate the length of each of the lines.

   a) $(2, 6)$ $(-2, 3)$ b) $(-10, -10)$ $(0, 14)$

5. Solve the following pairs of simultaneous equations graphically.

        a) $x + y = 6$ b) $x + 2y = 8$
           $x - y = 0$ $x - y = -1$

        c) $2x - y = -5$ d) $4x - 2y = -2$
           $x - 3y = 0$ $3x - y + 2 = 0$

ICT Section

You have seen in this chapter that the solution of two simultaneous equations, when plotted, represents the point at which both lines intersect.

The graph below shows the solution to the simultaneous equations:

$$x + y = 5$$
$$2x - y = 1$$

The solution occurs at $(2, 3)$ i.e $x = 2$ and $y = 3$

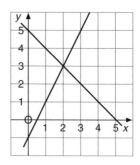

1. a) On a grid plot the point $(2, 2)$.
   b) Write down the equation of two straight lines that pass through the point $(2, 2)$.
   c) Check that these lines do pass through the point $(2, 2)$ using a graphing package.
   d) Solve your two equations simultaneously.

2. Repeat Q.1 above for the following points:
   i) $(3, 4)$
   ii) $(2, 0)$
   iii) $(0, 6)$
   iv) $(-5, 1)$
   v) $(-4, -\frac{1}{2})$
   vi) $(\frac{2}{5}, -3)$

9 FUNCTIONS

Use function notation, e.g. f(x) = 3x − 5, f: x → 3x − 5 to describe simple functions, and the notation f⁻¹(x) to describe their inverses; form composite functions as defined by gf(x) = g(f(x)).

An expression such as $4x - 9$, in which the variable is x, is called 'a function of x'. Its numerical value depends on the value of x. We sometimes write $f(x) = 4x - 9$, or $f: x \rightarrow 4x - 9$.

Worked examples **a)** For the function $f(x) = 3x - 5$, evaluate:

i) f(2)   ii) f(0)

$$f(2) = 3 \times 2 - 5 \qquad\qquad f(0) = 3 \times 0 - 5$$
$$= 6 - 5 \qquad\qquad\qquad = 0 - 5$$
$$= 1 \qquad\qquad\qquad\quad = -5$$

iii) f(−2)

$$f(-2) = 3 \times (-2) - 5$$
$$= -6 - 5$$
$$= -11$$

b) For the function $f: x \rightarrow \dfrac{2x + 6}{3}$, evaluate:

i) f(3)   ii) f(1.5)

$$f(3) = \frac{2 \times 3 + 6}{3} \qquad\qquad f(1.5) = \frac{2 \times 1.5 + 6}{3}$$
$$= \frac{6 + 6}{3} \qquad\qquad\qquad = \frac{3 + 6}{3}$$
$$= 4 \qquad\qquad\qquad\quad = 3$$

iii) f(−1)

$$f(-1) = \frac{2 \times (-1) + 6}{3}$$
$$= \frac{-2 + 6}{3}$$
$$= \frac{4}{3}$$

c) For the function $f(x) = x^2 + 4$, evaluate:

i) f(2)   ii) f(6)

$$f(2) = 2^2 + 4 \qquad\qquad f(6) = 6^2 + 4$$
$$= 4 + 4 \qquad\qquad\qquad = 36 + 4$$
$$= 8 \qquad\qquad\qquad\quad = 40$$

iii) f(−1)

$$f(-1) = -1^2 + 4$$
$$= 1 + 4$$
$$= 5$$

Exercise 9.1

1. If $f(x) = 2x + 2$, calculate:

a) $f(2)$ b) $f(4)$ c) $f(0.5)$ d) $f(1.5)$
e) $f(0)$ f) $f(-2)$ g) $f(-6)$ h) $f(-0.5)$

2. If $f(x) = 4x - 6$, calculate:

a) $f(4)$ b) $f(7)$ c) $f(3.5)$ d) $f(0.5)$
e) $f(0.25)$ f) $f(-3)$ g) $f(-4.25)$ h) $f(0)$

3. If $g(x) = -5x + 2$, calculate:

a) $g(0)$ b) $g(6)$ c) $g(4.5)$ d) $g(3.2)$
e) $g(0.1)$ f) $g(-2)$ g) $g(-6.5)$ h) $g(-2.3)$

4. If $h(x) = -3x - 7$, calculate:

a) $h(4)$ b) $h(6.5)$ c) $h(0)$ d) $h(0.4)$
e) $h(-9)$ f) $h(-5)$ g) $h(-2)$ h) $h(-3.5)$

Exercise 9.2

1. If $f(x) = \dfrac{3x + 2}{4}$, calculate:

a) $f(2)$ b) $f(8)$ c) $f(2.5)$ d) $f(0)$
e) $f(-0.5)$ f) $f(-6)$ g) $f(-4)$ h) $f(-1.6)$

2. If $g(x) = \dfrac{5x - 3}{5}$, calculate:

a) $g(3)$ b) $g(6)$ c) $g(0)$ d) $g(-3)$
e) $g(-1.5)$ f) $g(-9)$ g) $g(-0.2)$ h) $g(-0.1)$

3. If $h: x \rightarrow \dfrac{-6x + 8}{4}$, calculate:

a) $h(1)$ b) $h(0)$ c) $h(4)$ d) $h(1.5)$
e) $h(-2)$ f) $h(-0.5)$ g) $h(-22)$ h) $h(-1.5)$

4. If $f(x) = \dfrac{-5x - 7}{-8}$, calculate:

a) $f(5)$ b) $f(1)$ c) $f(3)$ d) $f(-1)$
e) $f(-7)$ f) $f(-\tfrac{3}{5})$ g) $f(-0.8)$ h) $f(0)$

Exercise 9.3

1. If $f(x) = x^2 + 3$, calculate:

a) $f(4)$ b) $f(7)$ c) $f(1)$ d) $f(0)$
e) $f(-1)$ f) $f(0.5)$ g) $f(-3)$ h) $f(\sqrt{2})$

2. If $f(x) = 3x^2 - 5$, calculate:

a) $f(5)$ b) $f(8)$ c) $f(1)$ d) $f(0)$
e) $f(-2)$ f) $f(\sqrt{3})$ g) $f(-\tfrac{1}{2})$ h) $f(-\tfrac{1}{3})$

3. If $g(x) = -2x^2 + 4$, calculate:

a) $g(3)$ b) $g(\tfrac{1}{2})$ c) $g(0)$ d) $g(1.5)$
e) $g(-4)$ f) $g(-1)$ g) $g(\sqrt{5})$ h) $g(-6)$

4. If $h(x) = \dfrac{-5x^2 + 15}{-2}$, calculate:

a) $h(1)$ b) $h(4)$ c) $h(\sqrt{3})$ d) $h(0.5)$

e) $h(0)$ f) $h(-3)$ g) $h\left(\dfrac{1}{\sqrt{5}}\right)$ h) $h(-2.5)$

5. If $f(x) = -6x(x - 4)$, calculate:
   a) $f(0)$ b) $f(2)$ c) $f(4)$ d) $f(0.5)$
   e) $f(-\frac{1}{2})$ f) $f(-\frac{1}{6})$ g) $f(-2.5)$ h) $f(\sqrt{2})$

6. If $g: x \rightarrow \dfrac{(x + 2)(x - 4)}{-x}$, calculate:
   a) $g(1)$ b) $g(4)$ c) $g(8)$ d) $g(0)$
   e) $g(-2)$ f) $g(-10)$ g) $g(-\frac{3}{2})$ h) $g(-8)$

Exercise 9.4

1. If $f(x) = 2x + 1$, write down the following in their simplest form:
   a) $f(x + 1)$ b) $f(2x - 3)$ c) $f(x^2)$
   d) $f\left(\dfrac{x}{2}\right)$ e) $f\left(\dfrac{x}{4} + 1\right)$ f) $f(x) - x$

2. If $g(x) = 3x^2 - 4$, write down the following in their simplest form:
   a) $g(2x)$ b) $g\left(\dfrac{x}{4}\right)$ c) $g(\sqrt{2x})$

   d) $g(3x) + 4$ e) $g(x - 1)$ f) $g(2x + 2)$

3. If $f(x) = 4x^2 + 3x - 2$, write down the following in their simplest form:
   a) $f(x) + 4$ b) $f(2x) + 2$ c) $f(x + 2) - 20$

   d) $f(x - 1) + 1$ e) $f\left(\dfrac{x}{2}\right)$ f) $f(3x + 2)$

Inverse functions

The **inverse** of a function is its reverse, i.e. it 'undoes' the function's effects. The inverse of the function $f(x)$ is written as $f^{-1}(x)$.

Worked examples   **a)** Find the inverse of each of the following functions:
   i) $f(x) = x + 2$ ii) $g(x) = 2x - 3$
      $f^{-1}(x) = x - 2$ $g^{-1}(x) = \dfrac{(x + 3)}{2}$

   b) If $f(x) = \dfrac{x - 3}{3}$ calculate:
   i) $f^{-1}(2)$ ii) $f^{-1}(-3)$
      $f^{-1}(x) = 3x + 3$ $f^{-1}(x) = 3x + 3$
      $f^{-1}(2) = 9$ $f^{-1}(-3) = -6$

Exercise 9.5   Find the inverse of each of the following functions:

1. a) $f(x) = x + 3$ b) $f(x) = x + 6$
   c) $f(x) = x - 5$ d) $g(x) = x$

   e) $h(x) = 2x$ f) $p(x) = \dfrac{x}{3}$

2. a) $f(x) = 4x$ b) $f(x) = 2x + 5$

    c) $f(x) = 3x - 6$ d) $f(x) = \dfrac{x + 4}{2}$

    e) $g(x) = \dfrac{3x - 2}{4}$ f) $g(x) = \dfrac{8x + 7}{5}$

3. a) $f(x) = \tfrac{1}{2}x + 3$ b) $g(x) = \tfrac{1}{4}x - 2$
    c) $h(x) = 4(3x - 6)$ d) $p(x) = 6(x + 3)$
    e) $q(x) = -2(-3x + 2)$ f) $f(x) = \tfrac{2}{3}(4x - 5)$

Exercise 9.6

1. If $f(x) = x - 4$, evaluate:
    a) $f^{-1}(2)$ b) $f^{-1}(0)$ c) $f^{-1}(-5)$

2. If $f(x) = 2x + 1$, evaluate:
    a) $f^{-1}(5)$ b) $f^{-1}(0)$ c) $f^{-1}(-11)$

3. If $g(x) = 6(x - 1)$, evaluate:
    a) $g^{-1}(12)$ b) $g^{-1}(3)$ c) $g^{-1}(6)$

4. If $g(x) = \dfrac{2x + 4}{3}$, evaluate:
    a) $g^{-1}(4)$ b) $g^{-1}(0)$ c) $g^{-1}(-6)$

5. If $h(x) = \tfrac{1}{3}x - 2$, evaluate:
    a) $h^{-1}(-\tfrac{1}{2})$ b) $h^{-1}(0)$ c) $h^{-1}(-2)$

6. If $f(x) = \dfrac{4x - 2}{5}$, evaluate:
    a) $f^{-1}(6)$ b) $f^{-1}(-2)$ c) $f^{-1}(0)$

▦ Composite functions

Worked examples   **a)** If $f(x) = x + 2$ and $g(x) = x + 3$, find $fg(x)$.

$$fg(x) = f(x + 3)$$
$$= (x + 3) + 2$$
$$= x + 5$$

b) If $f(x) = 2x - 1$ and $g(x) = x - 2$, find $fg(x)$.

$$fg(x) = f(x - 2)$$
$$= 2(x - 2) - 1$$
$$= 2x - 4 - 1$$
$$= 2x - 5$$

c) If $f(x) = 2x + 3$ and $g(x) = 2x$, evaluate $fg(3)$.

$$fg(x) = f(2x)$$
$$= 2(2x) + 3$$
$$= 4x + 3$$
$$fg(3) = 4 \times 3 + 3$$
$$= 15$$

Exercise 9.7

1. Write a formula for fg(x) in each of the following:
   a) $f(x) = x - 3$ $g(x) = x + 5$
   b) $f(x) = x + 4$ $g(x) = x - 1$
   c) $f(x) = x$ $g(x) = 2x$
   d) $f(x) = \dfrac{x}{2}$ $g(x) = 2x$

2. Write a formula for pq(x) in each of the following:
   a) $p(x) = 2x$ $q(x) = x + 4$
   b) $p(x) = 3x + 1$ $q(x) = 2x$
   c) $p(x) = 4x + 6$ $q(x) = 2x - 1$
   d) $p(x) = -x + 4$ $q(x) = x + 2$

3. Write a formula for jk(x) in each of the following:
   a) $j(x) = \dfrac{x - 2}{4}$ $k(x) = 4x$
   b) $j(x) = 3x + 2$ $k(x) = \dfrac{x - 3}{2}$
   c) $j(x) = \dfrac{2x + 5}{3}$ $k(x) = \frac{1}{2}x + 1$
   d) $j(x) = \frac{1}{4}(x - 3)$ $k(x) = \dfrac{8x + 2}{5}$

4. Evaluate fg(2) in each of the following:
   a) $f(x) = x - 4$ $g(x) = x + 3$
   b) $f(x) = 2x$ $g(x) = -x + 6$
   c) $f(x) = 3x$ $g(x) = 6x + 1$
   d) $f(x) = \dfrac{x}{2}$ $g(x) = -2x$

5. Evaluate gh(−4) in each of the following:
   a) $g(x) = 3x + 2$ $h(x) = -4x$
   b) $g(x) = \frac{1}{2}(3x - 1)$ $h(x) = \dfrac{2x}{5}$
   c) $g(x) = 4(-x + 2)$ $h(x) = \dfrac{2x + 6}{4}$
   d) $g(x) = \dfrac{4x + 4}{5}$ $h(x) = -\frac{1}{3}(-x + 5)$

Student Assessment 1

1. For the function $f(x) = 5x - 1$, evaluate:
   a) $f(2)$
   b) $f(0)$
   c) $f(-3)$

2. For the function $g: x \rightarrow \dfrac{3x - 2}{2}$, evaluate:

   a) $g(4)$
   b) $g(0)$
   c) $g(-3)$

3. For the function $f(x) = \dfrac{(x + 3)(x - 4)}{2}$, evaluate:

   a) $f(0)$
   b) $f(-3)$
   c) $f(-6)$

4. Find the inverse of each of the following functions:

   a) $f(x) = -x + 4$
   b) $g(x) = \dfrac{3(x - 6)}{2}$

5. If $h(x) = \frac{3}{2}(-x + 3)$, evaluate:
   a) $h^{-1}(-3)$
   b) $h^{-1}(\frac{3}{2})$

6. If $f(x) = 4x + 2$ and $g(x) = -x + 3$, find $fg(x)$.

Student Assessment 2

1. For the function $f(x) = 3x + 1$, evaluate:
   a) $f(4)$
   b) $f(-1)$
   c) $f(0)$

2. For the function $g: x \rightarrow \dfrac{-x - 2}{3}$, evaluate:

   a) $g(4)$
   b) $g(-5)$
   c) $g(1)$

3. For the function $f(x) = x^2 - 3x$, evaluate:
   a) $f(1)$
   b) $f(3)$
   c) $f(-3)$

4. Find the inverse of the following functions:

   a) $f(x) = -3x + 9$
   b) $g(x) = \dfrac{(x - 6)}{4}$

5. If $h(x) = -5(-2x + 4)$, evaluate:
   a) $h^{-1}(-10)$
   b) $h^{-1}(0)$

6. If $f(x) = 8x + 2$ and $g(x) = 4x - 1$, find $fg(x)$.

ICT Section

Using a graphing package for this activity, you will investigate the graphical relationship between a function and its inverse.

1. Calculate the inverse of each of the following linear functions.

   a) $f(x) = 3x$

   b) $f(x) = 2x - 2$

   c) $f(x) = \frac{1}{2}x + 1$

   d) $f(x) = \dfrac{2x + 1}{3}$

   e) $f(x) = \dfrac{4x - 3}{2}$

   f) $f(x) = \dfrac{3x + 5}{8}$

2. Using a graphing package, plot on the same axes each of the functions and their inverses in Q.1 above.

3. Comment on the graphical relationship between a linear function and its inverse.

10 GRAPHS OF FUNCTIONS

Core Section

> Construct tables of values for functions of the form $\pm ax^2 + bx + c$, a/x ($x \neq 0$) where a, b and c are integral constants; draw and interpret such graphs; solve linear and quadratic equations approximately by graphical methods.

You should already be familiar with the work covered on straight line graphs before embarking on this chapter.

Quadratic functions

The general expression for a quadratic function takes the form $ax^2 + bx + c$ where a, b and c are constants. Some examples of quadratic functions are given below.

$$y = 2x^2 + 3x - 12 \qquad y = x^2 - 5x + 6 \qquad y = 3x^2 + 2x - 3$$

If a graph of a quadratic function is plotted, the smooth curve produced is called a **parabola**, e.g.

$$y = x^2$$

x	-4	-3	-2	-1	0	1	2	3	4
y	16	9	4	1	0	1	4	9	16

$$y = -x^2$$

x	-4	-3	-2	-1	0	1	2	3	4
y	-16	-9	-4	-1	0	-1	-4	-9	-16

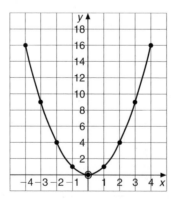
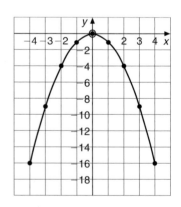

Worked examples **a)** Plot a graph of the function $y = x^2 - 5x + 6$ for $0 \leqslant x \leqslant 5$.

A table of values for x and y is given below:

x	0	1	2	3	4	5
y	6	2	0	0	2	6

This can then be plotted to give the graph:

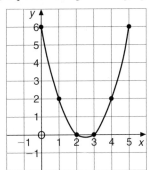

b) Plot a graph of the function $y = -x^2 + x + 2$ for $-3 \leqslant x \leqslant 4$.

Drawing up a table of values gives:

x	-3	-2	-1	0	1	2	3	4
y	-10	-4	0	2	2	0	-4	-10

The graph of the function is given below:

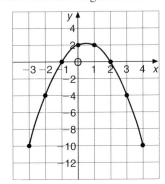

Exercise 10.1   For each of the following quadratic functions, construct a table of values and then draw the graph.

1. $y = x^2 + x - 2$, $-4 \leqslant x \leqslant 3$

2. $y = -x^2 + 2x + 3$, $-3 \leqslant x \leqslant 5$

3. $y = x^2 - 4x + 4$, $-1 \leqslant x \leqslant 5$

4. $y = -x^2 - 2x - 1$, $-4 \leqslant x \leqslant 2$

5. $y = x^2 - 2x - 15$, $-4 \leqslant x \leqslant 6$

6. $y = 2x^2 - 2x - 3$, $-2 \leqslant x \leqslant 3$

7. $y = -2x^2 + x + 6$, $-3 \leqslant x \leqslant 3$

8. $y = 3x^2 - 3x - 6$, $-2 \leqslant x \leqslant 3$

9. $y = 4x^2 - 7x - 4$, $-1 \leqslant x \leqslant 3$

10. $y = -4x^2 + 4x - 1$, $-2 \leqslant x \leqslant 3$

Graphical solution of a quadratic equation

Worked example i) Draw a graph of $y = x^2 - 4x + 3$ for $-2 \leqslant x \leqslant 5$.

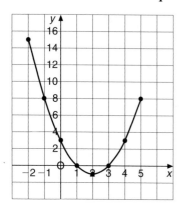

x	-2	-1	0	1	2	3	4	5
y	15	8	3	0	-1	0	3	8

ii) Use the graph to solve the equation $x^2 - 4x + 3 = 0$.

To solve the equation it is necessary to find the values of x when $y = 0$, i.e. where the graph crosses the x-axis. These points occur when $x = 1$ and $x = 3$ and are therefore the solutions.

Exercise 10.2   Solve each of the quadratic functions below by plotting a graph for the ranges of x stated.

1. $x^2 - x - 6 = 0$, $-4 \leqslant x \leqslant 4$

2. $-x^2 + 1 = 0$, $-4 \leqslant x \leqslant 4$

3. $x^2 - 6x + 9 = 0$, $0 \leqslant x \leqslant 6$

4. $-x^2 - x + 12 = 0$, $-5 \leqslant x \leqslant 4$

5. $x^2 - 4x + 4 = 0$, $-2 \leqslant x \leqslant 6$

6. $2x^2 - 7x + 3 = 0$, $-1 \leqslant x \leqslant 5$

7. $-2x^2 + 4x - 2 = 0$ $-2 \leqslant x \leqslant 4$

8. $3x^2 - 5x - 2 = 0$, $-1 \leqslant x \leqslant 3$

In the previous worked example, as $y = x^2 - 4x + 3$, a solution could be found to the equation $x^2 - 4x + 3 = 0$ by reading off where the graph crossed the x-axis. The graph can, however, also be used to solve other quadratic equations.

Worked example   Use the graph of $y = x^2 - 4x + 3$ to solve the equation $x^2 - 4x + 1 = 0$.

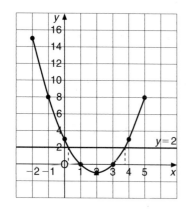

$x^2 - 4x + 1 = 0$ can be rearranged to give:

$$x^2 - 4x + 3 = 2$$

Using the graph of $y = x^2 - 4x + 3$ and plotting the line $y = 2$ on the same graph gives the graph shown on the left.

Where the curve and the line cross gives the solution to $x^2 - 4x + 3 = 2$ and hence also $x^2 - 4x + 1 = 0$.

Therefore the solutions to $x^2 - 4x + 1 = 0$ are $x \approx 0.3$ and 3.7.

Exercise 10.3   Using the graphs which you drew in Ex.10.2, solve the
following quadratic equations. Show your method clearly.

1. $x^2 - x - 4 = 0$

2. $-x^2 - 1 = 0$

3. $x^2 - 6x + 8 = 0$

4. $-x^2 - x + 9 = 0$

5. $x^2 - 4x + 1 = 0$

6. $2x^2 - 7x = 0$

7. $-2x^2 + 4x = -1$

8. $3x^2 = 2 + 5x$

▨ The reciprocal function

Worked example   Draw the graph of $y = \dfrac{2}{x}$ for $-4 \leqslant x \leqslant 4$.

x	-4	-3	-2	-1	0	1	2	3	4
y	-0.5	-0.7	-1	-2	—	2	1	0.7	0.5

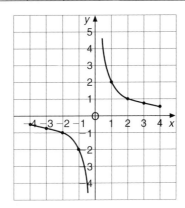

This is a reciprocal function giving a hyperbola.

Exercise 10.4   1.   Plot the graph of the function $y = \dfrac{1}{x}$ for $-4 \leqslant x \leqslant 4$.

2.   Plot the graph of the function $y = \dfrac{3}{x}$ for $-4 \leqslant x \leqslant 4$.

3.   Plot the graph of the function $y = \dfrac{5}{3x}$ for $-4 \leqslant x \leqslant 4$.

Core Section

Student Assessment 1

1. Sketch the graphs of the following functions:

   a) $y = x^2$

   b) $y = -x^2$

2. a) Copy and complete the table below for the function $y = x^2 + 8x + 15$

x	−7	−6	−5	−4	−3	−2	−1	0	1	2
y		3			3					

   b) Plot a graph of the function.

3. Plot a graph of each of the functions below between the given limits of x.

   a) $y = -x^2 - 2x - 1$, $\quad -3 \leqslant x \leqslant 3$

   b) $y = x^2 + 2x - 7$, $\qquad -5 \leqslant x \leqslant 3$

4. a) Plot the graph of the quadratic function $y = x^2 + 9x + 20$ for $-7 \leqslant x \leqslant -2$.

   b) Showing your method clearly, use your graph to solve the equation $x^2 = -9x - 14$.

5. a) Plot the graph of $y = \dfrac{1}{x}$ for $-4 \leqslant x \leqslant 4$.

   b) Showing your method clearly, use your graph to solve the equation $1 = -x^2 + 3x$.

Student Assessment 2

1. Sketch the graph of the function $y = \dfrac{1}{x}$.

2. a) Copy and complete the table below for the function $y = -x^2 - 7x - 12$.

x	−7	−6	−5	−4	−3	−2	−1	0	1	2
y		−6				−2				

   b) Plot a graph of the function.

3. Plot a graph of each of the functions below between the given limits of x.

   a) $y = x^2 - 3x - 10$, $\quad -3 \leqslant x \leqslant 6$

   b) $y = -x^2 - 4x - 4$, $\quad -5 \leqslant x \leqslant 1$

4. a) Plot the graph of the quadratic equation $y = -x^2 - x + 15$ for $-6 \leqslant x \leqslant 4$.

   b) Showing your method clearly, use your graph to solve the following equations:

   i) $10 = x^2 + x$

   ii) $x^2 = x + 5$

5. a) Plot the graph of $y = \dfrac{2}{x}$ for $-4 \leqslant x \leqslant 4$.

   b) Showing your method clearly, use your graph to solve the equation $x^2 + x = 2$.

GRAPHS OF FUNCTIONS

Extended Section

Construct tables of values and draw graphs for functions of the form ax^n where a is a rational constant and $n = -2, -1, 0, 1, 2, 3$ and simple sums of not more than three of these and for functions of the form a^x where a is a positive integer; estimate gradients of curves by drawing tangents; solve associated equations approximately by graphical methods.

▨ Types of graph

Graphs of functions of the form ax^n take different forms depending on the values of a and n. The different types of line produced also have different names, as described below.

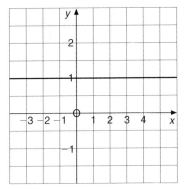

If $a = 1$ and $n = 0$, then $f(x) = x^0$. This is a **linear** function giving a **straight line**.

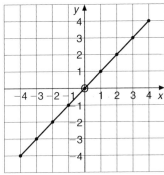

If $a = 1$ and $n = 1$, then $f(x) = x^1$. This is a **linear** function giving a **straight line**.

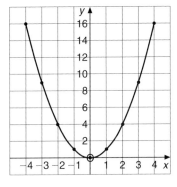

If $a = 1$ and $n = 2$, then $f(x) = x^2$. This is a **quadratic** function giving a **parabola**.

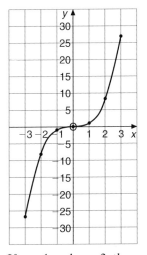

If $a = 1$ and $n = 3$, then $f(x) = x^3$. This is a **cubic** function giving a **cubic curve**.

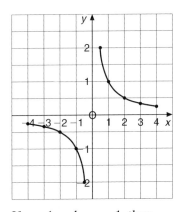

If $a = 1$ and $n = -1$, then $f(x) = x^{-1}$ or $f(x) = \dfrac{1}{x}$. This is a **reciprocal** function giving a **hyperbola**.

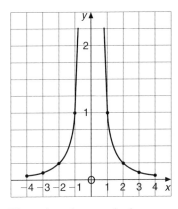

If $a = 1$ and $n = -2$, then $f(x) = x^{-2}$ or $f(x) = \dfrac{1}{x^2}$. This is a **reciprocal** function, shown on the graph above.

Worked example   Draw a graph of the function $y = 2x^2$ for $-3 \leqslant x \leqslant 3$.

x	−3	−2	−1	0	1	2	3
y	18	8	2	0	2	8	18

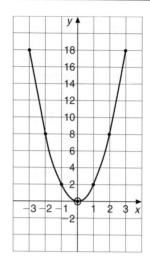

Exercise 10.5

For each of the functions given below:
i) draw up a table of values for x and $f(x)$,
ii) plot the graph of the function.

1. $f(x) = 3x + 2,$ $-3 \leqslant x \leqslant 3$

2. $f(x) = \frac{1}{2}x + 4,$ $-3 \leqslant x \leqslant 3$

3. $f(x) = -2x - 3,$ $-4 \leqslant x \leqslant 2$

4. $f(x) = 2x^2 - 1,$ $-3 \leqslant x \leqslant 3$

5. $f(x) = 0.5x^2 + x - 2,$ $-5 \leqslant x \leqslant 3$

6. $f(x) = 3x^2 - 2x + 1,$ $-2 \leqslant x \leqslant 2$

7. $f(x) = 2x^3,$ $-2 \leqslant x \leqslant 2$

8. $f(x) = -2x^3 + x,$ $-2 \leqslant x \leqslant 2$

9. $f(x) = \frac{1}{2}x^3 - 2x + 3,$ $-3 \leqslant x \leqslant 3$

10. $f(x) = 3x^{-1},$ $-3 \leqslant x \leqslant 3$

11. $f(x) = 2x^{-2},$ $-3 \leqslant x \leqslant 3$

12. $f(x) = \dfrac{1}{x^2} + 3x$ $-3 \leqslant x \leqslant 3$

Exponential functions

Functions of the form $y = a^x$ are known as **exponential** functions. Plotting an exponential function is done in the same way as for other functions.

Worked example   Plot the graph of the function $y = 2^x$ for $-3 \leqslant x \leqslant 3$.

x	−3	−2	−1	0	1	2	3
y	0.125	0.25	0.5	1	2	4	8

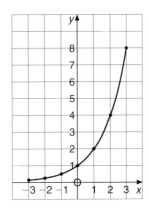

Exercise 10.6

For each of the functions below:
  i) draw up a table of values of x and $f(x)$,
  ii) plot a graph of the function.

1. $f(x) = 3^x$, $-3 \leqslant x \leqslant 3$

2. $f(x) = 1^x$, $-3 \leqslant x \leqslant 3$

3. $f(x) = 2^x + 3$, $-3 \leqslant x \leqslant 3$

4. $f(x) = 2^x + x$, $-3 \leqslant x \leqslant 3$

5. $f(x) = 2^x - x$, $-3 \leqslant x \leqslant 3$

6. $f(x) = 3^x - x^2$, $-3 \leqslant x \leqslant 3$

▨ Gradients of curves

The gradient of a straight line is constant and is calculated by considering the coordinates of two of the points on the line and then carrying out the calculation $\dfrac{y_2 - y_1}{x_2 - x_1}$ as shown below:

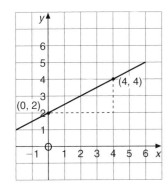

$$\text{Gradient} = \frac{4 - 2}{4 - 0}$$

$$= \frac{1}{2}$$

The gradient of a curve, however, is not constant: its slope changes. To calculate the gradient of a curve at a specific point, the following steps need to be taken:

■ draw a tangent to the curve at that point,
■ calculate the gradient of the tangent.

Worked example   For the function $y = 2x^2$, calculate the gradient of the curve at the point where $x = 1$.

The graph of the function $y = 2x^2$ is given on the left. Identifying the point on the curve where $x = 1$ and then drawing a tangent to that point gives:

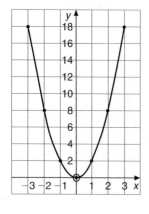

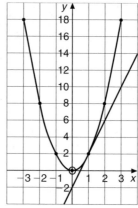

Two points on the tangent are identified in order to calculate its gradient.

$$\text{Gradient} = \frac{10 - (-2)}{3 - 0}$$

$$= \frac{12}{3}$$

$$= 4$$

Therefore the gradient of the function $y = 2x^2$ when $x = 1$ is 4.

Exercise 10.7   For each of the functions below:
  i)   plot a graph,
  ii) calculate the gradient of the function at the specified point.

1. $y = x^2$, $-4 \leqslant x \leqslant 4$,   gradient where $x = 1$
2. $y = \frac{1}{2}x^2$, $-4 \leqslant x \leqslant 4$,   gradient where $x = -2$
3. $y = x^3$, $-3 \leqslant x \leqslant 3$,   gradient where $x = 1$
4. $y = x^3 - 3x^2$, $-4 \leqslant x \leqslant 4$,   gradient where $x = -2$
5. $y = 4x^{-1}$, $-4 \leqslant x \leqslant 4$,   gradient where $x = -1$
6. $y = 2^x$, $-3 \leqslant x \leqslant 3$,   gradient where $x = 0$

Solving equations by graphical methods

As shown in the core section of this chapter, if a graph of a function is plotted, then it can be used to solve equations.

Worked examples

a) i) Plot a graph of $y = 3x^2 - x - 2$ for $-3 \leqslant x \leqslant 3$.

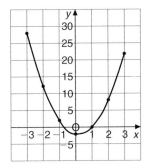

x	-3	-2	-1	0	1	2	3
y	28	12	2	-2	0	8	22

ii) Use the graph to solve the equation $3x^2 - x - 2 = 0$.

To solve the equation, $y = 0$. Therefore where the curve intersects the x-axis gives the solution to the equation.
i.e. $3x^2 - x - 2 = 0$ when $x = -0.7$ and 1

iii) Use the graph to solve the equation $3x^2 - 7 = 0$

To be able to use the original graph, this equation needs to be manipulated in such a way that one side of the equation becomes:
$$3x^2 - x - 2.$$
Manipulating $3x^2 - 7 = 0$ gives:
$$3x^2 - x - 2 = -x + 5 \quad \text{(subtracting } x \text{ from both sides, and adding 5 to both sides)}$$
Hence finding where the curve $y = 3x^2 - x - 2$ intersects the line $y = -x + 5$ gives the solution to the equation $3x^2 - 7 = 0$.
Therefore the solutions to $3x^2 - 7 = 0$ are $x \approx -1.5$ and 1.5.

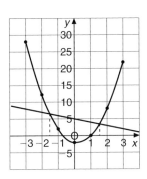

b) i) Plot a graph of $y = \dfrac{1}{x} + x$ for $-4 \leqslant x \leqslant 4$.

x	-4	-3	-2	-1	0	1	2	3	4
y	-4.25	-3.3	-2.5	-2	—	2	2.5	3.3	4.25

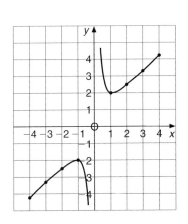

ii) Use the graph to explain why $\dfrac{1}{x} + x = 0$ has no solution.

For $\dfrac{1}{x} + x = 0$, the graph will need to intersect the x-axis. From the above plot, it can be seen that the graph does not intersect the x-axis and hence the equation $\dfrac{1}{x} + x = 0$ has no solution.

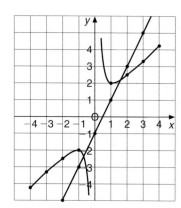

iii) Use the graph to find the solution to $x^2 - x = 1$.

This equation needs to be manipulated in such a way that one side becomes $\frac{1}{x} + x$.

Manipulating $x^2 - x = 1$ gives:

$$x - 1 = \frac{1}{x} \qquad \text{(dividing both sides by } x\text{)}$$

$$2x - 1 = \frac{1}{x} + x \qquad \text{(adding } x \text{ to both sides)}$$

Hence finding where the curve $y = \frac{1}{x} + x$ intersects the line $y = 2x - 1$ will give the solution to the equation $x^2 - x = 1$.

Therefore the solutions to the equation $x^2 - x = 1$ are $x \approx -0.6$ and 1.6.

Exercise 10.8

1. a) Plot the function $y = \frac{1}{2}x^2 + 1$ for $-4 \leqslant x \leqslant 4$.
   b) Showing your method clearly, use the graph to solve the equation $\frac{1}{2}x^2 = 4$.

2. a) Plot the function $y = x^3 + x - 2$ for $-3 \leqslant x \leqslant 3$.
   b) Showing your method clearly, use the graph to solve the equation $x^3 = 7 - x$.

3. a) Plot the function $y = 2x^3 - x^2 + 3$ for $-2 \leqslant x \leqslant 2$.
   b) Showing your method clearly, use the graph to solve the equation $2x^3 - 7 = 0$.

4. a) Plot the function $y = \frac{2}{x^2} - x$ for $-4 \leqslant x \leqslant 4$.
   b) Showing your method clearly, use the graph to solve the equation $4x^3 - 10x^2 + 2 = 0$.

5. a) Plot the function $y = 2^x - x$ for $-2 \leqslant x \leqslant 5$.
   b) Showing your method clearly, use the graph to solve the equation $2^x = 2x + 2$.

Extended Section

Student Assessment 1

1. a) Name the types of graph shown below:

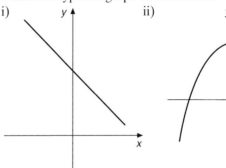

    b) Give a **possible** equation for each of the graphs drawn.

2. For each of the functions below:
    i) draw up a table of values,
    ii) plot a graph of the function.
    a) $f(x) = x^2 + 3x,\quad -5 \leqslant x \leqslant 2$
    b) $f(x) = \dfrac{1}{x} + 3x,\quad -3 \leqslant x \leqslant 3$

3. a) Plot the function $y = \frac{1}{2}x^3 + 2x^2$ for $-5 \leqslant x \leqslant 2$.
    b) Calculate the gradient of the curve when:
       i) $x = 1$ ii) $x = -1$

4. a) Plot a graph of the function $y = 2x^2 - 5x - 5$ for $-2 \leqslant x \leqslant 5$.
    b) Use the graph to solve the equation $2x^2 - 5x - 5 = 0$.
    c) Showing your method clearly, use the graph to solve the equation $2x^2 - 3x = 10$.

Student Assessment 2

1. a) Name the types of graph shown below:

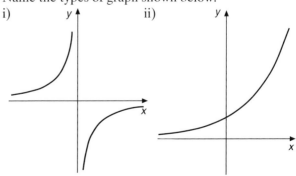

    b) Give a **possible** equation for each of the graphs drawn.

2. For each of the functions below:
   i) draw up a table of values,
   ii) plot a graph of the function.
      a) $f(x) = 2^x + x,$ $-3 \leqslant x \leqslant 3$
      b) $f(x) = 3^x - x^2,$ $-3 \leqslant x \leqslant 3$

3. a) Plot the function $y = -x^3 - 4x^2 + 5$ for $-5 \leqslant x \leqslant 2$.
   b) Calculate the gradient of the curve when:
      i) $x = 0$ ii) $x = -2$

4. a) Copy and complete the table below for the function

   $$y = \frac{1}{x^2} - 5.$$

x	-3	-2	-1	-0.5	-0.25	0	0.25	0.5	1	2	3
y				-1		—					

   b) Plot a graph of the function.
   c) Use the graph to solve the equation $\dfrac{1}{x^2} = 5$.
   d) Showing your method clearly, use your graph to solve

      the equation $\dfrac{1}{x^2} + x^2 = 7$.

ICT Section

In this investigation you will be using a graphing package to investigate the relationship between the function $y = x^2$ and the gradient at any point on it.

1. a) Using a graphing package such as Autograph, draw the graph $y = x^2$ for the range $-4 \leqslant x \leqslant 4$.
   b) Place a point on the curve where $x = 2$.
   c) Use the program to draw a tangent to the curve at that point (see diagram below).

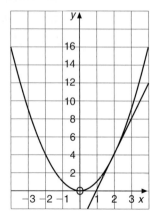

d) Either using the program, or manually, deduce the gradient of the tangent at $x = 2$ (and therefore the gradient of the curve when $x = 2$).

2. By calculating the gradient of tangents drawn at other integer values of x, copy and complete the table below.

x	−4	−3	−2	−1	0	1	2	3	4
Gradient									

3. By looking at the results in your table above, describe any pattern you see relating the function $y = x^2$ to the gradient at any point on it.

11 INDICES

Core Section

Use and interpret positive, negative and zero indices.

The index refers to the power to which a number is raised. In the example 5^3 the number 5 is raised to the power 3. The 3 is known as the **index**. Indices is the plural of index.

Worked examples **a)** $5^3 = 5 \times 5 \times 5$
$= 125$

b) $7^4 = 7 \times 7 \times 7 \times 7$
$= 2401$

c) $3^1 = 3$

Laws of indices

When working with numbers involving indices there are three basic laws which can be applied. These are:

(1) $a^m \times a^n = a^{m+n}$

(2) $a^m \div a^n$ or $\dfrac{a^m}{a^n} = a^{m-n}$

(3) $(a^m)^n = a^{mn}$

Positive indices

Worked examples **a)** Simplify $4^3 \times 4^2$.

$4^3 \times 4^2 = 4^{(3+2)}$
$= 4^5$

b) Simplify $2^5 \div 2^3$.

$2^5 \div 2^3 = 2^{(5-3)}$
$= 2^2$

c) Evaluate $3^3 \times 3^4$.

$3^3 \times 3^4 = 3^{(3+4)}$
$= 3^7$
$= 2187$

d) Evaluate $(4^2)^3$.

$(4^2)^3 = 4^{(2\times3)}$
$= 4^6$
$= 4096$

Exercise 11.1 **1.** Using indices, simplify the following expressions:
a) $3 \times 3 \times 3$
b) $2 \times 2 \times 2 \times 2 \times 2$
c) 4×4
d) $6 \times 6 \times 6 \times 6$
e) $8 \times 8 \times 8 \times 8 \times 8 \times 8$
f) 5

2. Simplify the following using indices:
a) $2 \times 2 \times 2 \times 3 \times 3$
b) $4 \times 4 \times 4 \times 4 \times 4 \times 5 \times 5$
c) $3 \times 3 \times 4 \times 4 \times 4 \times 5 \times 5$
d) $2 \times 7 \times 7 \times 7 \times 7$
e) $1 \times 1 \times 6 \times 6$
f) $3 \times 3 \times 3 \times 4 \times 4 \times 6 \times 6 \times 6 \times 6 \times 6$

3. Write out the following in full:
   a) 4^2
   b) 5^7
   c) 3^5
   d) $4^3 \times 6^3$
   e) $7^2 \times 2^7$
   f) $3^2 \times 4^3 \times 2^4$

4. Without a calculator work out the value of the following:
   a) 2^5
   b) 3^4
   c) 8^2
   d) 6^3
   e) 10^6
   f) 4^4
   g) $2^3 \times 3^2$
   h) $10^3 \times 5^3$

Exercise 11.2

1. Simplify the following using indices:
   a) $3^2 \times 3^4$
   b) $8^5 \times 8^2$
   c) $5^2 \times 5^4 \times 5^3$
   d) $4^3 \times 4^5 \times 4^2$
   e) $2^1 \times 2^3$
   f) $6^2 \times 3^2 \times 3^3 \times 6^4$
   g) $4^5 \times 4^3 \times 5^5 \times 5^4 \times 6^2$
   h) $2^4 \times 5^7 \times 5^3 \times 6^2 \times 6^6$

2. Simplify the following:
   a) $4^6 \div 4^2$
   b) $5^7 \div 5^4$
   c) $2^5 \div 2^4$
   d) $6^5 \div 6^2$
   e) $\dfrac{6^5}{6^2}$
   f) $\dfrac{8^6}{8^5}$
   g) $\dfrac{4^8}{4^5}$
   h) $\dfrac{3^9}{3^2}$

3. Simplify the following:
   a) $(5^2)^2$
   b) $(4^3)^4$
   c) $(10^2)^5$
   d) $(3^3)^5$
   e) $(6^2)^4$
   f) $(8^2)^3$

4. Simplify the following:
   a) $\dfrac{2^2 \times 2^4}{2^3}$
   b) $\dfrac{3^4 \times 3^2}{3^5}$
   c) $\dfrac{5^6 \times 5^7}{5^2 \times 5^8}$
   d) $\dfrac{(4^2)^5 \times 4^2}{4^7}$
   e) $\dfrac{4^4 \times 2^5 \times 4^2}{4^3 \times 2^3}$
   f) $\dfrac{6^3 \times 6^3 \times 8^5 \times 8^6}{8^6 \times 6^2}$
   g) $\dfrac{(5^5)^2 \times (4^4)^3}{5^8 \times 4^9}$
   h) $\dfrac{(6^3)^4 \times 6^3 \times 4^9}{6^8 \times (4^2)^4}$

5. Simplify the following:
   a) $c^5 \times c^3$
   b) $m^4 \div m^2$
   c) $(b^3)^5 \div b^6$
   d) $\dfrac{m^4 n^9}{mn^3}$
   e) $\dfrac{6a^6 b^4}{3a^2 b^3}$
   f) $\dfrac{12x^5 y^7}{4x^2 y^5}$
   g) $\dfrac{4u^3 v^6}{8u^2 v^3}$
   h) $\dfrac{3x^6 y^5 z^3}{9x^4 y^2 z}$

6. Simplify the following:
   a) $4a^2 \times 3a^3$
   b) $2a^2b \times 4a^3b^2$
   c) $(2p^2)^3$
   d) $(4m^2n^3)^2$
   e) $(5p^2)^2 \times (2p^3)^3$
   f) $(4m^2n^2) \times (2mn^3)^3$
   g) $\dfrac{(6x^2y^4)^2 \times (2xy)^3}{12x^6y^8}$
   h) $(ab)^d \times (ab)^e$

▨ The zero index

The zero index indicates that a number is raised to the power 0.
A number raised to the power 0 is equal to 1. This can be
explained by applying the laws of indices.

$$a^m \div a^n = a^{m-n} \qquad \text{therefore} \quad \frac{a^m}{a^m} = a^{m-m}$$
$$= a^0$$

However, $\qquad\qquad\qquad \dfrac{a^m}{a^m} = 1$

therefore $a^0 = 1$

▨ Negative indices

A negative index indicates that a number is being raised to a
negative power: e.g. 4^{-3}.

Another law of indices states that $a^{-m} = \dfrac{1}{a^m}$. It can be
proved as follows.

$$a^{-m} = a^{0-m}$$
$$= \frac{a^0}{a^m} \text{ (from the second law of indices)}$$
$$= \frac{1}{a^m}$$

therefore $a^{-m} = \dfrac{1}{a^m}$

Exercise 11.3 Without using a calculator, evaluate the following:

1. a) $2^3 \times 2^0$
   b) $5^2 \div 6^0$
   c) $5^2 \times 5^{-2}$
   d) $6^3 \times 6^{-3}$
   e) $(4^0)^2$
   f) $4^0 \div 2^2$

2. a) 4^{-1}
   b) 3^{-2}
   c) 6×10^{-2}
   d) 5×10^{-3}
   e) 100×10^{-2}
   f) 10^{-3}

3. a) 9×3^{-2}
   b) 16×2^{-3}
   c) 64×2^{-4}
   d) 4×2^{-3}
   e) 36×6^{-3}
   f) 100×10^{-1}

4. a) $\dfrac{3}{2^{-2}}$ b) $\dfrac{4}{2^{-3}}$

c) $\dfrac{9}{5^{-2}}$ d) $\dfrac{5}{4^{-2}}$

e) $\dfrac{7^{-3}}{7^{-4}}$ f) $\dfrac{8^{-6}}{8^{-8}}$

Exponential equations

Equations that involve indices as unknowns are known as **exponential equations**.

Worked examples **a)** Find the value of x if $2^x = 32$.

32 can be expressed as a power of 2,
$32 = 2^5$.

Therefore $2^x = 2^5$

$x = 5$

b) Find the value of m if $3^m = 81$.

81 can be expressed as a power of 3,
$81 = 3^4$.

Therefore $3^m = 3^4$

$m = 4$

Exercise 11.4 **1.** Find the value of x in each of the following:
a) $2^x = 4$ b) $2^x = 16$
c) $4^x = 64$ d) $10^x = 1000$
e) $5^x = 625$ f) $3^x = 1$

2. Find the value of z in each of the following:
a) $2^{(z-1)} = 8$ b) $3^{(z+2)} = 27$
c) $4^{2z} = 64$ d) $10^{(z+1)} = 1$
e) $3^z = 9^{(z-1)}$ f) $5^z = 125^z$

3. Find the value of n in each of the following:
a) $\left(\tfrac{1}{2}\right)^n = 8$ b) $\left(\tfrac{1}{3}\right)^n = 81$
c) $\left(\tfrac{1}{2}\right)^n = 32$ d) $\left(\tfrac{1}{2}\right)^n = 4^{(n+1)}$
e) $\left(\tfrac{1}{2}\right)^{(n+1)} = 2$ f) $\left(\tfrac{1}{16}\right)^n = 4$

4. Find the value of x in each of the following:
a) $3^{-x} = 27$ b) $2^{-x} = 128$
c) $2^{(-x+3)} = 64$ d) $4^{-x} = \tfrac{1}{16}$
e) $2^{-x} = \tfrac{1}{265}$ f) $3^{(-x+1)} = \tfrac{1}{81}$

Exercise 11.5 **1.** A tap is dripping at a constant rate into a container. The level (l cm) of the water in the container, is given by the equation $l = 2^t - 1$ where t hours is the time taken.
a) Calculate the level of the water after 3 hours.
b) Calculate the level of the water in the container at the start.

c) Calculate the time taken for the level of the water to reach 31 cm.

d) Plot a graph showing the level of the water over the first 6 hours.

e) From your graph, estimate the time taken for the water to reach a level of 45 cm.

2. Draw a graph of $y = 4^x$ for values of x between -1 and 3. Use your graph to find approximate solutions to the following equations:

a) $4^x = 30$

b) $4^x = \frac{1}{2}$

3. Draw a graph of $y = 2^x$ for values of x between -2 and 5. Use your graph to find approximate solutions to the following equations:

a) $2^x = 20$

b) $2^{(x + 2)} = 40$

c) $2^{-x} = 0.2$

4. Draw a graph of $y = 3^x$ for values of x between -1 and 3. Use your graph to find approximate solutions to the following equations:

a) $3^{(x + 2)} = 12$

b) $3^{(x - 3)} = 0.5$

Core Section

Student Assessment 1

1. Using indices, simplify the following:
   a) $2 \times 2 \times 2 \times 5 \times 5$ b) $2 \times 2 \times 3 \times 3 \times 3 \times 3 \times 3$

2. Write the following out in full:
   a) 4^3 b) 6^4

3. Work out the value of the following without using a calculator:
   a) $2^3 \times 10^2$ b) $1^4 \times 3^3$

4. Simplify the following using indices:
   a) $3^4 \times 3^3$ b) $6^3 \times 6^2 \times 3^4 \times 3^5$

   c) $\dfrac{4^5}{2^3}$ d) $\dfrac{(6^2)^3}{6^5}$

   e) $\dfrac{3^5 \times 4^2}{3^3 \times 4^0}$ f) $\dfrac{4^{-2} \times 2^6}{2^2}$

5. Without using a calculator, evaluate the following:

   a) $2^4 \times 2^{-2}$ b) $\dfrac{3^5}{3^3}$

   c) $\dfrac{5^{-5}}{5^{-6}}$ d) $\dfrac{2^5 \times 4^{-3}}{2^{-1}}$

6. Find the value of x in each of the following:

   a) $2^{(x-2)} = 32$ b) $\dfrac{1}{4^x} = 16$

   c) $5^{(-x+2)} = 125$ d) $8^{-x} = \frac{1}{2}$

Student Assessment 2

1. Using indices, simplify the following:
   a) $3 \times 2 \times 2 \times 3 \times 27$
   b) $2 \times 2 \times 4 \times 4 \times 4 \times 2 \times 32$

2. Write the following out in full:
   a) 6^5 b) 2^{-5}

3. Work out the value of the following without using a calculator:
   a) $3^3 \times 10^3$ b) $1^{-4} \times 5^3$

4. Simplify the following using indices:
   a) $2^4 \times 2^3$ b) $7^5 \times 7^2 \times 3^4 \times 3^8$

   c) $\dfrac{4^8}{2^{10}}$ d) $\dfrac{(3^3)^4}{27^3}$

   e) $\dfrac{7^6 \times 4^2}{4^3 \times 7^6}$ f) $\dfrac{8^{-2} \times 2^6}{2^{-2}}$

5. Without using a calculator, evaluate the following:

   a) $5^2 \times 5^{-1}$ b) $\dfrac{4^5}{4^3}$

   c) $\dfrac{7^{-5}}{7^{-7}}$ d) $\dfrac{3^{-5} \times 4^2}{3^{-6}}$

6. Find the value of x in each of the following:

   a) $2^{(2x + 2)} = 128$ b) $\dfrac{1}{4^{-x}} = \dfrac{1}{2}$

   c) $3^{(-x + 4)} = 81$ d) $8^{-3x} = \frac{1}{4}$

INDICES

Extended Section

Use and interpret fractional indices; calculate *x* when, for example, $32^x = 2$.

Fractional indices

$16^{\frac{1}{2}}$ can be written as $(4^2)^{\frac{1}{2}}$.

$$(4^2)^{\frac{1}{2}} = 4^{(2 \times \frac{1}{2})}$$
$$= 4^1$$
$$= 4$$

Therefore $16^{\frac{1}{2}} = 4$

but $\sqrt{16} = 4$

therefore $16^{\frac{1}{2}} = \sqrt{16}$

Similarly:

$27^{\frac{1}{3}}$ can be written as $(3^3)^{\frac{1}{3}}$

$$(3^3)^{\frac{1}{3}} = 3^{(3 \times \frac{1}{3})}$$
$$= 3^1$$
$$= 3$$

Therefore $27^{\frac{1}{3}} = 3$

but $\sqrt[3]{27} = 3$

therefore $27^{\frac{1}{3}} = \sqrt[3]{27}$

In general:

$$a^{\frac{1}{n}} = \sqrt[n]{a}$$

$$a^{\frac{m}{n}} = \sqrt[n]{(a^m)} \text{ or } (\sqrt[n]{a})^m$$

Worked examples   **a)**   Evaluate $16^{\frac{1}{4}}$ without the use of a calculator.

$$16^{\frac{1}{4}} = \sqrt[4]{16} \qquad \text{Alternatively:} \quad 16^{\frac{1}{4}} = (2^4)^{\frac{1}{4}}$$
$$= \sqrt[n]{(2^4)} \qquad\qquad\qquad\qquad = 2^1$$
$$= 2 \qquad\qquad\qquad\qquad\qquad = 2$$

b)   Evaluate $25^{\frac{3}{2}}$ without the use of a calculator.

$$25^{\frac{3}{2}} = (25^{\frac{1}{2}})^3 \qquad \text{Alternatively:} \quad 25^{\frac{3}{2}} = (5^2)^{\frac{3}{2}}$$
$$= (\sqrt{25})^3 \qquad\qquad\qquad\qquad = 5^3$$
$$= 5^3 \qquad\qquad\qquad\qquad\qquad = 125$$
$$= 125$$

c) Solve $32^x = 2$

32 is 2^5 so $\sqrt[5]{32} = 2$
or $\qquad 32^{\frac{1}{5}} = 2$
therefore $x = \frac{1}{5}$

d) Solve $125^x = 5$

125 is 5^3 so $\sqrt[3]{125} = 5$
$\qquad\qquad 125^{\frac{1}{3}} = 5$
or
therefore $x = \frac{1}{3}$

Exercise 11.6

Evaluate the following without the use of a calculator:

1. a) $16^{\frac{1}{2}}$ b) $25^{\frac{1}{2}}$ c) $100^{\frac{1}{2}}$
   d) $27^{\frac{1}{3}}$ e) $81^{\frac{1}{2}}$ f) $1000^{\frac{1}{3}}$

2. a) $16^{\frac{1}{4}}$ b) $81^{\frac{1}{4}}$ c) $32^{\frac{1}{5}}$
   d) $64^{\frac{1}{6}}$ e) $216^{\frac{1}{3}}$ f) $256^{\frac{1}{4}}$

3. a) $4^{\frac{3}{2}}$ b) $4^{\frac{5}{2}}$ c) $9^{\frac{3}{2}}$
   d) $16^{\frac{3}{2}}$ e) $1^{\frac{5}{2}}$ f) $27^{\frac{2}{3}}$

4. a) $125^{\frac{2}{3}}$ b) $32^{\frac{3}{5}}$ c) $65^{\frac{5}{6}}$
   d) $1000^{\frac{2}{3}}$ e) $16^{\frac{5}{4}}$ f) $81^{\frac{3}{4}}$

5. a) solve $16^x = 4$ b) solve $8^x = 2$
   c) solve $9^x = 3$ d) solve $27^x = 3$
   e) solve $100^x = 10$ f) solve $64^x = 2$

6. a) solve $1000^x = 10$ b) solve $49^x = 7$
   c) solve $81^x = 3$ d) solve $343^x = 7$
   e) solve $1\ 000\ 000^x = 10$ f) solve $216^x = 6$

Exercise 11.7

Evaluate the following without the use of a calculator:

1. a) $\dfrac{27^{\frac{2}{3}}}{3^2}$ b) $\dfrac{7^{\frac{3}{2}}}{\sqrt{7}}$ c) $\dfrac{4^{\frac{5}{2}}}{4^2}$

   d) $\dfrac{16^{\frac{3}{2}}}{2^6}$ e) $\dfrac{27^{\frac{5}{3}}}{\sqrt{9}}$ f) $\dfrac{6^{\frac{4}{3}}}{6^{\frac{1}{3}}}$

2. a) $5^{\frac{2}{3}} \times 5^{\frac{4}{3}}$ b) $4^{\frac{1}{4}} \times 4^{\frac{3}{4}}$ c) 8×2^{-2}
   d) $3^{\frac{4}{3}} \times 3^{\frac{5}{3}}$ e) $2^{-2} \times 16$ f) $8^{\frac{5}{3}} \times 8^{-\frac{4}{3}}$

3. a) $\dfrac{2^{\frac{1}{2}} \times 2^{\frac{5}{2}}}{2}$ b) $\dfrac{4^{\frac{5}{6}} \times 4^{\frac{1}{6}}}{4^{\frac{1}{2}}}$ c) $\dfrac{2^3 \times 8^{\frac{3}{2}}}{\sqrt{8}}$

   d) $\dfrac{(3^2)^{\frac{3}{2}} \times 3^{-\frac{1}{2}}}{3^{\frac{1}{2}}}$ e) $\dfrac{8^{\frac{1}{3}} + 7}{27^{\frac{1}{3}}}$ f) $\dfrac{9^{\frac{1}{2}} \times 3^{\frac{5}{2}}}{3^{\frac{2}{3}} \times 3^{-\frac{1}{6}}}$

Extended Section

Student Assessment 1

1. Evaluate the following without the use of a calculator:
   a) $81^{\frac{1}{2}}$
   b) $27^{\frac{1}{3}}$
   c) $9^{\frac{1}{2}}$
   d) $625^{\frac{3}{4}}$
   e) $343^{\frac{2}{3}}$
   f) $16^{-\frac{1}{4}}$
   g) $\dfrac{1}{25^{-\frac{1}{2}}}$
   h) $\dfrac{2}{16^{-\frac{3}{4}}}$

2. Evaluate the following without the use of a calculator:
   a) $\dfrac{16^{\frac{1}{2}}}{2^2}$
   b) $\dfrac{9^{\frac{5}{2}}}{3^3}$
   c) $\dfrac{8^{\frac{4}{3}}}{8^{\frac{2}{3}}}$
   d) $5^{\frac{6}{8}} \times 5^{\frac{4}{8}}$
   e) $4^{\frac{3}{2}} \times 2^{-\frac{1}{2}}$
   f) $\dfrac{27^{\frac{2}{3}} \times 3^{-2}}{4^{-\frac{3}{2}}}$
   g) $\dfrac{(4^3)^{-\frac{1}{2}} \times 2^{\frac{3}{2}}}{2^{-\frac{3}{2}}}$
   h) $\dfrac{(5^{\frac{1}{3}})^{\frac{1}{2}} \times 5^{\frac{2}{3}}}{3^{-2}}$

3. Draw a pair of axes with x from -4 to 4 and y from 0 to 10.
   a) Plot a graph of $y = 3^{\frac{x}{2}}$.
   b) Use your graph to estimate when $3^{\frac{x}{2}} = 5$.

Student Assessment 2

1. Evaluate the following without the use of a calculator:
   a) $64^{\frac{1}{6}}$
   b) $27^{\frac{4}{3}}$
   c) $9^{-\frac{1}{2}}$
   d) $512^{\frac{2}{3}}$
   e) $\sqrt[3]{27}$
   f) $\sqrt[4]{16}$
   g) $\dfrac{1}{36^{-\frac{1}{2}}}$
   h) $\dfrac{2}{64^{-\frac{2}{3}}}$

2. Evaluate the following without the use of a calculator:
   a) $\dfrac{25^{\frac{1}{2}}}{9^{-\frac{1}{2}}}$
   b) $\dfrac{4^{\frac{5}{2}}}{2^3}$
   c) $\dfrac{27^{\frac{4}{3}}}{3^3}$
   d) $25^{\frac{3}{2}} \times 5^2$
   e) $4^{\frac{6}{4}} \times 4^{-\frac{1}{2}}$
   f) $\dfrac{27^{\frac{2}{3}} \times 3^{-3}}{9^{-\frac{1}{2}}}$
   g) $\dfrac{(4^2)^{-\frac{1}{4}} \times 9^{\frac{3}{2}}}{\left(\dfrac{1}{4}\right)^{\frac{1}{2}}}$
   h) $\dfrac{(5^{\frac{1}{3}})^{\frac{1}{2}} \times 5^{\frac{5}{6}}}{4^{-\frac{1}{2}}}$

3. Draw a pair of axes with x from -4 to 4 and y from 0 to 18.
   a) Plot a graph of $y = 4^{-\frac{x}{2}}$.
   b) Use your graph to estimate when $4^{-\frac{x}{2}} = 6$.

ICT Section

You have seen that by applying the laws of indices, it is possible to solve some exponential equations.

Use a graphing package and appropriate graphs to solve the following exponential equations:

1. $4^x = 40$

2. $3^x = 17$

3. $5^{x-1} = 6$

4. $3^{-x} = 0.5$

12 LINEAR PROGRAMMING

Represent inequalities graphically and use this representation in the solution of simple linear programming problems (the conventions of using broken lines for strict inequalities and shading unwanted regions will be expected).

This chapter builds upon the previous work covered in Chapters 7 and 8. You may need to revise those chapters first.

▦ Revision

An understanding of the following symbols is necessary:

>   means 'is greater than'
⩾   means 'is greater than or equal to'
<   means 'is less than'
⩽   means 'is less than or equal to'

Exercise 12.1

1. Solve each of the following inequalities:

a) $15 + 3x < 21$ b) $18 \leqslant 7y + 4$

c) $19 - 4x \geqslant 27$ d) $2 \geqslant \dfrac{y}{3}$

e) $-4t + 1 < 1$ f) $1 \geqslant 3p + 10$

2. Solve each of the following inequalities:

a) $7 < 3y + 1 \leqslant 13$ b) $3 \leqslant 3p < 15$
c) $9 \leqslant 3(m - 2) < 15$ d) $20 < 8x - 4 < 28$

The solution to an inequality can also be illustrated on a graph.

Worked examples

a) On a pair of axes, shade the region which satisfies the inequality $x \geqslant 3$.

     To do this the line $x = 3$ is drawn.
The region to the right of $x = 3$ represents the inequality $x \geqslant 3$ and therefore is shaded as shown below.

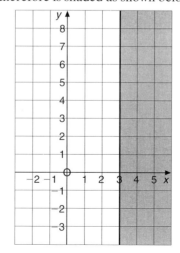

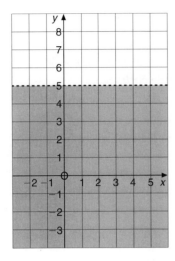

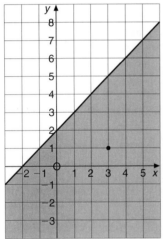

b) On a pair of axes, shade the region which satisfies the inequality $y < 5$.

The line $y = 5$ is drawn first (in this case it is drawn as a broken line).

The region below the line $y = 5$ represents the inequality $y < 5$ and therefore is shaded as shown (left).

Note that a broken (dashed) line shows $<$ or $>$ whilst a solid line shows $\leqslant$ or $\geqslant$.

c) On a pair of axes, shade the region which satisfies the inequality $y \leqslant x + 2$.

The line $y = x + 2$ is drawn first (since it is included, this line is solid).

To know which region satisfies the inequality, and hence to know which side of the line to shade, the following steps are taken.

- Choose a point at random which does not lie on the line e.g. (3, 1).
- Substitute those values of x and y into the inequality i.e. $1 \leqslant 3 + 2$
- If the inequality holds true, then the region in which the point lies satisfies the inequality and can therefore be shaded.

Note that in some questions the region which satisfies the inequality is left **unshaded** whilst in others it is **shaded**. You will therefore need to read the question carefully to see which is required.

Exercise 12.2

1. By drawing appropriate axes, shade the region which satisfies each of the following inequalities:

a) $y > 2$ b) $x < 3$ c) $y \leqslant 4$

d) $x \geqslant -1$ e) $y > 2x + 1$ f) $y \leqslant x - 3$

2. By drawing appropriate axes, leave unshaded the region which satisfies each of the following inequalities:

a) $y \geqslant -x$ b) $y \leqslant 2 - x$ c) $x \geqslant y - 3$

d) $x + y \geqslant 4$ e) $2x - y \geqslant 3$ f) $2y - x < 4$

Several inequalities can be graphed on the same set of axes. If the regions which satisfy each inequality are left unshaded then a solution can be found which satisfies all the inequalities, i.e. the region left unshaded by all the inequalities.

Worked example On the same pair of axes leave unshaded the regions which satisfy the following inequalities simultaneously:

$$x \leqslant 2 \qquad y > -1 \qquad y \leqslant 3 \qquad y \leqslant x + 2$$

Hence find the region which satisfies all four inequalities.

If the four inequalities are graphed on separate axes the solutions are as shown below:

$x \leqslant 2$

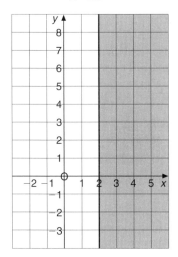

$y > -1$

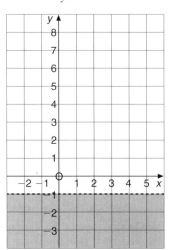

$y \leqslant 3$

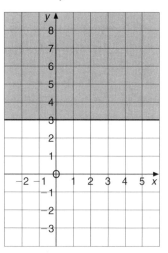

$y \leqslant x + 2$

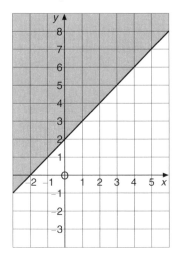

Combining all four on one pair of axes gives this diagram.

The unshaded region therefore gives a solution which satisfies all four inequalities.

Exercise 12.3

On the same pair of axes plot the following inequalities and leave unshaded the region which satisfies all of them simultaneously.

1. $y \leqslant x$ $y > 1$ $x \leqslant 5$

2. $x + y \leqslant 6$ $y < x$ $y \geqslant 1$

3. $y \geqslant 3x$ $y \leqslant 5$ $x + y > 4$

4. $2y \geqslant x + 4$ $y \leqslant 2x + 2$ $y < 4$ $x \leqslant 3$

▦ Linear programming

Linear programming is a way of finding a number of possible solutions to a problem given a number of constraints. But it is more than this – it is also a method for minimising a linear function in two (or more) variables.

Worked example The number of fields a farmer plants with wheat is w and the number of fields he plants with corn is c. There are, however, certain restrictions which govern how many fields he can plant of each. These are as follows.

- There must be at least two fields of corn.
- There must be at least two fields of wheat.
- Not more than 10 fields are to be sown with wheat or corn.

i)   Construct three inequalities from the information given above.

$$c \geqslant 2 \quad w \geqslant 2 \quad c + w \leqslant 10$$

ii) On one pair of axes, graph the three inequalities and leave unshaded the region which satisfies all three simultaneously.

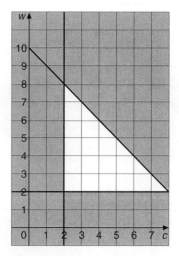

iii) Give one possible arrangement as to how the farmer should plant his fields.

Four fields of corn and four fields of wheat.

Exercise 12.4 In the problems below draw both axes numbered from 0 to 12. For each question:

a)   write an inequality for each statement,
b)   graph the inequalities leaving the region which satisfies the inequalities unshaded,
c)   using your graph, state one solution which satisfies all the inequalities simultaneously.

1. A taxi firm has at its disposal one morning a car and a minibus for hire. During the morning it makes *x* car trips and *y* minibus trips.

   ■ It makes at least five car trips.
   ■ It makes between two and eight minibus trips.
   ■ The total number of car and minibus trips does not exceed 12.

2. A woman is baking bread and cakes. She makes *p* loaves and *q* cakes. She bakes at least five loaves and at least two cakes but no more than ten loaves and cakes altogether.

3. A couple are buying curtains for their house. They buy *m* long curtains and *n* short curtains. They buy at least two long curtains. They also buy at least twice as many short curtains as long curtains. A maximum of 11 curtains are bought altogether.

4. A shop sells large and small oranges. A girl buys *L* large oranges and *S* small oranges. She buys at least three but fewer than nine large oranges. She also buys fewer than six small oranges. The maximum number of oranges she needs to buy is 10.

Student Assessment 1

1. Solve the following inequalities:

   a) $17 + 5x \leqslant 42$
   b) $3 \geqslant \dfrac{y}{3} + 2$

2. Find the range of values for which:
   a) $7 < 4y - 1 \leqslant 15$
   b) $18 < 3(p + 2) \leqslant 30$

3. A garage stocks two kinds of engine oil. They have *r* cases of regular oil and *s* cases of super oil.
   They have fewer than ten cases of regular oil in stock and between three and nine cases of super oil in stock. They also have fewer than 12 cases in stock altogether.
   a) Express the three constraints described above as inequalities.
   b) Draw an appropriate pair of axes and identify the region which satisfies all the inequalities by shading the unwanted regions.
   c) State two possible solutions for the number of each case in stock.

4. Students from Argentina and students from England are meeting on a cultural exchange.
   In total there will be between 12 and 20 students. There are fewer than 10 students from Argentina, whilst the number from England cannot be more than three greater than the number from Argentina.
   a) Write inequalities for the number (*A*) of students from Argentina and the number (*E*) of students from England.

b) On an appropriate pair of axes, graph the inequalities, leaving unshaded the region which satisfies all of them.
c) State two of the possible combinations of A and E which satisfy the given conditions.

Student Assessment 2

1. Solve the following inequalities:

   a) $5 + 6x \leqslant 47$ b) $4 \geqslant \dfrac{y + 3}{3}$

2. Find the range of values for which:
   a) $3 \leqslant 3p < 12$ b) $24 < 8(x - 1) \leqslant 48$

3. A breeder of dogs has x dogs and y bitches. She has fewer than four dogs and more than two bitches. She has enough room for a maximum of eight animals in total.
   a) Express the three conditions above as inequalities.
   b) Draw an appropriate pair of axes and leave unshaded the region which satisfies all the inequalities.
   c) State two of the possible combinations of dogs and bitches which she can have.

4. Antonio is employed by a company to do two jobs. He mends cars and also repairs electrical goods. His terms of employment are listed below.

   ■ He is employed for a maximum of 40 hours.
   ■ He must spend at least 16 hours mending cars.
   ■ He must spend at least 5 hours repairing electrical goods.
   ■ He must spend more than twice as much time mending cars as repairing electrical goods.

   a) Express the conditions above as inequalities, using c to represent the number of hours spent mending cars, and e to represent the number of hours spent mending electrical goods.
   b) On an appropriate pair of axes, graph the inequalities, leaving unshaded the region which satisfies all four.
   c) State two of the possible combinations which satisfy the above conditions.

ICT Section

Using a graphing package, plot on the same pair of axes the following inequalities. Leave unshaded the region which satisfies all of them simultaneously.

1. $y \leqslant x$ $y > 0$ $x \leqslant 3$

2. $x + y > 3$ $y \leqslant 4$ $y - x > 2$

3. $2y + x \leqslant 5$ $y - 3x - 6 < 0$ $2y - x > 3$

SHAPE AND SPACE

■ Essential Revision ■

LOCI

Use the following loci and the method of intersecting loci for sets of points in two dimensions:
- which are at a given distance from a given point,
- which are at a given distance from a given straight line,
- which are equidistant from two given points,
- which are equidistant from two given intersecting straight lines.

NB: All diagrams are not drawn to scale.

A **locus** (plural **loci**) refers to all the points which fit a particular description. These points can either belong to a region, a line or both. The principal types of loci are explained below.

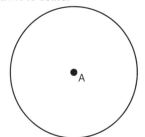

■ The locus of the points which are at a given distance from a given point

In the diagram (left) it can be seen that the locus of all the points equidistant from a point A is the circumference of a circle centre A. This is due to the fact that all points on the circumference of a circle are equidistant from the centre.

■ The locus of the points which are at a given distance from a given straight line

In the diagram (left) it can be seen that the locus of the points equidistant from a straight line AB runs parallel to that straight line. It is important to note that the distance of the locus from the straight line is measured at right angles to the line. This diagram, however, excludes the ends of the line. If these two points are taken into consideration then the locus takes the form shown in the next diagram (left).

■ The locus of the points which are equidistant from two given points

The locus of the points equidistant from points X and Y lies on the perpendicular bisector of the line XY.

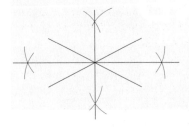

▧ The locus of the points which are equidistant from two given intersecting straight lines

The locus in this case lies on the bisectors of both pairs of opposite angles as shown below.

The application of the above cases will enable you to tackle problems involving loci at this level.

Worked example

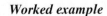
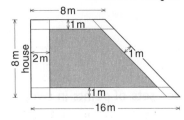

The diagram (left) shows a trapezoidal garden. Three of its sides are enclosed by a fence, and the fourth is adjacent to a house.

i) Grass is to be planted in the garden. However, it must be at least 2 m away from the house and at least 1 m away from the fence. Shade the region in which the grass can be planted.

The shaded region is therefore the locus of all the points which are both at least 2 m away from the house and at least 1 m away from the surrounding fence. Note that the boundary of the region also forms part of the locus of the points.

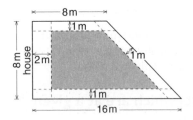

ii) Using the same garden as before, grass must now be planted according to the following conditions: it must be **more than** 2 m away from the house and **more than** 1 m away from the fence. Shade the region in which the grass can be planted.

The shape of the region is the same as in the first case; however, in this instance the boundary is not included in the locus of the points as the grass cannot be exactly 2 m away from the house or exactly 1 m away from the fence.

Note: If the boundary is included in the locus points, it is represented by a **solid** line. If it is not included then it is represented by a **dashed** (broken) line.

Exercise A

Questions 1–4 are about a rectangular garden measuring 8 m by 6 m. For each question draw a scale diagram of the garden and identify the locus of the points which fit the criteria.

1. Draw the locus of all the points at least 1 m from the edge of the garden.

2. Draw the locus of all the points at least 2 m from each corner of the garden.

3. Draw the locus of all the points more than 3 m from the centre of the garden.

4. Draw the locus of all the points equidistant from the longer sides of the garden.

5. A port has two radar stations at P and Q which are 20 km apart. The radar at P is set to a range of 20 km, whilst the radar at Q is set to a range of 15 km.
a) Draw a scale diagram to show the above information.
b) Shade the region in which a ship must be sailing if it is only picked up by radar P. Label this region 'a'.
c) Shade the region in which a ship must be sailing if it is only picked up by radar Q. Label this region 'b'.
d) Identify the region in which a ship must be sailing if it is picked up by both radars. Label this region 'c'.

6. X and Y are two ship-to-shore radio receivers. They are 25 km apart.
A ship sends out a distress signal. The signal is picked up by both X and Y. The radio receiver at X indicates that the ship is within a 30 km radius of X, whilst the radio receiver at Y indicates that the ship is within 20 km of Y. Draw a scale diagram and identify the region in which the ship must lie.

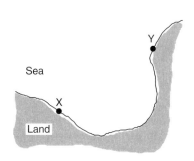

7. a) Mark three points L, M and N not in a straight line. By construction find the point which is equidistant from L, M and N.
b) What would happen if L, M and N were on the same straight line?

8. Draw a line AB 8 cm long. What is the locus of a point C such that the angle ACB is always a right angle?

9. Draw a circle by drawing round a circular object (do not use a pair of compasses). By construction determine the position of the centre of the circle.

10. Three lionesses L_1, L_2 and L_3 have surrounded a gazelle. The three lionesses are equidistant from the gazelle. Draw a diagram with the lionesses in similar positions to those shown (left) and by construction determine the position (g) of the gazelle.

Exercise B

1. Three girls are playing hide and seek. Ayshe and Belinda are at the positions shown (left) and are trying to find Cristina. Cristina is on the opposite side of a wall PQ to her two friends.
Assuming Ayshe and Belinda cannot see over the wall identify, by copying the diagram, the locus of points where Cristina could be if:
a) Cristina can only be seen by Ayshe,
b) Cristina can only be seen by Belinda,
c) Cristina can not be seen by either of her two friends,
d) Cristina can be seen by both of her friends.

2. A security guard S is inside a building in the position shown. The building is inside a rectangular compound. If the building has three windows as shown, identify the locus of points in the compound which can be seen by the security guard.

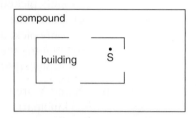

3. The circular cage shown (left) houses a snake. Inside the cage are three obstacles.
A rodent is placed inside the cage at R. From where it is lying, the snake can see the rodent.
Trace the diagram and identify the regions in which the snake could be lying.

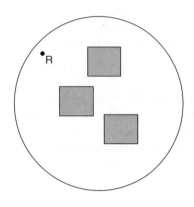

Exercise C

1. A coin is rolled in a straight line on a flat surface as shown below.

Draw the locus of the centre of the coin O as the coin rolls along the surface.

2. The diameter of the disc (left) is the same as the width and height of each of the steps shown.
Copy the diagram and draw the locus of the centre of the disc as it rolls down the steps.

3. A stone is thrown vertically upwards. Draw the locus of its trajectory from the moment it leaves the person's hand to the moment it is caught again.

4. A stone is thrown at an angle of elevation of 45°. Sketch the locus of its trajectory.

5. X and Y are two fixed posts in the ground. The ends of a rope are tied to X and Y. A goat is attached to the rope by a ring on its collar which enables it to move freely along the rope's length.
Copy the diagram (left) and sketch the locus of points in which the goat is able to graze.

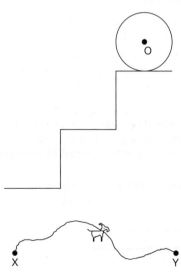

NB: All diagrams are not drawn to scale.

P
●

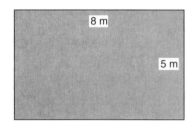

building

●
S

8 m

5 m

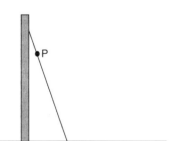

●⎯⎯⎯⎯⎯⎯⎯⎯⎯⎯⎯●
A   80 km   B

●P

Student Assessment 1

1. Pedro and Sara are on opposite sides of a building as shown (left).
   Their friend Raul is standing in a place such that he cannot be seen by either Pedro or Sara. Copy the above diagram and identify the locus of points at which Raul could be standing.

2. A rectangular garden measures 10 m by 6 m. A tree stands in the centre of the garden. Grass is to be planted according to the following conditions:

   ■ it must be at least 1 m from the edge of the garden,
   ■ it must be more than 2 m away from the centre of the tree.

   a) Make a scale drawing of the garden.
   b) Draw the locus of points in which the grass can be planted.

3. A rectangular rose bed in a park measures 8 m by 5 m as shown (left).
   The park keeper puts a low fence around the rose bed. The fence is at a constant distance of 2 m from the rose bed.
   a) Make a scale drawing of the rose bed.
   b) Draw the position of the fence.

4. A and B are two radio beacons 80 km apart, either side of a shipping channel. A ship sails in such a way that it is always equidistant from A and B.
   Showing your method of construction clearly, draw the path of the ship.

5. A ladder 10 m long is propped up against a wall as shown. A point P on the ladder is 2 m from the top.
   Make a scale drawing to show the locus of point P if the ladder were to slide down the wall. Note: several positions of the ladder will need to be shown.

6. The equilateral triangle PQR is rolled along the line shown. At first, corner Q acts as the pivot point until P reaches the line, then P acts as the pivot point until R reaches the line, and so on.

P
●
╱ ╲
╱   ╲
R   Q

Showing your method clearly, draw the locus of point P as the triangle makes one full rotation, assuming there is no slipping.

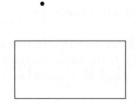

J
•

• K

•
L

Student Assessment 2

1. Jose, Katrina and Luis are standing at different points around a building as shown (left).
    Trace the diagram and show whether any of the three friends can see each other or not.

2. A rectangular courtyard measures 20 m by 12 m. A horse is tethered in the centre with a rope 7 m long. Another horse is tethered, by a rope 5 m long, to a rail which runs along the whole of the left-hand (shorter) side of the courtyard. This rope is able to run freely along the length of the rail. Draw a scale diagram of the courtyard and draw the locus of points which can be reached by both horses.

3. The view in the diagram (left) is of two walls which form part of an obstacle course. A girl decides to ride her bicycle in between the two walls in such a way that she is always equidistant from them.
    Copy the diagram and, showing your construction clearly, draw the locus of her path.

4. A ball is rolling along the line shown in the diagram (below). Copy the diagram and draw the locus of the centre, O, of the ball as it rolls.

5. A square ABCD is 'rolled' along the flat surface shown below. Initially corner C acts as a pivot point until B touches the surface, then B acts as a pivot point until A touches the surface, and so on.

    Assuming there is no slipping, draw the locus of point A as the square makes one complete rotation. Show your method clearly.

ICT Section

In this chapter you will have found that it is possible to draw a circle through any three points as long as they do not lie in a straight line.

In this activity you should use either a dynamic geometry package such as Cabri or a graphing package such as Autograph to demonstrate that it is possible to draw a circle through three non-collinear points.

1. a) Using an appropriate software package, plot three points on a pair of axes.
    b) Draw line segments connecting each of the three points.
    c) Construct the perpendicular bisectors of each of the three line segments.

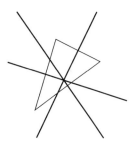

    d) Construct a circle with its centre at the point of intersection of the three perpendicular bisectors and with its circumference passing through the three original points.

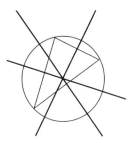

2. Click and drag one of the points to change its position. Describe what happens to the perpendicular bisectors and the circle.

3. Describe what happens when the points are dragged so that they become collinear.

13 GEOMETRICAL RELATIONSHIPS

Use the relationships between the areas of similar triangles, with corresponding results for similar figures and extension to volumes of similar solids.

NB: All diagrams are not drawn to scale.

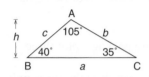

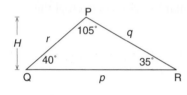

Similar triangles

Two polygons are said to be **similar** if (a) they are equi-angular, and (b) corresponding sides are in proportion. For triangles (a) ⇔ (b).

In the diagrams (left) △ ABC and △ PQR are similar.

For similar figures the ratios of the lengths of the sides are the same and represent the **scale factor**, i.e.

$$\frac{p}{a} = \frac{q}{b} = \frac{r}{c} = k \text{ (where } k \text{ is the scale factor of enlargement)}$$

The heights of similar triangles are proportional also:

$$\frac{H}{h} = \frac{p}{a} = \frac{q}{b} = \frac{r}{c} = k$$

The ratio of the areas of similar triangles (the **area factor**) is equal to the square of the scale factor.

$$\frac{\text{Area of } \triangle \text{ PQR}}{\text{Area of } \triangle \text{ ABC}} = \frac{\frac{1}{2}H \times p}{\frac{1}{2}h \times a} = \frac{H}{h} \times \frac{p}{a} = k^2$$

$$\frac{\text{Area of } \triangle \text{ PQR}}{\text{Area of } \triangle \text{ ABC}} = \left(\frac{H}{h}\right)^2 = \left(\frac{p}{a}\right)^2 = \left(\frac{q}{b}\right)^2 = \left(\frac{r}{c}\right)^2 = k^2$$

Exercise 13.1

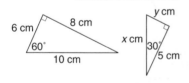

1. a) Explain why the two triangles (left) are similar.
    b) Calculate the scale factor which reduces the larger triangle to the smaller one.
    c) Calculate the value of x and the value of y.

2. Which of the triangles below are similar?

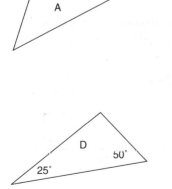

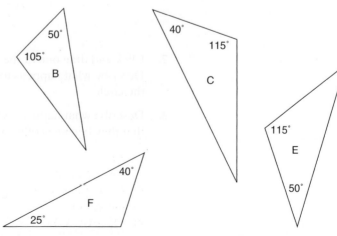

3. The triangles below are similar.

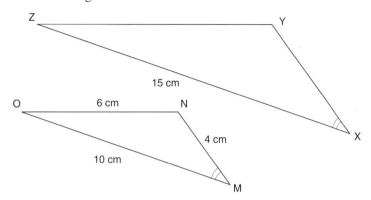

   a) Calculate the length XY.
   b) Calculate the length YZ.

4. In the triangle below calculate the lengths of sides *p*, *q* and *r*.

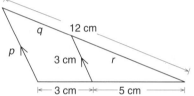

5. In the triangle below calculate the lengths of sides *e* and *f*.

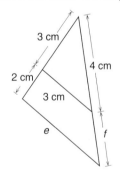

6. The triangles PQR and LMN are similar.

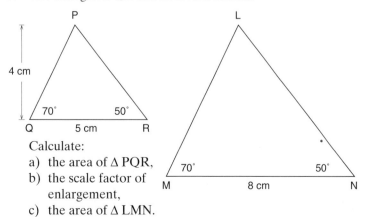

   Calculate:
   a) the area of △ PQR,
   b) the scale factor of enlargement,
   c) the area of △ LMN.

7. The triangles ABC and XYZ below are similar.

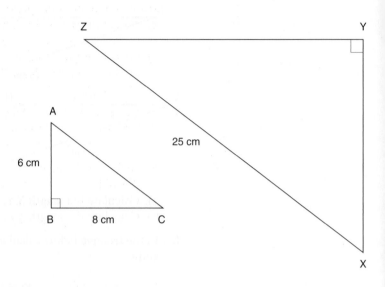

a) Using Pythagoras' theorem, calculate the length AC.
b) Calculate the scale factor of enlargement.
c) Calculate the area of $\triangle$ XYZ.

8. a) Calculate the area of each triangle shown below.

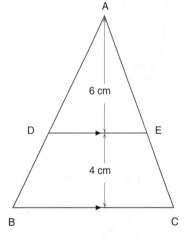

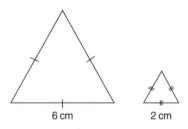

b) Show that the ratio of their areas is equal to the square of the ratio of their sides.

9. The triangle ADE shown (left) has an area of 12 cm^2.
a) Calculate the area of $\triangle$ ABC.
b) Calculate the length BC.

Exercise 13.2

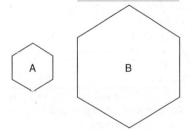

1. In the hexagons (left), hexagon B is an enlargement of hexagon A by a scale factor of 2.5.
If the area of A is 8 cm^2, calculate the area of B.

2. P and Q are two regular pentagons. Q is an enlargement of P by a scale factor of 3. If the area of pentagon Q is 90 cm^2, calculate the area of P.

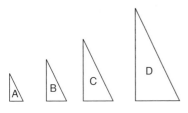

3. On the left is a row of four triangles A, B, C and D. Each is an enlargement of the previous one by a scale factor of 1.5.
 a) If the area of C is 202.5 cm² calculate the area of:
  i) triangle D, ii) triangle B, iii) triangle A.
 b) If the triangles were to continue in this sequence, which letter triangle would be the first to have an area greater than 15 000 cm²?

4. A square is enlarged by increasing the length of its sides by 10%. If the length of its sides was originally 6 cm, calculate the area of the enlarged square.

5. A square of side length 4 cm is enlarged by increasing the lengths of its sides by 25% and then increasing them by a further 50%. Calculate the area of the final square.

6. An equilateral triangle has an area of 25 cm². If the lengths of its sides are reduced by 15%, calculate the area of the reduced triangle.

▓ **Area and volume of similar shapes**

Earlier in the chapter we found the following relationship between the scale factor and the area factor of enlargement:

$$\text{Area factor} = (\text{scale factor})^2$$

A similar relationship can be stated for volumes of similar shapes:

$$\text{i.e. Volume factor} = (\text{scale factor})^3$$

Exercise 13.3

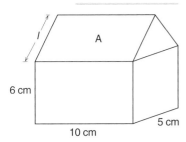

1. The diagram (left) is of a scale model of a garage. Its width is 5 cm, its length 10 cm and the height of its walls 6 cm.
 a) If the width of the real garage is 4 m, calculate:
  i) the length of the real garage,
  ii) the real height of the garage wall.
 b) If the apex of the roof of the real garage is 2 m above the top of the walls, use Pythagoras' theorem to find the real slant length l.
 c) What is the actual area of the roof section marked A?

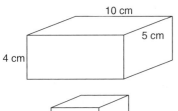

2. The cuboid (left) has dimensions as shown.
 If the cuboid is enlarged by a scale factor of 2.5, calculate:
 a) the total surface area of the original cuboid,
 b) the total surface area of the enlarged cuboid,
 c) the volume of the original cuboid,
 d) the volume of the enlarged cuboid.

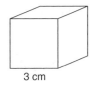

3. On the left is a cube of side length 3 cm.
 a) Calculate its total surface area.
 b) If the cube is enlarged and has a total surface area of 486 cm², calculate the scale factor of enlargement.
 c) Calculate the volume of the enlarged cube.

4. Two cubes P and Q are of different sizes. If n is the ratio of their corresponding sides, express in terms of n:
   a) the ratio of their surface areas,
   b) the ratio of their volumes.

5. The cuboids A and B shown below are similar.

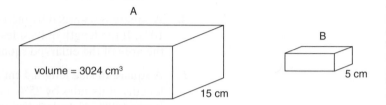

Calculate the volume of cuboid B.

6. Two similar troughs X and Y are shown below.

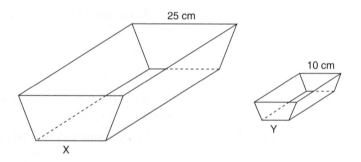

If the capacity of X is 10 litres, calculate the capacity of Y.

Exercise 13.4

1. The two cylinders L and M shown below are similar.

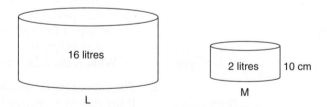

If the height of cylinder M is 10 cm, calculate the height of cylinder L.

2. A square-based pyramid is cut into two shapes by a cut running parallel to the base and made half-way up.
   a) Calculate the ratio of the volume of the smaller pyramid to that of the original one.
   b) Calculate the ratio of the volume of the small pyramid to that of the truncated base.

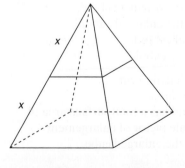

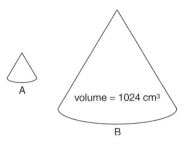

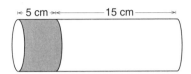

3. The two cones A and B shown (left) are similar. Cone B is an enlargement of A by a scale factor of 4.
If the volume of cone B is 1024 cm³, calculate the volume of cone A.

4. a) Stating your reasons clearly, decide whether the two cylinders shown below are similar or not.

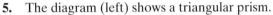

b) What is the ratio of the curved surface area of the shaded cylinder to that of the unshaded cylinder?

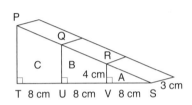

5. The diagram (left) shows a triangular prism.
a) Calculate the area of Δ RSV.
b) Calculate the area of Δ QSU.
c) Calculate the area of Δ PST.
d) Calculate the volume of each of the sections A, B and C.

6. The area of an island on a map is 30 cm². The scale used on the map is 1 : 100 000.
a) Calculate the area in square kilometres of the real island.
b) An airport on the island is on a rectangular piece of land measuring 3 km by 2 km. Calculate the area of the airport on the map in cm³.

7. The two packs of cheese X and Y shown below are similar. The total surface area of pack Y is four times that of pack X.

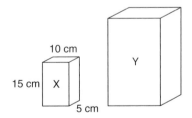

Calculate:
a) the dimensions of pack Y,
b) the mass of pack X if pack Y has a mass of 800 g.

NB: All diagrams are not drawn to scale.

Student Assessment I

1. Which of the triangles below are similar?

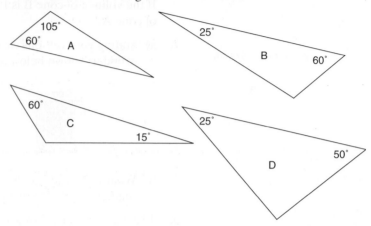

2. Triangles P and Q are similar.

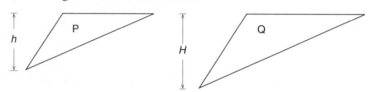

Express the ratio of their areas in the form, area of P : area of Q.

3. Using the triangle (left),
a) explain whether △ ABC and △ PBQ are similar,
b) calculate the length QB,
c) calculate the length BC,
d) calculate the length AP.

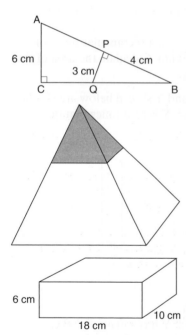

4. The vertical height of the large square-based solid pyramid shown (left) is 30 m. Its mass is 16 000 tonnes.
If the mass of the smaller (shaded) pyramid is 2000 tonnes, calculate its vertical height.

5. The cuboid shown (below left) undergoes a reduction by a scale factor of 0.6.
a) Draw a sketch of the reduced cuboid labelling its dimensions clearly.
b) What is the volume of the new cuboid?
c) What is the total surface area of the new cuboid?

6. Cuboids V and W (below) are similar.

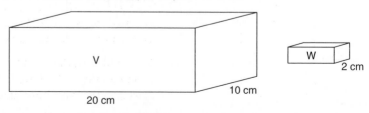

If the volume of cuboid V is 1600 cm³, calculate:
a) the volume of cuboid W,
b) the total surface area of cuboid V,
c) the total surface area of cuboid W.

7. An island has an area of 50 km². What would be its area on a map of scale 1 : 20 000?

8. A box in the shape of a cube has a surface area of 2400 cm². What would be the volume of a similar box enlarged by a scale factor of 1.5?

Student Assessment 2

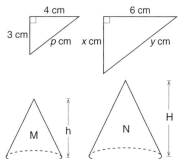

1. The two triangles (left) are similar.
a) Using Pythagoras' theorem, calculate the value of p.
b) Calculate the values of x and y.

2. Cones M and N are similar.
a) Express the ratio of their surface areas in the form, area of M : area of N.
b) Express the ratio of their volumes in the form, volume of M : volume of N.

3. Calculate the values of x, y and z in the triangle below.

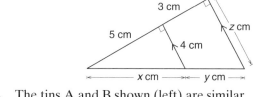

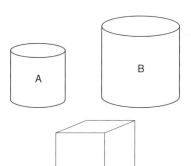

4. The tins A and B shown (left) are similar. The capacity of tin B is three times that of tin A. If the label on tin A has an area of 75 cm², calculate the area of the label on tin B.

5. The cube shown (left) is enlarged by a scale factor of 2.5.
a) Calculate the volume of the enlarged cube.
b) Calculate the surface area of the enlarged cube.

6. The two troughs X and Y shown (left) are similar. The scale factor of enlargement from Y to X is 4. If the capacity of trough X is 1200 cm³, calculate:
a) the depth of trough X,
b) the capacity of trough Y.

7. The rectangular floor plan of a house measures 8 cm by 6 cm. If the scale of the plan is 1 : 50, calculate:
a) the dimensions of the actual floor,
b) the area of the actual floor in m².

8. The volume of the cylinder shown (left) is 400 cm³. Calculate the surface area of a similar cylinder formed by enlarging the one shown above by a scale factor 2.

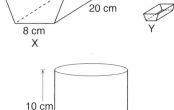

ICT Section

In this activity you will be using a dynamic geometry package such as Cabri to demonstrate that for the triangle below

$$\frac{AB}{ED} = \frac{AC}{EC} = \frac{BC}{DC}$$

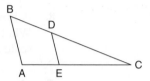

1. a) Using the geometry package construct the triangle ABC.
   b) Construct the line segment ED such that it is parallel to AB. (You will need to construct a line parallel to AB first and then attach the line segment ED to it.)
   c) Using a 'measurement' tool, measure each of the lengths AB, AC, BC, ED, EC and DC.
   d) Using a 'calculator' tool, calculate the ratios
   $$\frac{AB}{ED}, \frac{AC}{EC}, \frac{BC}{DC}$$

2. Comment on your answers to Q.1(d) above.

3. a) Grab vertex B and move it to a new position. What happens to the ratios you calculated in Q.1(d)?
   b) Grab each of the vertices A and C in turn and move them to new positions. What happens to the ratios? Explain why this happens.

4. Grab point D and move it to a new position along the side BC. Explain, giving reasons, what happens to the ratios.

14 ANGLE PROPERTIES

Core Section

Calculate unknown angles using the following geometrical properties:
- angle properties of regular polygons,
- angle in semi-circle,
- angle between tangent and radius of a circle.

NB: All diagrams are not drawn to scale.

A **polygon** is a closed two-dimensional shape bounded by straight lines. Examples of polygons include triangles, quadrilaterals, pentagons and hexagons. Hence the shapes below all belong to the polygon family:

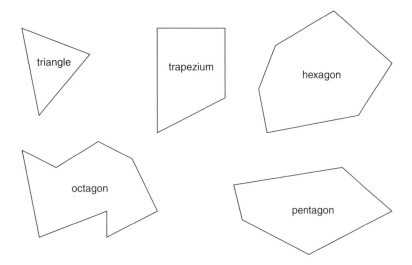

A **regular polygon** is distinctive in that all its sides are of equal length and all its angles are of equal size. Below are some examples of regular polygons.

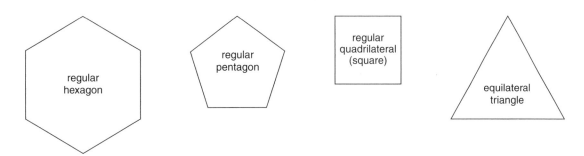

The name of each polygon is derived from the number of angles it contains. The following list identifies some of these polygons.

3 angles = triangle
4 angles = quadrilateral (tetragon)
5 angles = pentagon
6 angles = hexagon
7 angles = heptagon
8 angles = octagon
9 angles = nonagon
10 angles = decagon
12 angles = dodecagon

The sum of the interior angles of a polygon

In the polygons below a straight line is drawn from each vertex to vertex A.

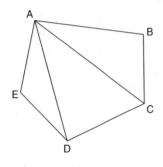 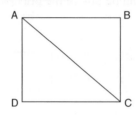 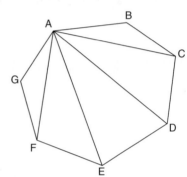

As can be seen, the number of triangles is always two less than the number of sides the polygon has, i.e. if there are *n* sides, there will be (*n* − 2) triangles.

Since the angles of a triangle add up to 180°, the sum of the interior angles of a polygon is therefore 180(*n* − 2) degrees.

Worked example   Find the sum of the interior angles of a regular pentagon and hence the size of each interior angle.

For a pentagon, *n* = 5.

Therefore the sum of the interior angles = 180(5 − 2)°
$$= 180 \times 3°$$
$$= 540°$$

For a regular pentagon the interior angles are of equal size.

Therefore each angle $= \dfrac{540°}{5} = 108°$.

The sum of the exterior angles of a polygon

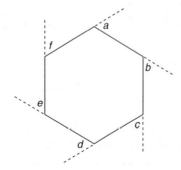

The angles marked *a*, *b*, *c*, *d*, *e* and *f* (left) represent the exterior angles of the regular hexagon drawn.

For any convex polygon the sum of the exterior angles is 360°.
If the polygon is regular and has *n* sides, then each exterior

angle $= \dfrac{360°}{n}$.

Worked examples   **a)**   Find the size of an exterior angle of a regular nonagon.

$$\frac{360°}{9} = 40°$$

b)   Calculate the number of sides a regular polygon has if each exterior angle is 15°.

$$n = \frac{360}{15}$$

$$= 24$$

The polygon has 24 sides.

Exercise 14.1   **1.**   Find the sum of the interior angles of the following polygons:
a) a hexagon   b) a nonagon   c) a heptagon

2.   Find the value of each interior angle of the following regular polygons:
a) an octagon   b) a square
c) a decagon   d) a dodecagon

3.   Find the size of each exterior angle of the following regular polygons:
a) a pentagon   b) a dodecagon   c) a heptagon

4.   The exterior angles of regular polygons are given below. In each case calculate the number of sides the polygon has.
a) 20°   b) 36°   c) 10°
d) 45°   e) 18°   f) 3°

5.   The interior angles of regular polygons are given below. In each case calculate the number of sides the polygon has.
a) 108°   b) 150°   c) 162°
d) 156°   e) 171°   f) 179°

6.   Calculate the number of sides a regular polygon has if an interior angle is five times the size of an exterior angle.

7.   Copy and complete the table below for regular polygons:

Number of sides	Name	Sum of exterior angles	Size of an exterior angle	Sum of interior angles	Size of an interior angle
3					
4					
5					
6					
7					
8					
9					
10					
12					

▧ The angle in a semi-circle

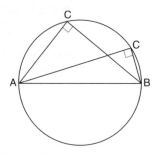

If AB represents the diameter of the circle, then the angle at C is 90°.

Exercise 14.2 In each of the following diagrams, O marks the centre of the circle. Calculate the value of *x* in each case.

1.

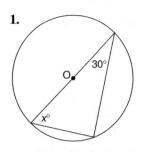

2.

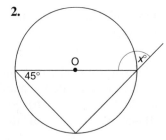

3.

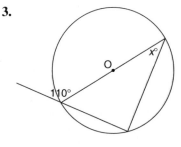

4.

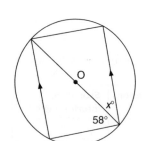

5.

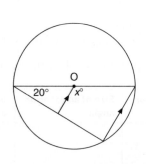

6.

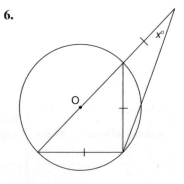

▧ The angle between a tangent and a radius of a circle

The angle between a tangent at a point and the radius to the same point on the circle is a right angle.

Triangles OAC and OBC are congruent as ∠OAC and ∠OBC are right angles, OA = OB because they are both radii and OC is common to both triangles.

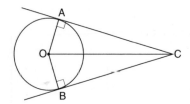

Exercise 14.3   In each of the following diagrams, O marks the centre of the circle. Calculate the value of x in each case.

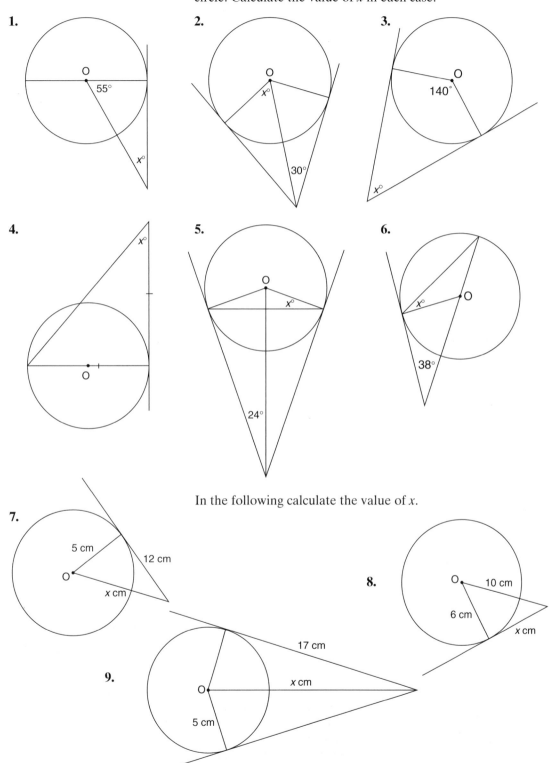

In the following calculate the value of x.

Core Section

Student Assessment I

1. Draw a diagram of an octagon to help illustrate the fact that the sum of the internal angles of an octagon is given by $180 \times (8 - 2)°$.

2. Find the size of each interior angle of a twenty-sided regular polygon.

3. What is the sum of the interior angles of a nonagon?

4. What is the sum of the exterior angles of a polygon?

5. What is the size of the exterior angle of a regular pentagon?

6. If AB is the diameter of the circle and AC = 5 cm and BC = 12 cm, calculate:
    a) the size of angle ACB,
    b) the length of the radius of the circle.

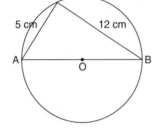

In Q.7–10, O marks the centre of the circle.

Calculate the size of the angle marked *x* in each case.

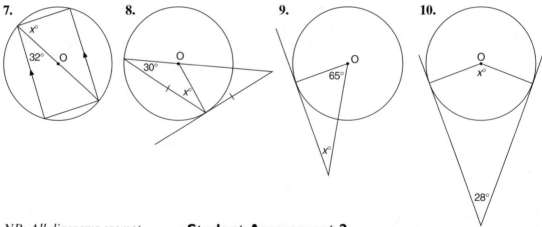

7.   8.   9.   10.

Student Assessment 2

1. Draw a diagram of a hexagon to help illustrate the fact that the sum of the internal angles of a hexagon is given by $180 \times (6 - 2)°$.

2. Find the value of each interior angle of a regular polygon with 24 sides.

3. What is the sum of the interior angles of a regular dodecagon?

4. What is the size of an exterior angle of a regular dodecagon?

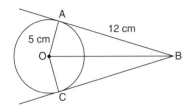

5. AB and BC are both tangents to the circle centre O. If
    OA = 5 cm and AB = 12 cm calculate:
    a) the size of angle OAB,
    b) the length OB.

6. If OA is a radius of the circle and PB the tangent to the
    circle at A, calculate angle ABO.

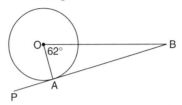

In Q.7–10, O marks the centre of the circle. Calculate the size
of the angle marked *x* in each case.

7.   8.   9.   10.

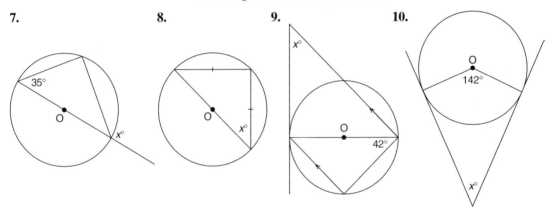

ANGLE PROPERTIES

Extended Section

Use the following geometrical properties:
- angle properties of irregular polygons,
- angle at the centre of a circle is twice the angle at the circumference,
- angles in the same segment are equal,
- angles in opposite segments are supplementary.

NB: All diagrams are not drawn to scale.

Angle properties of irregular polygons

As explained in the core section of this chapter, the sum of the interior angles of a polygon is given by $180(n - 2)°$, where n represents the number of sides the polygon has. The sum of the exterior angles of any polygon is $360°$.

Both of these rules also apply to irregular polygons, i.e. those where the lengths of the sides and the sizes of the interior angles are not all equal.

Exercise 14.4

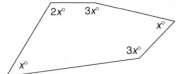

1. For the pentagon (left):
   a) calculate the value of x,
   b) calculate the size of each of the angles.

2. Find the size of each angle in the octagon (right).

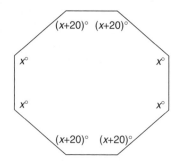

3. Calculate the value of x for the pentagon shown (right).

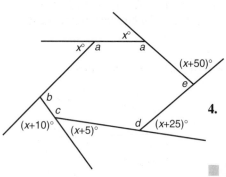

4. Calculate the size of each of the angles a, b, c, d and e in the hexagon (left).

Angle at the centre of a circle

The angle subtended at the centre of a circle by an arc is twice the size of the angle on the circumference subtended by the same arc.

Both diagrams below illustrate this theorem.

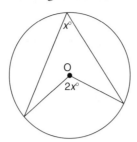 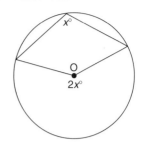

Exercise 14.5 In each of the following diagrams, O marks the centre of the circle. Calculate the size of the marked angles:

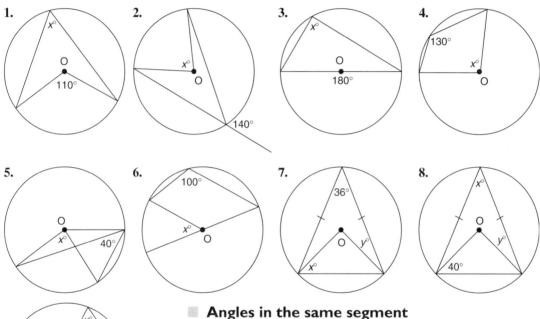

Angles in the same segment

Angles in the same segment of a circle are equal.

This can be explained simply by using the theorem that the angle subtended at the centre is twice the angle on the circumference. Looking at the diagram (left), if the angle at the centre is $2x°$, then each of the angles at the circumference must be equal to $x°$.

Exercise 14.6 Calculate the marked angles in the following diagrams:

1. 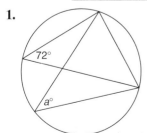 **2.** 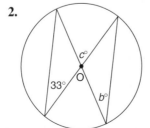 **3.**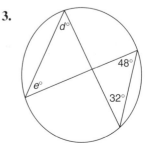

4.

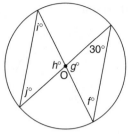

5.

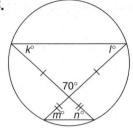

6.

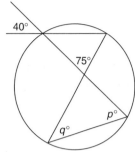

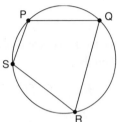

Angles in opposite segments

Points P, Q, R and S all lie on the circumference of the circle (left). They are called concyclic points. Joining the points P, Q, R and S produces a cyclic quadrilateral.

The opposite angles are **supplementary**, i.e. they add up to 180°.

Since $p° + r° = 180°$ (supplementary angles) and $r° + t° = 180°$ (angles on a straight line) it follows that $p° = t°$.

Therefore the exterior angle of a cyclic quadrilateral is equal to the interior opposite angle.

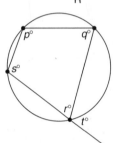

Exercise 14.7   Calculate the size of the marked angles in each of the following:

1.

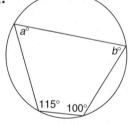

2.

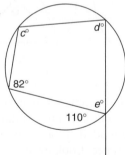

3.

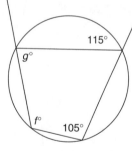

4.

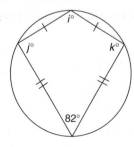

5.

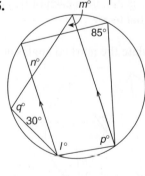

6.

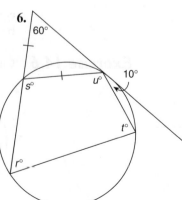

Extended Section

Student Assessment I

NB: All diagrams are not drawn to scale.

1. In the following diagrams, O marks the centre of the circle. Identify which angles are:
   i) supplementary angles,
   ii) right angles,
   iii) equal.

a)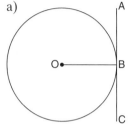

b)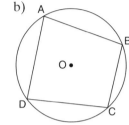

c)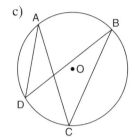

d)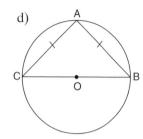

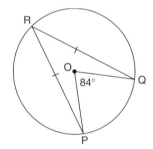

2. If ∠POQ = 84° and O marks the centre of the circle in the diagram (left), calculate the following:
   a) ∠PRQ   b) ∠OQR

3. Calculate ∠DAB and ∠ABC in the diagram below.

4. If DC is a diameter, and O marks the centre of the circle, calculate angles BDC and DAB.

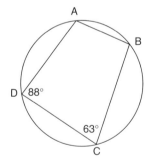

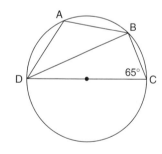

5. Calculate as many angles as possible in the diagram below. O marks the centre of the circle.

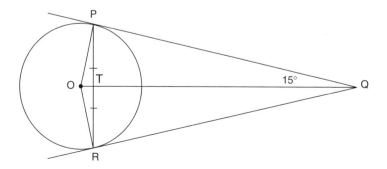

6. Calculate the values of *c*, *d* and *e*.

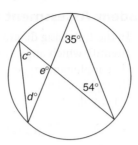

7. Calculate the values of *f*, *g*, *h* and *i*.

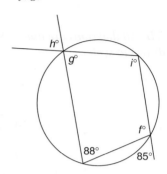

NB: All diagrams are not drawn to scale.

Student Assessment 2

1. In the following diagrams, O is the centre of the circle. Identify which angles are:
i) supplementary angles,
ii) right angles,
iii) equal.

a)
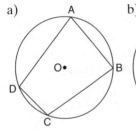

b)
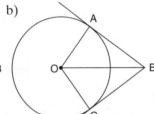

c)
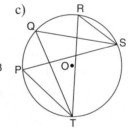

d)
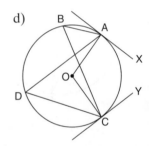

2. If ∠AOC is 72°, calculate ∠ABC.

3. If ∠AOB = 130° calculate ∠ABC, ∠OAB and ∠CAO.

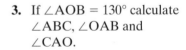

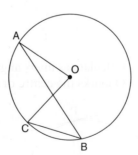

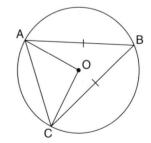

4. Show that ABCD is a cyclic quadrilateral.

5. Calculate f and g.

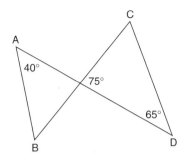

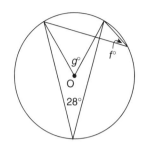

6. If $y° = 22.5°$ calculate the value of x.

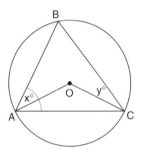

ICT Section

Using a geometry package, such as Cabri, demonstrate the following angle properties of a circle:

i)   The angle subtended at the centre of a circle by an arc is twice the size of the angle on the circumference subtended by the same arc.

     e.g. The diagram below demonstrates the first of these:

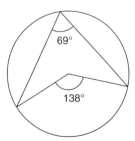

ii) The angles in the same segment of a circle are equal.

iii) The exterior angle of a cyclic quadrilateral is equal to the interior opposite angle.

15 TRIGONOMETRY
Core Section

> Apply Pythagoras' theorem and the sine, cosine and tangent ratios for acute angles to the calculation of a side or of an angle of a right-angled triangle.

NB: All diagrams are not drawn to scale.

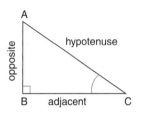

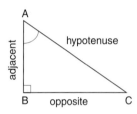

There are three basic trigonometric ratios: sine, cosine and tangent.

Each of these relates an angle of a right-angled triangle to a ratio of the lengths of two of its sides.

The sides of the triangle have names, two of which are dependent on their position in relation to a specific angle.

The longest side (always opposite the right angle) is called the **hypotenuse**. The side opposite the angle is called the **opposite** side and the side next to the angle is called the **adjacent** side.

Note that, when the chosen angle is at A, the sides labelled opposite and adjacent change (below left).

■ Tangent

$$\tan C = \frac{\text{length of opposite side}}{\text{length of adjacent side}}$$

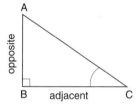

Worked examples

a) Calculate the size of angle BAC in each of the following triangles.

i) $\tan x° = \dfrac{\text{opposite}}{\text{adjacent}} = \dfrac{4}{5}$

$x = \tan^{-1}\left(\dfrac{4}{5}\right)$

$x = 38.7$ (3 s.f.)

$\angle \text{BAC} = 38.7°$ (3 s.f.)

ii) $\tan x° = \dfrac{8}{3}$

$x = \tan^{-1}\left(\dfrac{8}{3}\right)$

$x = 69.4$ (3 s.f.)

$\angle \text{BAC} = 69.4°$ (3 s.f.)

b) Calculate the length of the opposite side QR.

$\tan 42° = \dfrac{p}{6}$

$6 \times \tan 42° = p$

$p = 5.40$ (3 s.f.)

$\text{QR} = 5.40$ cm (3 s.f.)

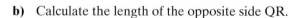

c) Calculate the length of the adjacent side XY.

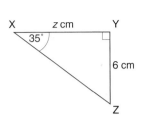

$$\tan 35° = \frac{6}{z}$$

$$z \times \tan 35° = 6$$

$$z = \frac{6}{\tan 35°}$$

$$z = 8.57 \ (3 \text{ s.f.})$$
$$XY = 8.57 \text{ cm} \ (3 \text{ s.f.})$$

Exercise 15.1

Calculate the length of the side marked x cm in each of the diagrams in questions 1 and 2. Give your answers to 1 d.p.

1. a)

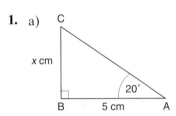

b)

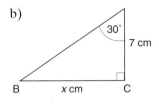

c)

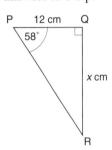

d)

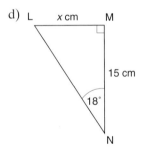

e)

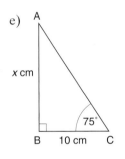

f)

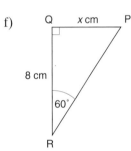

2. a)

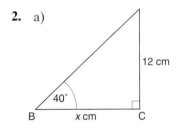

b)

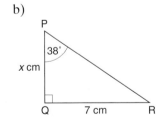

c)

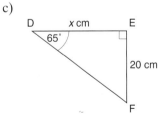

d)

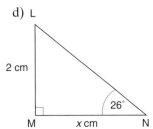

e)

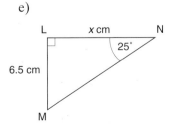

f)

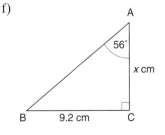

3. Calculate the size of the marked angle $x°$ in each of the following diagrams. Give your answer to 1 d.p.

a)
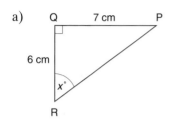

b)
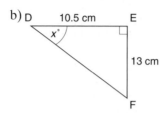

c)
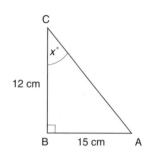

d)
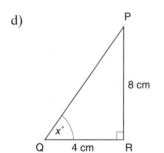

e)
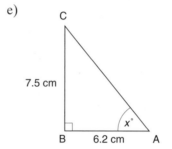

f)
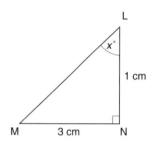

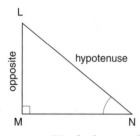

Sine

$$\sin N = \frac{\text{length of opposite side}}{\text{length of hypotenuse}}$$

Worked examples

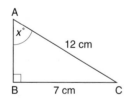

a) Calculate the size of angle BAC.

$$\sin x = \frac{\text{opposite}}{\text{hypotenuse}} = \frac{7}{12}$$

$$x = \sin^{-1}\left(\frac{7}{12}\right)$$

$$x = 35.7 \text{ (1 d.p.)}$$
$$\angle \text{BAC} = 35.7° \text{ (1 d.p.)}$$

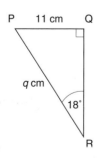

b) Calculate the length of the hypotenuse PR.

$$\sin 18° = \frac{11}{q}$$

$$q \times \sin 18° = 11$$

$$q = \frac{11}{\sin 18°}$$

$$q = 35.6 \text{ (1 d.p.)}$$

$$\text{PR} = 35.6 \text{ cm (1 d.p.)}$$

Exercise 15.2   **1.**   Calculate the length of the marked side in each of the following diagrams. Give your answers to 1 d.p.

a)
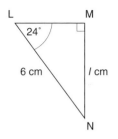

b)
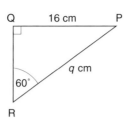

c)
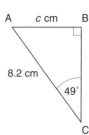

d)
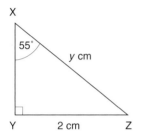

e)
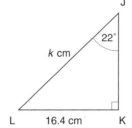

f)
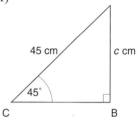

2.   Calculate the size of the angle marked x in each of the following diagrams. Give your answers to 1 d.p.

a)
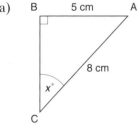

b)
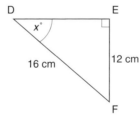

c)
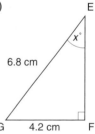

d)
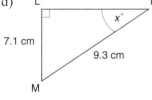

e)
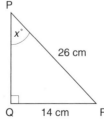

f)
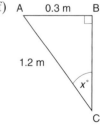

▨ Cosine

$$\cos Z = \frac{\text{length of adjacent side}}{\text{length of hypotenuse}}$$

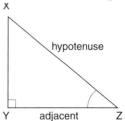

Worked examples **a)** Calculate the length XY.

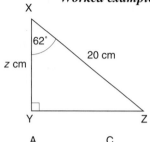

$$\cos 62° = \frac{\text{adjacent}}{\text{hypotenuse}} = \frac{z}{20}$$

$z = 20 \times \cos 62°$
$z = 9.4 \ (1 \text{ d.p.})$
$XY = 9.4 \text{ cm} \ (1 \text{ d.p.})$

b) Calculate the size of angle ABC.

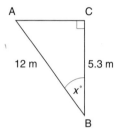

$$\cos x = \frac{5.3}{12}$$

$$x = \cos^{-1}\left(\frac{5.3}{12}\right)$$

$x = 63.8 \ (1 \text{ d.p.})$
$\angle ABC = 63.8° \ (1 \text{ d.p.})$

Exercise 15.3 **1.** Calculate either the marked side or angle in each of the following diagrams. Give your answers to 1 d.p.

a)
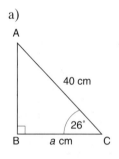

b)
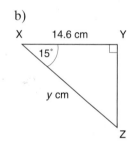

c) d)
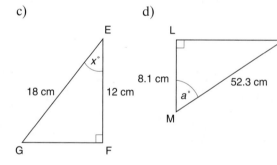

e)
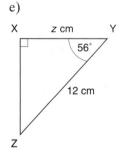

f)
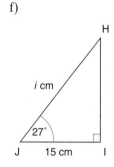

g)
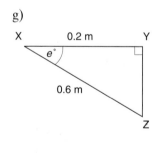

h)
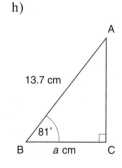

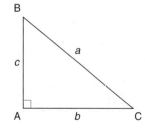

Pythagoras' theorem

Pythagoras' theorem states the relationship between the lengths of the three sides of a right-angled triangle.
Pythagoras' theorem states that:

$$a^2 = b^2 + c^2$$

Worked examples **a)** Calculate the length of the side BC.

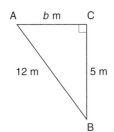

Using Pythagoras:

$$a^2 = b^2 + c^2$$
$$a^2 = 8^2 + 6^2$$
$$a^2 = 64 + 36 = 100$$
$$a = \sqrt{100}$$
$$a = 10$$
$$BC = 10 \text{ m}$$

b) Calculate the length of the side AC.

Using Pythagoras:
$$a^2 = b^2 + c^2$$
$$a^2 - c^2 = b^2$$
$$b^2 = 144 - 25 = 119$$
$$b = \sqrt{119}$$
$$b = 10.9 \text{ (1 d.p.)}$$
$$AC = 10.9 \text{ m (1 d.p.)}$$

Exercise 15.4 In each of the diagrams in Q.1 and 2, use Pythagoras' theorem to calculate the length of the marked side.

1. a) b) c) d)

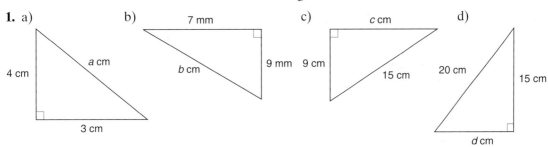

2. a) b) c)

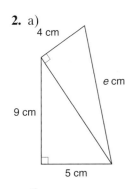

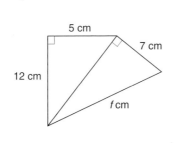

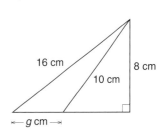

d) e) f)

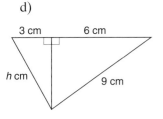

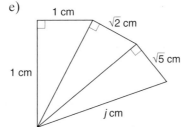

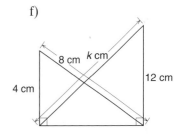

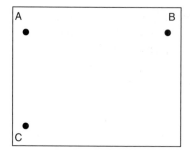

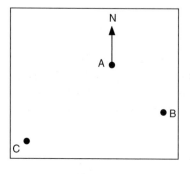

3. Villages A, B and C lie on the edge of the Namib desert. Village A is 30 km due North of village C. Village B is 65 km due East of A.
Calculate the shortest distance between villages C and B, giving your answer to the nearest 0.1 km.

4. Town X is 54 km due West of town Y. The shortest distance between town Y and town Z is 86 km. If town Z is due South of X calculate the distance between X and Z, giving your answer to the nearest kilometre.

5. Village B is on a bearing of 135° and at a distance of 40 km from village A. Village C is on a bearing of 225° and a distance of 62 km from village A.
   a) Show that triangle ABC is right-angled.
   b) Calculate the distance from B to C, giving your answer to the nearest 0.1 km.

6. Two boats set off from X at the same time. Boat A sets off on a bearing of 325° and with a velocity of 14 km/h. Boat B sets off on a bearing of 235° with a velocity of 18 km/h. Calculate the distance between the boats after they have been travelling for 2.5 hours. Give your answer to the nearest metre.

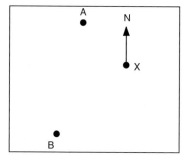

7. A boat sets off on a trip from S. It heads towards B, a point 6 km away and due North. At B it changes direction and heads towards point C, also 6 km away and due East of B. At C it changes direction once again and heads on a bearing of 135° towards D which is 13 km from C.
   a) Calculate the distance between S and C to the nearest 0.1 km.
   b) Calculate the distance the boat will have to travel if it is to return to S from D.

8. Two trees are standing on flat ground.
The height of the smaller tree is 7 m. The distance between the top of the smaller tree and the base of the taller tree is 15 m.
The distance between the top of the taller tree and the base of the smaller tree is 20 m.
   a) Calculate the horizontal distance between the two trees.
   b) Calculate the height of the taller tree.

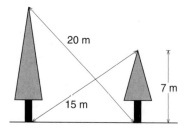

Exercise 15.5

1. By using Pythagoras' theorem, trigonometry or both, calculate the marked value in each of the following diagrams. In each case give your answer to 1 d.p.

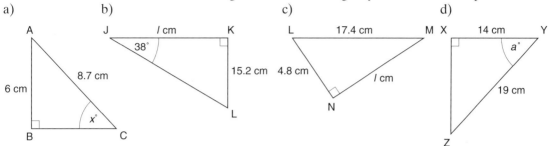

a)   b)   c)   d)

2. a) A sailing boat sets off from a point X and heads towards Y, a point 17 km North. At point Y it changes direction and heads towards point Z, a point 12 km away on a bearing of 090°. Once at Z the crew want to sail back to X. Calculate:
   i) the distance ZX,
   ii) the bearing of X from Z.

   b) An aeroplane sets off from G (left) on a bearing of 024° towards H, a point 250 km away. At H it changes course and heads towards J on a bearing of 055° and a distance of 180 km away.
   i) How far is H to the North of G?
   ii) How far is H to the East of G?
   iii) How far is J to the North of H?
   iv) How far is J to the East of H?
   v) What is the shortest distance between G and J?
   vi) What is the bearing of G from J?

   c) Two trees are standing on flat ground. The angle of elevation of their tops from a point X on the ground is 40°. If the horizontal distance between X and the small tree is 8 m and the distance between the tops of the two trees is 20 m, calculate:
   i) the height of the small tree,
   ii) the height of the tall tree,
   iii) the horizontal distance between the trees.

   d) PQRS is a quadrilateral. The sides RS and QR are the same length. The sides QP and RS are parallel. Calculate:
   i) angle SQR,
   ii) angle PSQ,
   iii) length PQ,
   iv) length PS,
   v) the area of PQRS.

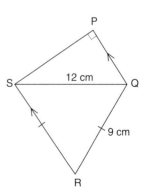

Core Section

Student Assessment 1

NB: All diagrams are not drawn to scale.

1. Calculate the length of the side marked *x* cm in each of the following. Give your answers correct to 1 d.p.

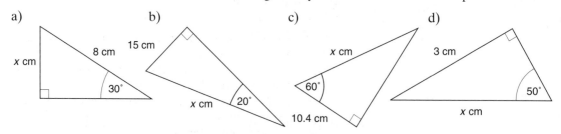

a)   b)   c)   d)

2. Calculate the angle marked $\theta°$ in each of the following. Give your answers correct to the nearest degree.

a)   b)   c)

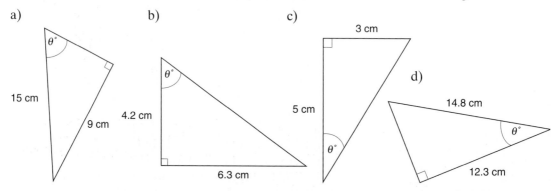

d)

3. Calculate the length of the side marked *q* cm in each of the following. Give your answers correct to 1 d.p.

a)   b)

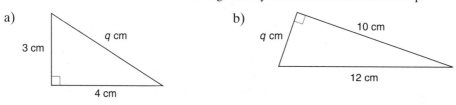

c)   d)

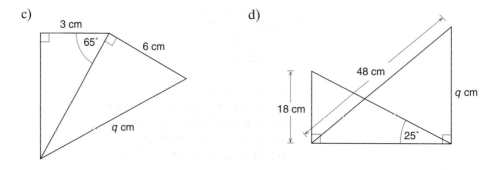

NB: All diagrams are not drawn to scale.

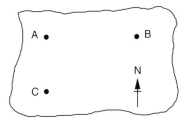

Student Assessment 2

1. A map shows three towns A, B and C. Town A is due North of C. Town B is due East of A. The distance AC is 75 km and the bearing of C from B is 245°. Calculate, giving your answers to the nearest 100 m:
a) the distance AB,
b) the distance BC.

2. Two trees stand 16 m apart. Their tops make an angle of $\theta°$ at point A on the ground.
a) Express $\theta°$ in terms of the height of the shorter tree and its distance x metres from point A.
b) Express $\theta°$ in terms of the height of the taller tree and its distance from A.
c) Form an equation in terms of x.
d) Calculate the value of x.
e) Calculate the value θ.

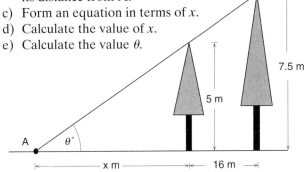

3. Two boats X and Y, sailing in a race, are shown in the diagram (left). Boat X is 145 m due North of a buoy B. Boat Y is due East of buoy B. Boats X and Y are 320 m apart. Calculate:
a) the distance BY,
b) the bearing of Y from X,
c) the bearing of X from Y.

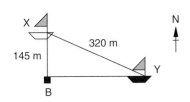

4. Two hawks P and Q are flying vertically above one another. Hawk Q is 250 m above hawk P. They both spot a snake at R.

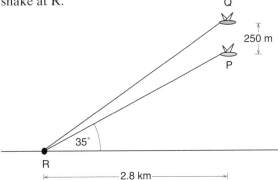

Using the information given, calculate:
a) the height of P above the ground,
b) the distance between P and R,
c) the distance between Q and R.

TRIGONOMETRY

Extended Section

Solve trigonometrical problems in two dimensions involving angles of elevation and depression, extend the sine and cosine functions to angles between 90° and 180°, solve problems using the sine and cosine rules for any triangle and the formula area of triangle $= \frac{1}{2}ab$ sin C; solve simple trigonometrical problems in three dimensions including angle between a line and a plane.

NB: All diagrams are not drawn to scale.

▨ Angles of elevation and depression

The **angle of elevation** is the angle above the horizontal through which a line of view is raised. The **angle of depression** is the angle below the horizontal through which a line of view is lowered.

Worked examples **a)** The base of a tower is 60 m away from a point X on the ground. If the angle of elevation of the top of the tower from X is 40° calculate the height of the tower.

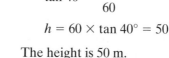

Give your answer to the nearest metre.

$$\tan 40° = \frac{h}{60}$$

$$h = 60 \times \tan 40° = 50$$

The height is 50 m.

b) An aeroplane receives a signal from a point X on the ground. If the angle of depression of point X from the aeroplane is 30° calculate the height at which the plane is flying.

Give your answer to the nearest 0.1 km.

$$\sin 30° = \frac{h}{6}$$

$$h = 6 \times \sin 30° = 3.0$$

The height is 3.0 km.

Exercise 15.6

1. **a)** A and B are two villages. If the horizontal distance between them is 12 km and the vertical distance between them is 2 km calculate:
   i) the shortest distance between the two villages,
   ii) the angle of elevation of B from A.

b) X and Y are two towns. If the horizontal distance between them is 10 km and the angle of depression of Y from X is 7° calculate:
   i) the shortest distance between the two towns,
   ii) the vertical height between the two towns.

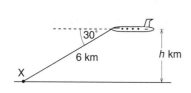

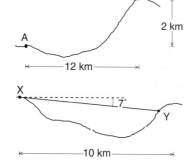

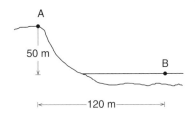

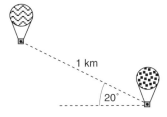

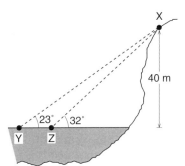

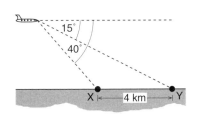

c) A girl standing on a hill at A, overlooking a lake, can see a small boat at a point B on the lake. If the girl is at a height of 50 m above B and at a horizontal distance of 120 m away from B, calculate:
   i) the angle of depression of the boat from the girl,
   ii) the shortest distance between the girl and the boat.

d) Two hot air balloons are 1 km apart in the air. If the angle of elevation of the higher from the lower balloon is 20°, calculate, giving your answers to the nearest metre:
   i) the vertical height between the two balloons,
   ii) the horizontal distance between the two balloons.

2. a) A boy X can be seen by two of his friends Y and Z, who are swimming in the sea. If the angle of elevation of X from Y is 23° and from Z is 32°, and the height of X above Y and Z is 40 m calculate:
   i) the horizontal distance between X and Z,
   ii) the horizontal distance between Y and Z.
   Note: XYZ is a vertical plane

   b) A plane is flying at an altitude of 6 km directly over the line AB. It spots two boats A and B, on the sea. If the angles of depression of A and B from the plane are 60°

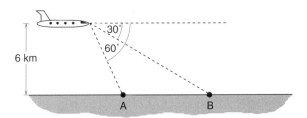

   and 30° respectively, calculate the horizontal distance between A and B.

   c) A plane is flying at a constant altitude over the sea directly over the line XY. It can see two boats X and Y which are 4 km apart. If the angles of depression of X and Y from the plane are 40° and 15° respectively, calculate:
   i) the horizontal distance between Y and the plane,
   ii) the altitude at which the plane is flying.

   d) Two planes are flying directly above each other. A person standing at P can see both of them. The horizontal distance between the two planes and the person is 2 km. If the angles of elevation of the planes from the person are 65° and 75° calculate:
   i) the altitude at which the higher plane is flying,
   ii) the vertical distance between the two planes.

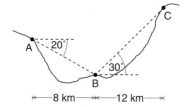

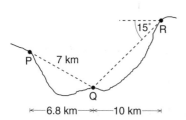

3. a) Three villages A, B and C can see each other across a valley. The horizontal distance between A and B is 8 km, and the horizontal distance between B and C is 12 km. The angle of depression of B from A is 20° and the angle of elevation of C from B is 30°. Calculate, giving all answers to 1 d.p.:
   i) the vertical height between A and B,
   ii) the vertical height between B and C,
   iii) the angle of elevation of C from A,
   iv) the shortest distance between A and C.
   Note: A, B and C are in the same vertical plane.

b) Using binoculars, three people P, Q and R can see each other across a valley. The horizontal distance between P and Q is 6.8 km and the horizontal distance between Q and R is 10 km. If the shortest distance between P and Q is 7 km and the angle of depression of Q from R is 15°, calculate, giving your answers to 1 d.p.:
   i) the vertical height between Q and R,
   ii) the vertical height between P and R,
   iii) the angle of elevation of R from P,
   iv) the shortest distance between P and R.
   Note: P, Q and R are in the same vertical plane.

c) Two people A and B are standing either side of a transmission mast. A is 130 m away from the mast and B is 200 m away. If the angle of elevation of the top of the mast from A is 60°, calculate to the nearest metre:
   i) the height of the mast,
   ii) the angle of elevation of the top of the mast from B.

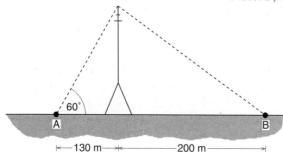

Angles between 0° and 180°

When calculating the size of angles using trigonometry, there are often two solutions. Most calculators, however, will only give the first solution. To be able to calculate the value of the second possible solution, an understanding of the shape of trigonometrical graphs is needed.

The sine curve

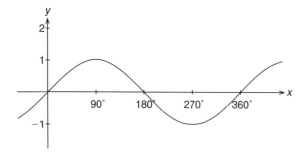

The graph of $y = \sin x$ is plotted (left), where x is the size of the angle in degrees.

The graph of $y = \sin x$ has

■ a period of 360° (i.e. it repeats itself every 360°),
■ a maximum value of 1, (90°)
■ a minimum value of −1 (270°).

Worked examples **a)** sin 30° = 0.5. Which other angle between 0° and 180° has a sine of 0.5?

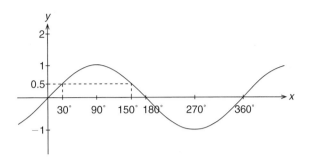

From the graph above it can be seen that sin 150° = 0.5.

sin x = sin (180° − x)

Exercise 15.7

1. Express each of the following in terms of the sine of another angle between 0° and 180°:

a) sin 60°   b) sin 80°   c) sin 115°
d) sin 140°   e) sin 128°   f) sin 167°

2. Express each of the following in terms of the sine of another angle between 0° and 180°:

a) sin 35°   b) sin 50°   c) sin 30°
d) sin 48°   e) sin 104°   f) sin 127°

3. Find the two angles between 0° and 180° which have the following sine. Give each angle to the nearest degree.

a) 0.33   b) 0.99   c) 0.09
d) 0.95   e) 0.22   f) 0.47

4. Find the two angles between 0° and 180° which have the following sine. Give each angle to the nearest degree.

a) 0.94   b) 0.16   c) 0.80
d) 0.56   e) 0.28   f) 0.33

▨ The cosine curve

The graph of $y = \cos x$ is plotted below, where x is the size of the angle in degrees.

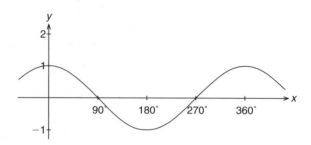

As with the sine curve, the graph of $y = \cos x$ has

■ a period of 360°,
■ a maximum value of 1,
■ a minimum value of −1.

Note $\cos x° = -\cos(180 - x)°$

Worked examples **a)** $\cos 60° = 0.5$. Which other angle between 0° and 180° has a cosine of −0.5?

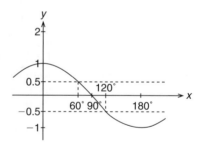

From the graph above it can be seen that $\cos 120° = -0.5$.

b) The cosine of which other angle between 0° and 180° is equal to the negative of $\cos 50°$?

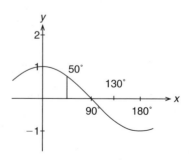

$\cos 130° = -\cos 50°$

Exercise 15.8

1. Express each of the following in terms of the cosine of another angle between 0° and 180°:
   a) cos 20° b) cos 85° c) cos 32°
   d) cos 95° e) cos 147° f) cos 106°

2. Express each of the following in terms of the cosine of another angle between 0° and 180°:
   a) cos 98° b) cos 144° c) cos 160°
   d) cos 143° e) cos 171° f) cos 123°

3. Express each of the following in terms of the cosine of another angle between 0° and 180°:
   a) −cos 100° b) cos 90° c) −cos 110°
   d) −cos 45° e) −cos 122° f) −cos 25°

4. The cosine of which acute angle has the same value as:
   a) cos 125° b) cos 107° c) −cos 120°
   d) −cos 98° e) −cos 92° f) −cos 110°?

The sine rule

With right-angled triangles we can use the basic trigonometric ratios of sine, cosine and tangent. The **sine rule** is a relationship which can be used with non right-angled triangles.

The sine rule states that:

$$\frac{a}{\sin A} = \frac{b}{\sin B} = \frac{c}{\sin C}$$

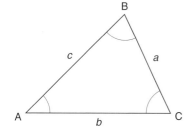

Worked examples **a)** Calculate the length of side BC.

Using the sine rule:

$$\frac{a}{\sin A} = \frac{b}{\sin B}$$

$$\frac{a}{\sin 40°} = \frac{6}{\sin 30°}$$

$$a = \frac{6 \times \sin 40°}{\sin 30°}$$

$$a = 7.7 \ (1 \ \text{d.p.})$$

$$BC = 7.7 \ \text{cm} \ (1 \ \text{d.p.})$$

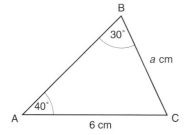

b) Calculate the size of angle C.

Using the sine rule:

$$\frac{\sin A}{a} = \frac{\sin C}{c}$$

$$\sin C = \frac{6.5 \times \sin 60°}{6}$$

$$C = \sin^{-1}(0.94)$$

$$C = 69.8° \ (1 \ \text{d.p.})$$

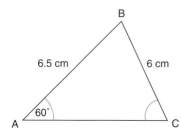

Exercise 15.9

1. Calculate the length of the side marked *x* in each of the following. Give your answers to 1 d.p.

a)   b)   c)   d)

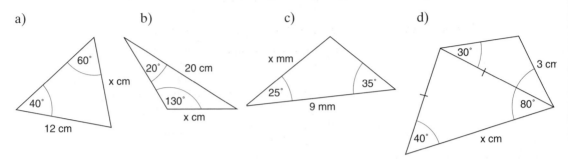

2. Calculate the size of the angle marked $\theta°$ in each of the following. Give your answers to 1 d.p.

a)   b)   c)   d)

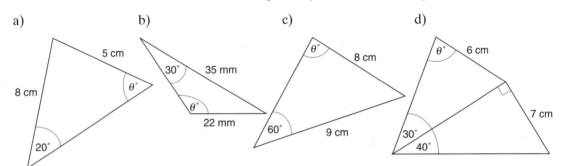

3. Δ ABC has the following dimensions:

AC = 10 cm, AB = 8 cm and ∠ ACB = 20°.

a) Calculate the two possible values for ∠CAB.
b) Sketch and label the two possible shapes for Δ ABC.

4. Δ PQR has the following dimensions:

PQ = 6 cm, PR = 4 cm and ∠PQR = 40°.

a) Calculate the two possible values for ∠QRP.
b) Sketch and label the two possible shapes for Δ PQR.

The cosine rule

The **cosine rule** is another relationship which can be used with non right-angled triangles.

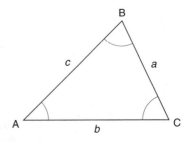

The cosine rule states that:

$$a^2 = b^2 + c^2 - 2bc \cos A$$

Worked examples

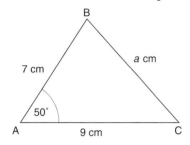

a) Calculate the length of the side BC.

Using the cosine rule:

$$a^2 = b^2 + c^2 - 2bc \cos A$$
$$a^2 = 9^2 + 7^2 - (2 \times 9 \times 7 \times \cos 50°)$$
$$= 81 + 49 - (126 \times \cos 50°) = 49.0$$
$$a = \sqrt{49.0}$$
$$a = 7.0 \text{ (1 d.p.)}$$
$$BC = 7.0 \text{ cm (1 d.p.)}$$

b) Calculate the size of angle A.

Using the cosine rule:

$$a^2 = b^2 + c^2 - 2bc \cos A.$$

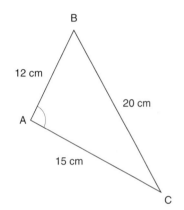

Rearranging the equation gives:

$$\cos A = \frac{b^2 + c^2 - a^2}{2bc}$$

$$\cos A = \frac{15^2 + 12^2 - 20^2}{2 \times 15 \times 12} = -0.086$$

$$A = \cos^{-1}(-0.086)$$
$$A = 94.9° \text{ (1 d.p.)}$$

Exercise 15.10

1. Calculate the length of the side marked x in each of the following. Give your answers to 1 d.p.

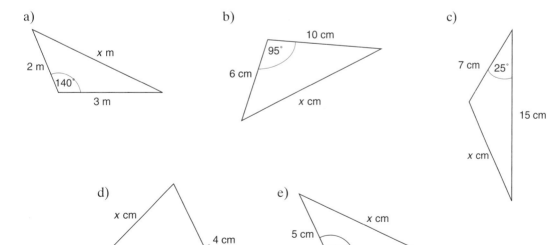

2. Calculate the angle marked $\theta°$ in each of the following. Give your answers to 1 d.p.

a)

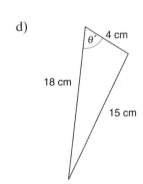

40 m
20 m
$\theta°$
25 m

b)

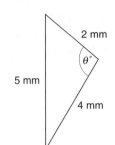

2 mm
$\theta°$
5 mm
4 mm

c)

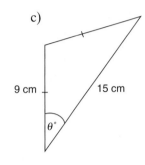

9 cm
15 cm
$\theta°$

d)

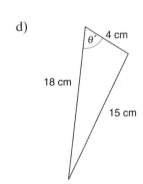

$\theta°$ 4 cm
18 cm
15 cm

e)

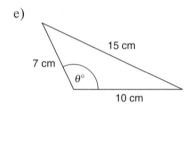

15 cm
7 cm
$\theta°$
10 cm

Exercise 15.11

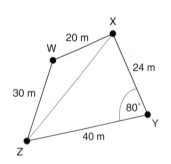

X
20 m
W
24 m
30 m
80°
Y
40 m
Z

1. Four players W, X, Y and Z are on a rugby pitch. The diagram (left) shows a plan view of their relative positions. Calculate:
a) the distance between players X and Z,
b) ∠ZWX,
c) ∠WZX,
d) ∠YZX,
e) the distance between players W and Y.

2. Three yachts A, B and C are racing off the 'Cape'. Their relative positions are shown below.
   Calculate the distance between B and C to the nearest 10 m.

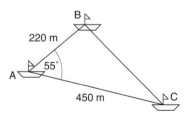

B
220 m
55°
A
450 m
C

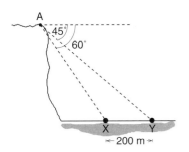

3. A girl standing on a cliff top at A can see two buoys X and Y, 200 m apart, floating on the sea. The angle of depression of Y from A is 45°, and the angle of depression of X from A is 60° (see diagram on left).

If AXY are in the same vertical plane, calculate:

a) the distance AY,

b) the distance AX,

c) the vertical height of the cliff.

4. There are two trees standing on one side of a river bank. On the opposite side is a boy standing at X.

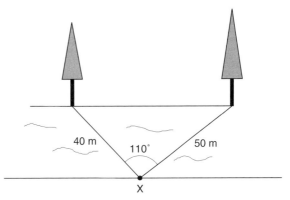

Using the information given, calculate the distance between the two trees.

The area of a triangle

$$\text{Area} = \tfrac{1}{2}bh$$

Also:

$$\sin C = \frac{h}{a}$$

Rearranging:

$$h = a \sin C$$

Therefore

$$\text{area} = \tfrac{1}{2}ab \sin C$$

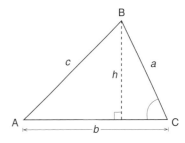

Exercise 15.12 **1.** Calculate the area of the following triangles. Give your answers to 1 d.p.

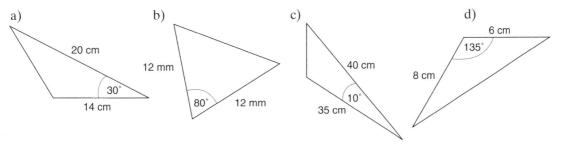

2. Calculate the value of x in each of the following. Give your answers correct to 1 d.p.

a)

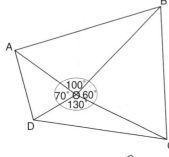

b)

c)

d)

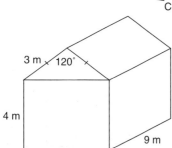

3. ABCD is a school playing field (left). The following lengths are known:

$$OA = 83 \text{ m}, \quad OB = 122 \text{ m}, \quad OC = 106 \text{ m}, \quad OD = 78 \text{ m}$$

Calculate the area of the school playing field to the nearest 100 m².

4. The roof of a garage has a slanting length of 3 m and makes an angle of 120° at its vertex (left). The height of the garage is 4 m and its depth is 9m.
Calculate:
a) the cross-sectional area of the roof,
b) the volume occupied by the whole garage.

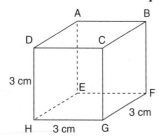

Trigonometry in three dimensions

The diagram (left) shows a cube of edge length 3 cm.
i) Calculate the length EG.

Triangle EHG (below) is right angled. Use Pythagoras' theorem to calculate the length EG.

Worked example

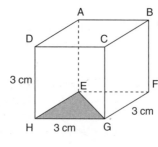

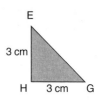

$$EG^2 = EH^2 + HG^2$$
$$EG^2 = 3^2 + 3^2 = 18$$
$$EG = \sqrt{18} \text{ cm}$$

ii) Calculate the length AG.

Triangle AEG (below) is right angled. Use Pythagoras' theorem to calculate the length AG.

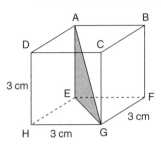

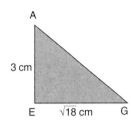

$$AG^2 = AE^2 + EG^2$$
$$AG^2 = 3^2 + (\sqrt{18})^2$$
$$AG^2 = 9 + 18$$
$$AG = \sqrt{27} \text{ cm}$$

iii) Calculate the angle EGA.

To calculate angle EGA we use the triangle EGA:

$$\tan G = \frac{3}{\sqrt{18}}$$

$$G = 35.3° \text{ (1 d.p.)}$$

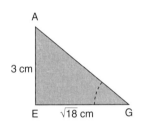

Exercise 15.13 **1.** a) Calculate the length HF.
b) Calculate the length of HB.
c) Calculate the angle BHG.

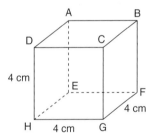

2. a) Calculate the length CA.
b) Calculate the length CE.
c) Calculate the angle ACE.

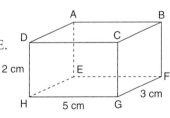

3. In the cuboid (left):
a) Calculate the length EG.
b) Calculate the length AG.
c) Calculate the angle AGE.

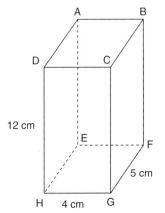

4.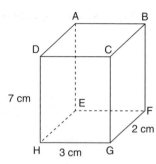
a) Calculate the angle BCE.
b) Calculate the angle GFH.

5. The diagram (left) shows a right pyramid where A is vertically above X.
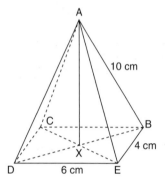
a) i) Calculate the length DB.
   ii) Calculate the angle DAX.
b) i) Calculate the angle CED.
   ii) Calculate the angle DBA.

6. The diagram (right) shows a right pyramid where A is vertically above X.
a) i) Calculate the length CE.
   ii) Calculate the angle CAX.
b) i) Calculate the angle BDE.
   ii) Calculate the angle ADB.

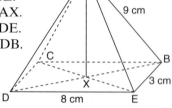

7. In this cone (left) the angle YXZ = 60°; calculate:
a) the length XY,
b) the length YZ,
c) the circumference of the base.

8. In this cone (right) the angle XZY = 40°; calculate:
a) the length XZ,
b) the length XY.

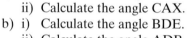

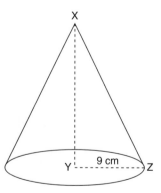

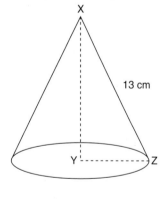

9. One corner of this cuboid has been sliced off along the plane QTU. WU = 4 cm.
a) Calculate the length of the three sides of the triangle QTU.
b) Calculate the three angles P, Q and T in triangle PQT.
c) Calculate the area of triangle PQT.

▨ The angle between a line and a plane

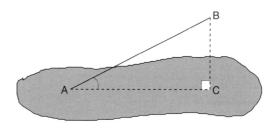

To calculate the size of the angle between the line AB and the shaded plane, drop a perpendicular from B. It meets the shaded plane at C. Then join AC.

The angle between the lines AB and AC represents the angle between the line AB and the shaded plane.

The line AC is the projection of the line AB on the shaded plane.

Worked example   i)   Calculate the length CE.

First use Pythagoras' theorem to calculate the length EG:

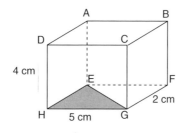

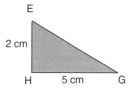

$$EG^2 = EH^2 + HG^2$$
$$EG^2 = 2^2 + 5^2$$
$$EG^2 = 29$$
$$EG = \sqrt{29} \text{ cm}$$

Now use Pythagoras' theorem to calculate CE:

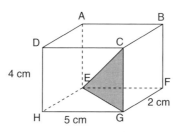

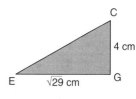

$$EC^2 = EG^2 + CG^2$$
$$EC^2 = (\sqrt{29})^2 + 4^2$$
$$EC^2 = 29 + 16$$
$$EC = \sqrt{45} \text{ cm}$$

ii) Calculate the angle between the line CE and the plane ADHE.

To calculate the angle between the line CE and the plane ADHE use the right-angled triangle CED and calculate the angle CED.

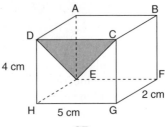
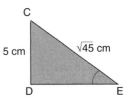

$$\sin E = \frac{CD}{CE}$$

$$\sin E = \frac{5}{\sqrt{45}}$$

$$E = \sin^{-1}\left(\frac{5}{\sqrt{45}}\right)$$

$$E = 48.2° \text{ (1 d.p.)}$$

Exercise 15.14

1. Name the projection of each line onto the given plane:
   a) TR onto RSWV
   b) TR onto PQUT
   c) SU onto PQRS
   d) SU onto TUVW
   e) PV onto QRVU
   f) PV onto RSWV

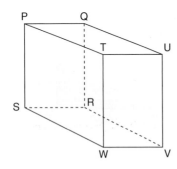

2. Name the projection of each line onto the given plane:
   a) KM onto IJNM
   b) KM onto JKON
   c) KM onto HIML
   d) IO onto HLOK
   e) IO onto JKON
   f) IO onto LMNO

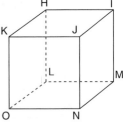

3. Name the angle between the given line and plane:
   a) PT and PQRS
   b) PU and PQRS
   c) SV and PSWT
   d) RT and TUVW
   e) SU and QRVU
   f) PV and PSWT

4. a) Calculate the length BH.
   b) Calculate the angle between the line BH and the plane EFGH.

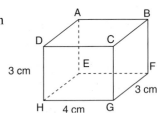

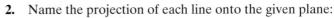

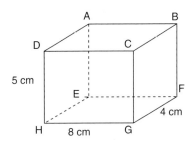

5. a) Calculate the length AG.
   b) Calculate the angle between the line AG and the plane EFGH.
   c) Calculate the angle between the line AG and the plane ADHE.

6. The diagram (right) shows a right pyramid where A is vertically above X.
   a) Calculate the length BD.
   b) Calculate the angle between AB and CBED.

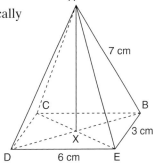

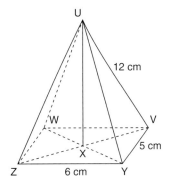

7. The diagram (left) shows a right pyramid where U is vertically above X.
   a) Calculate the length WY.
   b) Calculate the length UX.
   c) Calculate the angle between UX and UZY.

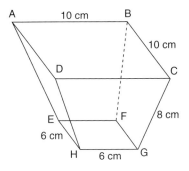

8. ABCD and EFGH are square faces lying parallel to each other.
   Calculate:
   a) the length DB,
   b) the length HF,
   c) the vertical height of the object,
   d) the angle DH makes with the plane ABCD.

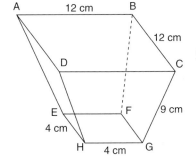

9. ABCD and EFGH are square faces lying parallel to each other.
   Calculate:
   a) the length AC,
   b) the length EG,
   c) the vertical height of the object,
   d) the angle CG makes with the plane EFGH.

NB: All diagrams are not drawn to scale.

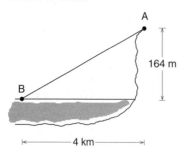

Extended Section

Student Assessment 1

1. A boy standing on a cliff top at A can see a boat sailing in the sea at B. The vertical height of the boy above sea level is 164 m, and the horizontal distance between the boat and the boy is 4 km. Calculate:
   a) the distance AB to the nearest metre,
   b) the angle of depression of the boat from the boy.

2. Draw the graph of $y = \sin x°$ for $0° \leqslant x° \leqslant 180°$. Mark on the graph the angles $0°, 90°, 180°$, and also the maximum and minimum values of y.

3. Express each of the following in terms of another angle between $0°$ and $180°$.
   a) $\sin 50°$ b) $\sin 150°$
   c) $\cos 45°$ d) $\cos 120°$

4. Calculate the size of the obtuse angle marked $\theta°$ in the triangle (left).

5. For the cuboid (left), calculate:
   a) the length EG,
   b) the length EC,
   c) $\angle$BEC.

Student Assessment 2

1. Using the triangular prism (left), calculate:
   a) the length AD,
   b) the length AC,
   c) the angle AC makes with the plane CDEF,
   d) the angle AC makes with the plane ABFE.

2. Draw a graph of $y = \cos \theta°$, for $0° \leqslant \theta° \leqslant 180°$. Mark on the angles $0°, 90°, 180°$, and also the maximum and minimum values of y.

3. The cosine of which other angle between 0 and $180°$ has the same value as
   a) $\cos 128°$ b) $-\cos 80°$?

4. For the triangle (left), calculate:
   a) the length PS,
   b) $\angle$QRS,
   c) the length SR.

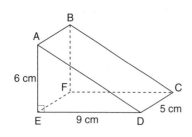

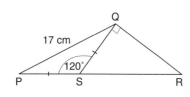

5.

The Great Pyramid at Giza is 146 m high. Two people A and B are looking at the top of the pyramid. The angle of elevation of the top of the pyramid from B is 12°. The distance between A and B is 25 m. If both A and B are 1.8 m tall, calculate:

a) the distance from B to the centre of the base of the pyramid,

b) the angle of elevation θ of the top of the pyramid from A,

c) the distance between A and the top of the pyramid.

Note: A, B and the top of the pyramid are in the same vertical plane.

Student Assessment 3

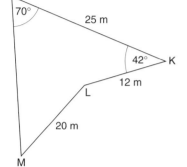

1. For the quadrilateral (left), calculate:
   a) the length JL,
   b) $\angle$KJL,
   c) the length JM,
   d) the area of JKLM.

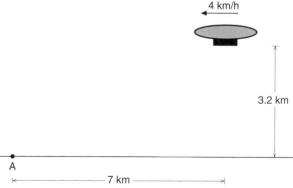

2. For the square-based right pyramid (left), calculate:
   a) the length BD,
   b) $\angle$ABD,
   c) the area of $\triangle$ ABD,
   d) the vertical height of the pyramid.

3. Find two angles between 0° and 360° which have the following cosine. Give each angle to the nearest degree.
   a) 0.79   b) −0.28

4.

An airship is travelling in a horizontal direction as shown, at a speed of 4 km/h. Its vertical height above the ground is 3.2 km. At 3.00 p.m. its horizontal distance from A is 7 km. A is under the flight path of the airship. Calculate:

a) the angle of elevation of the airship from A at 3.00 p.m.,

b) the angle of elevation of the airship from A at 3.30 p.m.,

c) the distance between A and the airship at 3.30 p.m.,

d) at what time, to the nearest minute, the angle of elevation of the airship from A will be 85°.

Student Assessment 4

1. Two hot air balloons A and B are travelling in the same horizontal direction as shown in the diagram below. A is travelling at 2 m/s and B at 3 m/s. Their heights above the ground are 1.6 km and 1 km, respectively.

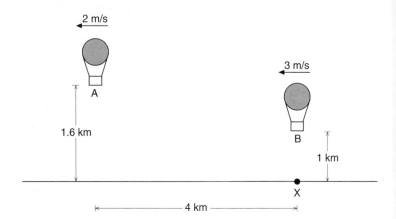

At midday, their horizontal distance apart is 4 km and balloon B is directly above a point X on the ground. Calculate:
a) the angle of elevation of A from X at midday,
b) the angle of depression of B from A at midday,
c) their horizontal distance apart at 12.30 p.m.,
d) the angle of elevation of B from X at 12.30 p.m.,
e) the angle of elevation of A from B at 12.30 p.m.,
f) how much closer A and B are at 12.30 p.m. compared with midday.

2. a) On one diagram plot the graph of $y = \sin \theta°$ and the graph of $y = \cos \theta°$, for $0° \leqslant \theta° \leqslant 360°$.
   b) Use your graph to find the angles for which $\sin \theta° = \cos° \theta°$.

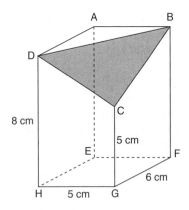

3. The cuboid (left) has one of its corners removed to leave a flat triangle BDC.
   Calculate:
   a) length DC,
   b) length BC,
   c) length DB,
   d) $\angle$CBD,
   e) the area of $\triangle$ BDC,
   f) the angle AC makes with the plane AEHD.

ICT Section

In this activity you will be using a graphing package such as Autograph to investigate the relationship between different trigonometric ratios.

1. a) Using a graphing package, plot the graph of $y = \sin x$ for $0° \leqslant x \leqslant 360°$.

    The graph should look similar to the one shown below:

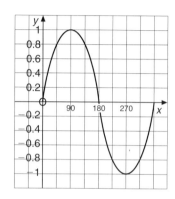

    b) Using the graph evaluate the following:
        i) $\sin 90°$
        ii) $\sin 180°$
        iii) $\sin 270°$
    c) Referring to the graph explain why $\sin^{-1} 0.7$ has two values between 0° and 180°.
    d) Use to graph to evaluate $\sin^{-1} 0.5$

2. a) On the same axes superimpose the plot of $y = \cos x$.
    b) How many solutions are there to the equation $\sin x = \cos x$ between 0° and 360°?
    c) What is the solution to the equation $\sin x = \cos x$ between 0° and 180°?

3. By plotting appropriate graphs solve the following for $0° \leqslant x \leqslant 180°$.
    a) $\sin x = \tan x$
    b) $\cos x = \tan x$

16 MENSURATION

Core Section

Carry out calculations involving the circumference and area of a circle, the area of a parallelogram and a trapezium, the volume of a prism and cylinder and the surface area of a cuboid and a cylinder.

NB: All diagrams are not drawn to scale.

■ The circumference and area of a circle

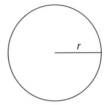 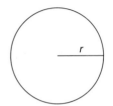

The circumference is $2\pi r$.

$$C = 2\pi r$$

The area is πr^2.

$$A = \pi r^2$$

Worked examples

3 cm

a) Calculate the circumference of this circle, giving your answer to 2 d.p.

$$C = 2\pi r$$
$$= 2\pi \times 3$$
$$= 18.85$$

The circumference is 18.85 cm.

b) If the circumference of this circle is 12 cm, calculate the radius, giving your answer to 2 d.p.

$$C = 2\pi r$$
$$r = \frac{C}{2\pi}$$
$$r = \frac{12}{2\pi}$$
$$= 1.91$$

The radius is 1.91 cm.

c) Calculate the area of this circle, giving your answer to 2 d.p.

$$A = \pi r^2$$
$$= \pi \times 5^2$$
$$= 78.54$$

The area is 78.54 cm^2.

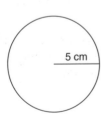

5 cm

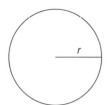

d) If the area of this circle is 34 cm², calculate the radius, giving your answer to 2 d.p.

$$A - \pi r^2$$

$$r = \sqrt{\frac{A}{\pi}}$$

$$r = \sqrt{\frac{34}{\pi}}$$

$$= 3.29$$

The radius is 3.29 cm.

Exercise 16.1

1. Calculate the circumference of each circle, giving your answer to 2 d.p.

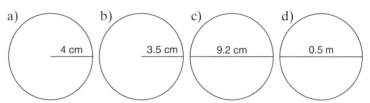

a) 4 cm   b) 3.5 cm   c) 9.2 cm   d) 0.5 m

2. Calculate the area of each of the circles in question 1. Give your answers to 2 d.p.

3. Calculate the radius of a circle when the circumference is:
a) 15 cm
b) π cm
c) 4 m
d) 8 mm

4. Calculate the diameter of a circle when the area is:
a) 16 cm²
b) 9π cm²
c) 8.2 m²
d) 14.6 mm²

Exercise 16.2

1. The wheel of a car has an outer radius of 25 cm. Calculate:
i) how far the car has travelled after one complete turn of the wheel,
ii) how many times the wheel turns for a journey of 1 km.

2. If the wheel of a bicycle has a diameter of 60 cm, calculate how far a cyclist will have travelled after the wheel has rotated 100 times.

3. A circular ring has a cross-section as shown (left). If the outer radius is 22 mm and the inner radius 20 mm, calculate the cross-sectional area of the ring.

4.

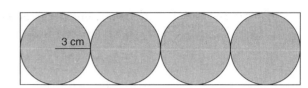

Four circles are drawn in a line and enclosed by a rectangle as shown. If the radius of each circle is 3 cm, calculate:
a) the area of the rectangle,
b) the area of each circle,
c) the unshaded area within the rectangle.

5. A garden is made up of a rectangular patch of grass and two semi-circular vegetable patches. If the length and width of the rectangular patch are 16 m and 8 m respectively, calculate:
a) the perimeter of the garden,
b) the total area of the garden.

The area of parallelograms and trapeziums

A **parallelogram** can be rearranged to form a rectangle in the way shown to the left:

Therefore: area of parallelogram
= base length × perpendicular height.
    A **trapezium** can be visualised as being split into two triangles as shown on the left:

Area of triangle $A = \frac{1}{2} \times a \times h$
Area of triangle $B = \frac{1}{2} \times b \times h$
Area of the trapezium
    = area of triangle A + area of triangle B
    $= \frac{1}{2}ah + \frac{1}{2}bh$
    $= \frac{1}{2}h(a + b)$

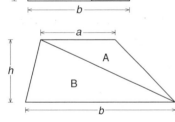

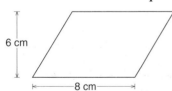

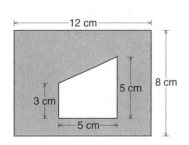

Worked examples **a)** Calculate the area of the parallelogram (left):

Area = base length × perpendicular height
        $= 8 \times 6$
        $= 48 \text{ cm}^2$

b) Calculate the shaded area in the shape (left):

Area of rectangle $= 12 \times 8$
                    $= 96 \text{ cm}^2$

Area of trapezium $= \frac{1}{2} \times 5(3 + 5)$
                    $= 2.5 \times 8$
                    $= 20 \text{ cm}^2$
Shaded area $= 96 - 20$
                    $= 76 \text{ cm}^2$

Exercise 16.3   Find the area of each of the following shapes:

1.

2.
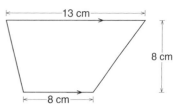

3.   **4.**
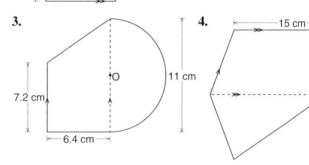

Exercise 16.4   **1.**   Calculate *a*.

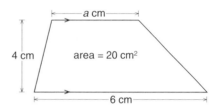

2.   If the areas of this trapezium and parallelogram are equal, calculate *x*.

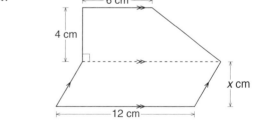

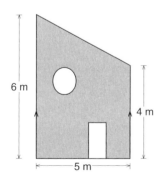

3.   The end view of a house is as shown in the diagram (left). If the door has a width and height of 0.75 m and 2 m respectively and the circular window has a diameter of 0.8 m, calculate the area of brickwork.

4.   A garden in the shape of a trapezium is split into three parts: flower beds in the shape of a triangle and a parallelogram; and a section of grass in the shape of a trapezium. The area of the grass is two and a half times the total area of flower beds. Calculate:
a) the area of each flower bed,
b) the area of grass,
c) the value of *x*.

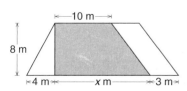

The surface area of a cuboid and cylinder

To calculate the surface area of a **cuboid** start by looking at its individual faces. These are either squares or rectangles. The surface area of a cuboid is the sum of the areas of its faces.

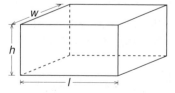

Area of top = wl Area of bottom = wl
Area of front = lh Area of back = lh
Area of one side = wh Area of other side = wh
Total surface area
$$= 2wl + 2lh + 2wh$$
$$= 2(wl + lh + wh)$$

For the surface area of a **cylinder** it is best to visualise the net of the solid: it is made up of one rectangular piece and two circular pieces.

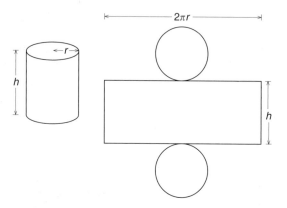

Area of circular pieces = $2 \times \pi r^2$
Area of rectangular piece = $2\pi r \times h$
Total surface area = $2\pi r^2 + 2\pi rh$
$$= 2\pi r(r + h)$$

Worked examples **a)** Calculate the surface area of the cuboid shown (left).

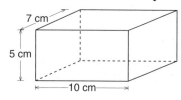

Total area of top and bottom = $2 \times 7 \times 10 = 140$ cm^2
Total area of front and back = $2 \times 5 \times 10 = 100$ cm^2
Total area of both sides   = $2 \times 5 \times 7$ = 70 cm^2

Total surface area   = 310 cm^2

b) If the height of a cylinder is 7 cm and the radius of its circular top is 3 cm, calculate its surface area.

Total surface area = $2\pi r(r + h)$
$$= 2\pi \times 3 \times (3 + 7)$$
$$= 6\pi \times 10$$
$$= 60\pi$$
$$= 188.50 \text{ cm}^2 \text{ (2 d.p.)}$$

The total surface area is 188.50 cm^2.

Exercise 16.5

1. Calculate the surface area of each of the following cuboids if:
   a) $l = 12$ cm, $w = 10$ cm, $h = 5$ cm
   b) $l = 4$ cm, $w = 6$ cm, $h = 8$ cm
   c) $l = 4.2$ cm, $w = 7.1$ cm, $h = 3.9$ cm
   d) $l = 5.2$ cm, $w = 2.1$ cm, $h = 0.8$ cm

2. Calculate the height of each of the following cuboids if:
   a) $l = 5$ cm, $w = 6$ cm, surface area = 104 cm^2
   b) $l = 2$ cm, $w = 8$ cm, surface area = 112 cm^2
   c) $l = 3.5$ cm, $w = 4$ cm, surface area = 118 cm^2
   d) $l = 4.2$ cm, $w = 10$ cm, surface area = 226 cm^2

3. Calculate the surface area of each of the following cylinders if:
   a) $r = 2$ cm, $h = 6$ cm   b) $r = 4$ cm, $h = 7$ cm
   c) $r = 3.5$ cm, $h = 9.2$ cm   d) $r = 0.8$ cm, $h = 4.3$ cm

4. Calculate the height of each of the following cylinders. Give your answers to 1 d.p.
   a) $r = 2.0$ cm, surface area = 40 cm^2
   b) $r = 3.5$ cm, surface area = 88 cm^2
   c) $r = 5.5$ cm, surface area = 250 cm^2
   d) $r = 3.0$ cm, surface area = 189 cm^2

Exercise 16.6

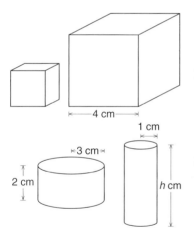

1. Two cubes (left) are placed next to each other. The length of each of the edges of the larger cube is 4 cm. If the ratio of their surface areas is 1 : 4, calculate:
   a) the surface area of the small cube,
   b) the length of an edge of the small cube.

2. A cube and a cylinder have the same surface area. If the cube has an edge length of 6 cm and the cylinder a radius of 2 cm calculate:
   a) the surface area of the cube,
   b) the height of the cylinder.

3. Two cylinders (left) have the same surface area. The shorter of the two has a radius of 3 cm and a height of 2 cm, and the taller cylinder has a radius of 1 cm. Calculate:
   a) the surface area of one of the cylinders,
   b) the height of the taller cylinder.

4. Two cuboids have the same surface area. The dimensions of one of them are: length = 3 cm, width = 4 cm and height = 2 cm.
   Calculate the height of the other cuboid if its length is 1 cm and width is 4 cm.

The volume of prisms

A prism is any three-dimensional object which has a constant cross-sectional area.

Below are a few examples of some of the more common types of prism:

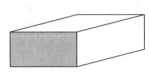

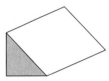

| Rectangular prism (cuboid) | Circular prism (cylinder) | Triangular prism |

When each of the shapes is cut parallel to the shaded face, the cross-section is constant and the shape is therefore classified as a prism.

Volume of a prism = area of cross-section × length

Worked examples   **a)**   Calculate the volume of the cylinder shown in the diagram:

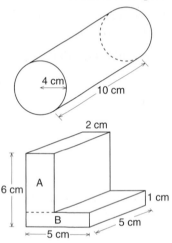

Volume = cross-sectional area × length
$$= \pi \times 4^2 \times 10$$
Volume = 502.7 cm³ (1 d.p.)

b)   Calculate the volume of the 'L' shaped prism shown in the diagram (left):

The cross-sectional area can be split into two rectangles:

Area of rectangle A = 5 × 2
= 10 cm²
Area of rectangle B = 5 × 1
= 5 cm²

Total cross-sectional area = (10 cm² + 5 cm²) = 15 cm²
Volume of prism = 15 × 5
= 75 cm³

Exercise 16.7   **1.**   Calculate the volume of each of the following cuboids, where *w*, *l* and *h* represent the width, length and height respectively.
a) *w* = 2 cm,   *l* = 3 cm,   *h* = 4 cm
b) *w* = 6 cm,   *l* = 1 cm,   *h* = 3 cm
c) *w* = 6 cm,   *l* = 23 mm,   *h* = 2 cm
d) *w* = 42 mm,   *l* = 3 cm,   *h* = 0.007 m

2.   Calculate the volume of each of the following cylinders, where *r* represents the radius of the circular face and *h* the height of the cylinder.
a) *r* = 4 cm,   *h* = 9 cm
b) *r* = 3.5 cm,   *h* = 7.2 cm
c) *r* = 25 mm,   *h* = 10 cm
d) *r* = 0.3 cm,   *h* = 17 mm

3.   Calculate the volume of each of the following triangular prisms, where *b* represents the base length of the triangular face, *h* its perpendicular height and *l* the length of the prism.
a) *b* = 6 cm,   *h* = 3 cm,   *l* = 12 cm
b) *b* = 4 cm,   *h* = 7 cm,   *l* = 10 cm
c) *b* = 5 cm,   *h* = 24 mm,   *l* = 7 cm
d) *b* = 62 mm,   *h* = 2 cm,   *l* = 0.01 m

4. Calculate the volume of each of the following prisms. All dimensions are given in centimetres.

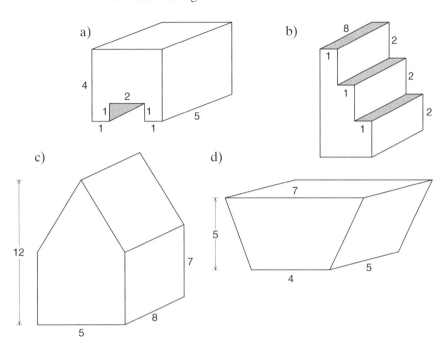

a)

b)

c)

d)

Exercise 16.8

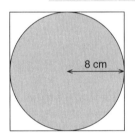

1. The diagram shows a plan view of a cylinder inside a box the shape of a cube. If the radius of the cylinder is 8 cm, calculate:
   a) the height of the cube,
   b) the volume of the cube,
   c) the volume of the cylinder,
   d) the percentage volume of the cube not occupied by the cylinder.

2. A chocolate bar is made in the shape of a triangular prism. The triangular face of the prism is equilateral and has an edge length of 4 cm and a perpendicular height of 3.5 cm. The manufacturer also sells these in special packs of six bars arranged as a hexagonal prism.
   If the prisms are 20 cm long, calculate:
   a) the cross-sectional area of the pack,
   b) the volume of the pack.

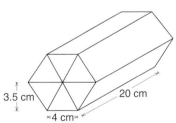

3. A cuboid and a cylinder have the same volume. The radius and height of the cylinder are 2.5 cm and 8 cm respectively. If the length and width of the cuboid are each 5 cm, calculate its height to 1 d.p.

4. A section of steel pipe is shown in the diagram. The inner radius is 35 cm and the outer radius 36 cm. Calculate the volume of steel used in making the pipe if it has a length of 130 m.

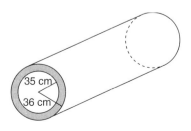

Core Section

Student Assessment 1

NB: All diagrams are not drawn to scale.

1. Calculate the circumference and area of each of the following circles. Give your answers to 1 d.p.
   a)   b)

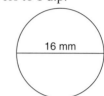

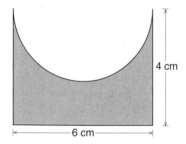

2. A semi-circular shape is cut out of the side of a rectangle as shown. Calculate the shaded area to 1 d.p.

3. For the diagram below, calculate the area of:
   a) the semi-circle,
   b) the trapezium,
   c) the whole shape.

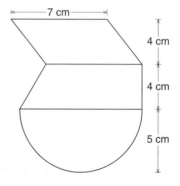

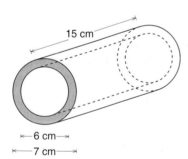

4. A cylindrical tube has an inner diameter of 6 cm, an outer diameter of 7 cm and a length of 15 cm. Calculate the following to 1 d.p.:
   a) the surface area of the shaded end,
   b) the inside surface area of the tube,
   c) the total surface area of the tube.

5. Calculate the volume of each of the following cylinders:
   a)   b)

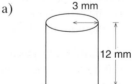

Student Assessment 2

1. Calculate the circumference and area of each of the following circles. Give your answers to 1 d.p.

   a)

   4.3 cm

   b)

   15 mm

2. A rectangle of length 32 cm and width 20 cm has a semi-circle cut out of two of its sides as shown (left). Calculate the shaded area to 1 d.p.

   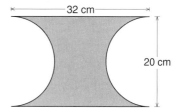

3. Calculate the area of:
   a) the semi-circle,
   b) the parallelogram,
   c) the whole shape.

   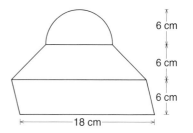

4. A prism in the shape of a hollowed-out cuboid has dimensions as shown. If the end is square, calculate the volume of the prism.

   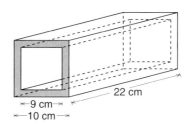

5. Calculate the surface area of each of the following cylinders:

   a)
   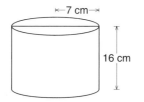

   b)
   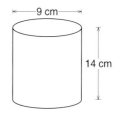

MENSURATION

Extended Section

Solve problems involving the arc length and sector area as fractions of the circumference and area of a circle, the surface area and volume of a sphere, pyramid and cone (given the formulae for the sphere, pyramid and cone).

NB: All diagrams are not drawn to scale.

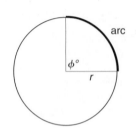

Arc length

An **arc** is part of the circumference of a circle between two radii.

Its length is proportional to the size of the angle ϕ between the two radii. The length of the arc as a fraction of the circumference of the whole circle is therefore equal to the fraction that ϕ is of 360°.

$$\text{Arc length} = \frac{\phi}{360} \times 2\pi r$$

Worked examples

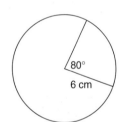

a) Find the length of the minor arc in the circle (left). Give your answer to 1 d.p.

$$\text{Arc length} = \frac{80}{360} \times 2 \times \pi \times 6$$

$$= 8.4 \text{ cm}$$

b) In this circle, the length of the minor arc is 12.4 cm and the radius is 7 cm.

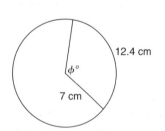

i) Calculate the angle $\phi°$.

$$\text{Arc length} = \frac{\phi}{360} \times 2\pi r$$

$$12.4 = \frac{\phi}{360} \times 2 \times \pi \times 7$$

$$\frac{12.4}{2 \times \pi \times 7} \quad \frac{360}{} = \phi$$

$$\phi = 101.5 \text{ (1 d.p.)}$$

ii) Calculate the length of the major arc.

$$C = 2\pi r$$
$$= 2 \times \pi \times 7$$
$$= 44.0 \text{ cm (1 d.p.)}$$
$$\text{Major arc} = \text{circumference} - \text{minor arc}$$
$$= (44.0 - 12.4) \text{ cm}$$
$$= 31.6 \text{ cm}$$

Exercise 16.9

1. For each of the following, give the length of the arc to 1 d.p. O is the centre of the circle.

a)

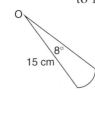

b)

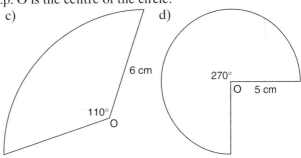

c)

d)

2. A sector is the region of a circle enclosed by two radii and an arc. Calculate the angle ϕ for each of the following sectors. The radius r and arc length a are given in each case.
   a) $r = 14$ cm, $a = 8$ cm
   b) $r = 4$ cm, $a = 16$ cm
   c) $r = 7.5$ cm, $a = 7.5$ cm
   d) $r = 6.8$ cm, $a = 13.6$ cm

3. Calculate the radius r for each of the following sectors. The angle ϕ and arc length a are given in each case.
   a) $\phi = 75°$, $a = 16$ cm
   b) $\phi = 300°$, $a = 24$ cm
   c) $\phi = 20°$, $a = 6.5$ cm
   d) $\phi = 243°$, $a = 17$ cm

Exercise 16.10

1. Calculate the perimeter of each of these shapes.

a)

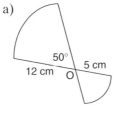

b)

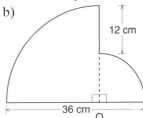

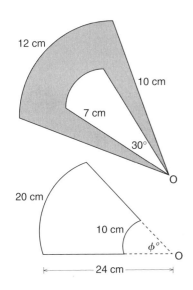

2. A shape is made from two sectors arranged in such a way that they share the same centre. The radius of the smaller sector is 7 cm and the radius of the larger sector is 10 cm. If the angle at the centre of the smaller sector is 30° and the arc length of the larger sector is 12 cm, calculate:
   a) the arc length of the smaller sector,
   b) the total perimeter of the two sectors,
   c) the angle at the centre of the larger sector.

3. For the diagram on the left, calculate:
   a) the radius of the smaller sector,
   b) the perimeter of the shape,
   c) the angle $\phi°$.

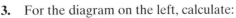

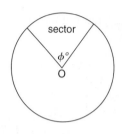

sector

$\phi°$

O

The area of a sector

A **sector** is the region of a circle enclosed by two radii and an arc. Its area is proportional to the size of the angle $\phi°$ between the two radii. The area of the sector as a fraction of the area of the whole circle is therefore equal to the fraction that $\phi°$ is of 360°.

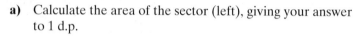

$$\text{Area of sector} = \frac{\phi}{360} \times \pi r^2$$

Worked examples   **a)**   Calculate the area of the sector (left), giving your answer to 1 d.p.

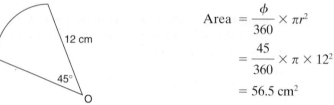

12 cm

45°

O

$$\text{Area} = \frac{\phi}{360} \times \pi r^2$$

$$= \frac{45}{360} \times \pi \times 12^2$$

$$= 56.5 \text{ cm}^2$$

b)   Calculate the radius of the sector (left), giving your answer to 1 d.p.

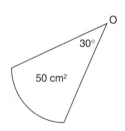

O

30°

50 cm²

$$\text{Area} = \frac{\phi}{360} \times \pi r^2$$

$$50 = \frac{30}{360} \times \pi \times r^2$$

$$\frac{50 \times 360}{30\pi} = r^2$$

$$r = 13.8$$

The radius is 13.8 cm.

Exercise 16.11   **1.**   Calculate the area of each of the following sectors, using the values of the angles ϕ and radius r in each case.
   a) $\phi = 60°$, $r = 8$ cm
   b) $\phi = 120°$, $r = 14$ cm
   c) $\phi = 2°$, $r = 18$ cm
   d) $\phi = 320°$, $r = 4$ cm

2.   Calculate the radius for each of the following sectors, using the values of the angle ϕ and the area A in each case.
   a) $\phi = 40°$, $A = 120$ cm²
   b) $\phi = 12°$, $A = 42$ cm²
   c) $\phi = 150°$, $A = 4$ cm²
   d) $\phi = 300°$, $A = 400$ cm²

3.   Calculate the value of the angle ϕ for each of the following sectors, using the values of A and r in each case.
   a) $r = 12$ cm, $A = 60$ cm²
   b) $r = 26$ cm, $A = 0.02$ m²
   c) $r = 0.32$ m, $A = 180$ cm²
   d) $r = 38$ mm, $A = 16$ cm²

Exercise 16.12

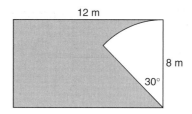

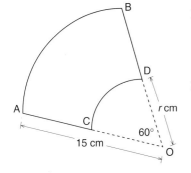

1. A rotating sprinkler is placed in one corner of a garden (above). If it has a reach of 8 m and rotates through an angle of 30°, calculate the area of garden not being watered.

2. Two sectors AOB and COD share the same centre O. The area of AOB is three times the area of COD. Calculate:
   a) the area of sector AOB,
   b) the area of sector COD,
   c) the radius r cm of sector COD.

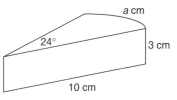

3. A circular cake is cut. One of the slices is shown. Calculate:
   a) the length a cm of the arc,
   b) the total surface area of all the sides of the slice,
   c) the volume of the slice.

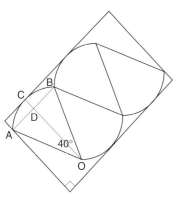

4. The diagram shows a plan view of four tiles in the shape of sectors placed in the bottom of a box. C is the midpoint of the arc AB and intersects the chord AB at point D. If the length ADB is 8 cm and the length OB is 10 cm, calculate:
   a) the length OD,
   b) the length CD,
   c) the area of the sector AOB,
   d) the length and width of the box,
   e) the area of the base of the box not covered by the tiles.

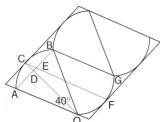

5. The tiles in Q.4 are repackaged and are now placed in a box, the base of which is a parallelogram. Given that C and F are the midpoints of arcs AB and OG respectively, calculate:
   a) the angle OCF,
   b) the length CE,
   c) the length of the sides of the box,
   d) the area of the base of the box not covered by the tiles.

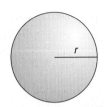

Volume of a sphere

Volume of sphere $= \frac{4}{3}\pi r^3$

Worked examples

a) Calculate the volume of the following sphere, giving your answer to 1 d.p.

$$\begin{aligned}
\text{Volume of sphere} &= \tfrac{4}{3}\pi r^3 \\
&= \tfrac{4}{3} \times \pi \times 3^3 \\
&= 113.1
\end{aligned}$$

The volume is 113.1 cm³.

b) Given that the volume of a sphere is 150 cm³, calculate its radius to 1 d.p.

$$V = \tfrac{4}{3}\pi r^3$$

$$r^3 = \frac{3V}{4\pi}$$

$$r^3 = \frac{3 \times 150}{4 \times \pi}$$

$$r = \sqrt[3]{35.8} = 3.3$$

The radius is 3.3 cm.

Exercise 16.13

1. Calculate the volume of each of the following spheres. The radius r is given in each case.
   a) $r = 6$ cm
   b) $r = 9.5$ cm
   c) $r = 8.2$ cm
   d) $r = 0.7$ cm

2. Calculate the radius of each of the following spheres. Give your answers in centimetres and to 1 d.p. The volume V is given in each case.
   a) $V = 130$ cm³
   b) $V = 720$ cm³
   c) $V = 0.2$ m³
   d) $V = 1000$ mm³

Exercise 16.14

1. Given that sphere B has twice the volume of sphere A, calculate the radius of sphere B. Give your answer to 1 d.p.

A   B

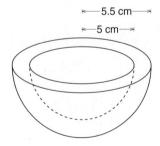

2. Calculate the volume of material used to make the hemispherical bowl on the left, if the inner radius of the bowl is 5 cm and its outer radius 5.5 cm.

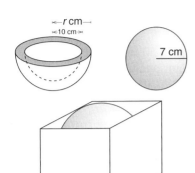

3. The volume of the material used to make the sphere and hemispherical bowl (left) are the same. Given that the radius of the sphere is 7 cm and the inner radius of the bowl is 10 cm, calculate, to 1 d.p., the outer radius *r* cm of the bowl.

4. A ball is placed inside a box into which it will fit tightly. If the radius of the ball is 10 cm, calculate:
   a) the volume of the ball,
   b) the volume of the box,
   c) the percentage volume of the box not occupied by the ball.

5. A steel ball is melted down to make eight smaller identical balls. If the radius of the original steel ball was 20 cm, calculate to the nearest millimetre the radius of each of the smaller balls.

6. A steel ball of volume 600 cm³ is melted down and made into three smaller balls A, B and C. If the volumes of A, B and C are in the ratio 7 : 5 : 3, calculate to 1 d.p. the radius of each of A, B and C.

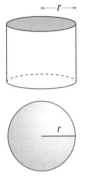

7. The cylinder and sphere shown (left) have the same radius and the same height. Calculate the ratio of their volumes, giving your answer in the form,
   volume of cylinder : volume of sphere.

The surface area of a sphere

Surface area of sphere $= 4\pi r^2$

Exercise 16.15
1. Calculate the surface area of each of the following spheres when:
   a) $r = 6$ cm   b) $r = 4.5$ cm

   c) $r = 12.25$ cm   d) $r = \dfrac{1}{\sqrt{\pi}}$ cm

2. Calculate the radius of each of the following spheres, given the surface area in each case.
   a) $A = 50$ cm²   b) $A = 16.5$ cm²
   c) $A = 120$ mm²   d) $A = \pi$ cm²

3. Sphere A has a radius of 8 cm and sphere B has a radius of 16 cm. Calculate the ratio of their surface areas in the form 1 : *n*.

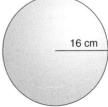

A

B

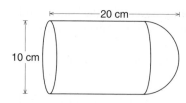

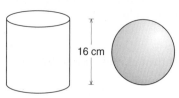

4. A hemisphere of diameter 10 cm is attached to a cylinder of equal diameter as shown.
If the total length of the shape is 20 cm, calculate:
a) the surface area of the hemisphere,
b) the length of the cylinder,
c) the surface area of the whole shape.

5. A sphere and a cylinder both have the same surface area and the same height of 16 cm.
Calculate:
a) the surface area of the sphere,
b) the radius of the cylinder.

▨ The volume of a pyramid

A pyramid is a three-dimensional shape in which each of its faces must be plane. A pyramid has a polygon for its base and the other faces are triangles with a common vertex, known as the **apex**. Its individual name is taken from the shape of the base.

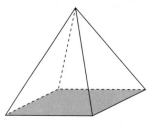

Square-based pyramid

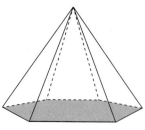

Hexagonal-based pyramid

Volume of any pyramid
$= \frac{1}{3} \times$ area of base $\times$ perpendicular height

Worked examples **a)** A rectangular-based pyramid has a perpendicular height of 5 cm and base dimensions as shown. Calculate the volume of the pyramid.

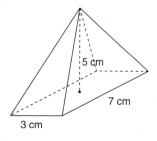

$$\text{Volume} = \frac{1}{3} \times \text{base area} \times \text{height}$$
$$= \frac{1}{3} \times 3 \times 7 \times 5$$
$$= 35$$

The volume is 35 cm³.

b) The pyramid shown has a volume of 60 cm³. Calculate its perpendicular height h cm.

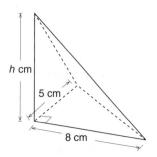

$$\text{Volume} = \frac{1}{3} \times \text{base area} \times \text{height}$$

$$\text{Height} = \frac{3 \times \text{volume}}{\text{base area}}$$

$$h = \frac{3 \times 60}{\frac{1}{2} \times 8 \times 5}$$

$$h = 9$$

The height is 9 cm.

Exercise 16.16   Find the volume of each of the following pyramids:

1.

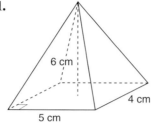

2.

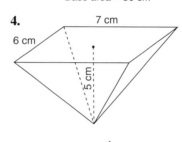

Base area = 50 cm²

3.

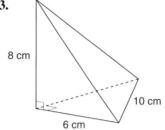

4.

Exercise 16.17

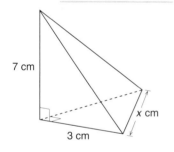

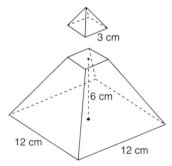

1. Calculate the perpendicular height *h* cm for the pyramid (right), given that it has a volume of 168 cm³.

2. Calculate the length of the edge marked *x* cm, given that the volume of the pyramid (left) is 14 cm³.

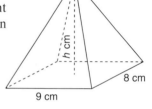

3. The top of a square-based pyramid (left) is cut off. The cut is made parallel to the base. If the base of the smaller pyramid has a side length of 3 cm and the vertical height of the truncated pyramid is 6 cm, calculate:
   a) the height of the original pyramid,
   b) the volume of the original pyramid,
   c) the volume of the truncated pyramid.

4. The top of a triangular-based pyramid (tetrahedron) is cut off. The cut is made parallel to the base. If the vertical height of the top is 6 cm, calculate:
   a) the height of the truncated piece,
   b) the volume of the small pyramid,
   c) the volume of the original pyramid.

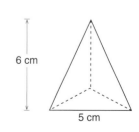

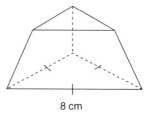

The surface area of a pyramid

The surface area of a pyramid is found simply by adding together the areas of all of its faces.

Exercise 16.18

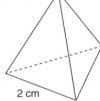

1. Calculate the surface area of a regular tetrahedron with edge length 2 cm.

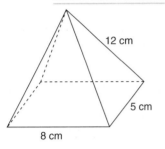

2. The rectangular-based pyramid shown (left) has a sloping edge length of 12 cm. Calculate its surface area.

3. Two square-based pyramids are glued together as shown. Given that all the triangular faces are identical, calculate the surface area of the whole shape.

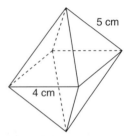

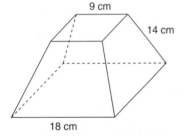

4. Calculate the surface area of the truncated square-based pyramid shown (left). Assume that all the sloping faces are identical.

5. The two pyramids shown below have the same surface area.

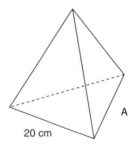

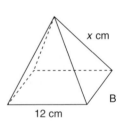

Calculate:
a) the surface area of the tetrahedron,
b) the area of one of the triangular faces on the square-based pyramid,
c) the value of x.

The volume of a cone

A cone is a pyramid with a circular base. The formula for its volume is therefore the same as for any other pyramid.

Volume $= \frac{1}{3} \times$ base area $\times$ height
$\qquad = \frac{1}{3}\pi r^2 h$

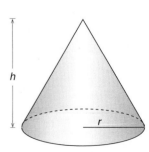

Worked examples **a)** Calculate the volume of the cone (left).

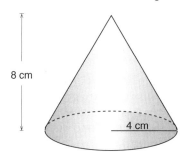

8 cm

4 cm

$$\text{Volume} = \tfrac{1}{3}\pi r^2 h$$
$$= \tfrac{1}{3} \times \pi \times 4^2 \times 8$$
$$= 134.0 \ (1 \ \text{d.p.})$$

The volume is 134.0 cm³.

b) The sector below is assembled to form a cone as shown.

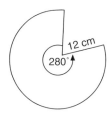

i)   Calculate the base circumference of the cone.
     The base circumference of the cone is equal to the
arc length of the sector.

$$\text{Sector arc length} = \frac{\phi}{360} \times 2\pi r$$

$$= \frac{280}{360} \times 2\pi \times 12 = 58.6 \ (1 \ \text{d.p.})$$

So the base circumference is 58.6 cm.

ii) Calculate the base radius of the cone.
     The base of a cone is circular, therefore:

$$C = 2\pi r$$

$$r = \frac{C}{2\pi} = \frac{58.6}{2\pi}$$

$$= 9.3 \ (1 \ \text{d.p.})$$

So the radius is 9.3 cm.

iii) Calculate the vertical height of the cone.
     The vertical height of the cone can be calculated
using Pythagoras' theorem on the right-angled triangle
enclosed by the base radius, vertical height and the
sloping face, as shown below.
     Note that the length of the sloping face is equal to
the radius of the sector.

$$12^2 = h^2 + 9.3^2$$
$$h^2 = 12^2 - 9.3^2$$
$$h^2 = 56.9$$
$$h = 7.5 \ (1 \ \text{d.p.})$$

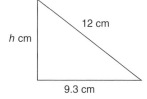

12 cm

h cm

9.3 cm

So the height is 7.5 cm.

iv) Calculate the volume of the cone.

$$\text{Volume} = \tfrac{1}{3} \times \pi r^2 h$$
$$= \tfrac{1}{3} \times \pi \times 9.3^2 \times 7.5$$
$$= 688.0 \ (1 \text{ d.p.})$$

So the volume is 688.0 cm³.

It is important to note that, although answers were given to 1 d.p. in each case, where the answer was needed in a subsequent calculation the exact value was used and not the rounded one. By doing this we avoid introducing rounding errors into the calculations.

Exercise 16.19

1. Calculate the volume of each of the following cones. Use the values for the base radius *r* and the vertical height *h* given in each case.
   a) $r = 3$ cm, $\qquad h = 6$ cm
   b) $r = 6$ cm, $\qquad h = 7$ cm
   c) $r = 8$ mm, $\qquad h = 2$ cm
   d) $r = 6$ cm, $\qquad h = 44$ mm

2. Calculate the base radius of each of the following cones. Use the values for the volume *V* and the vertical height *h* given in each case.
   a) $V = 600$ cm³, $\quad h = 12$ cm
   b) $V = 225$ cm³, $\quad h = 18$ mm
   c) $V = 1400$ mm³, $\ h = 2$ cm
   d) $V = 0.04$ m³, $\quad h = 145$ mm

3. The base circumference *C* and the length of the sloping face *l* is given for each of the following cones. Calculate i) the base radius, ii) the vertical height, iii) the volume in each case. Give all answers to 1 d.p.
   a) $C = 50$ cm, $\qquad l = 15$ cm
   b) $C = 100$ cm, $\qquad l = 18$ cm
   c) $C = 0.4$ m, $\qquad l = 75$ mm
   d) $C = 240$ mm, $\qquad l = 6$ cm

Exercise 16.20

1. The two cones A and B shown below have the same volume. Using the dimensions shown and given that the base circumference of cone B is 60 cm, calculate the height *h* cm.

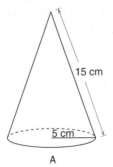

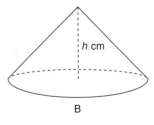

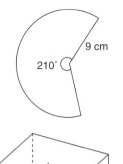

2. The sector shown is assembled to form a cone. Calculate:
   a) the base circumference of the cone,
   b) the base radius of the cone,
   c) the vertical height of the cone,
   d) the volume of the cone,
   e) the curved surface area of the cone.

3. A cone is placed inside a cuboid as shown (bottom left). If the base diameter of the cone is 12 cm and the height of the cuboid is 16 cm, calculate:
   a) the volume of the cuboid,
   b) the volume of the cone,
   c) the volume of the cuboid not occupied by the cone.

4. Two similar sectors are assembled into cones (below). Calculate:
   a) the volume of the smaller cone,
   b) the volume of the larger cone,

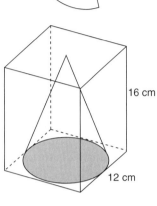

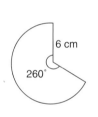

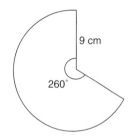

   c) the ratio of their volumes.

Exercise 16.21

1. An ice cream consists of a hemisphere and a cone. Calculate its total volume.

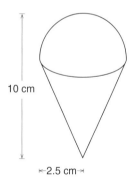

2. A cone is placed on top of a cylinder. Using the dimensions given, calculate the total volume of the shape.

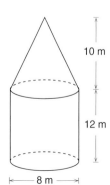

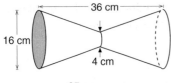

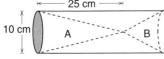

3. Two identical truncated cones are placed end to end as shown.

Calculate the total volume of the shape.

4. Two cones A and B are placed either end of a cylindrical tube as shown.

Given that the volumes of A and B are in the ratio 2 : 1, calculate:
a) the volume of cone A,
b) the height of cone B,
c) the volume of the cylinder.

■ The surface area of a cone

The surface area of a cone comprises the area of the circular base and the area of the curved face. The area of the curved face is equal to the area of the sector from which it is formed.

Worked example Calculate the total surface area of the cone shown (left).

$$\text{Surface area of base} = \pi r^2$$
$$= 25\pi \text{ cm}^2$$

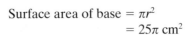

The curved surface area can best be visualised if drawn as a sector as shown in the diagram:

The radius of the sector is equivalent to the slant height of the cone. The curved perimeter of the sector is equivalent to the base circumference of the cone.

$$\frac{\phi}{360} = \frac{10\pi}{24\pi}$$

Therefore $\phi = 150°$

$$\text{Area of sector} = \frac{150}{360} \times \pi \times 12^2 = 60\pi \text{ cm}^2$$

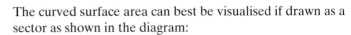

$$\text{Total surface area} = 60\pi + 25\pi$$
$$= 85\pi$$
$$= 267 \text{ (3 s.f.)}$$

The total surface area is 267 cm².

Exercise 16.22

1. Calculate the surface area of each of the following cones:

a)

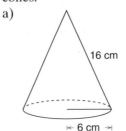

b)

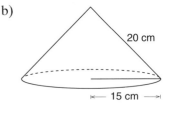

2. Two cones with the same base radius are stuck together as shown. Calculate the surface area of the shape.

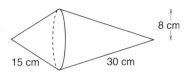

3. Two cones have the same total surface area.

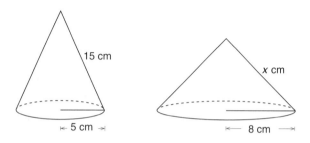

Calculate:
a) the total surface area of each cone,
b) the value of *x*.

Extended Section

Student Assessment I

NB: All diagrams are not drawn to scale.

1. Calculate the arc length of each of the following sectors. The angle ϕ and radius r are given in each case.
   a) $\phi = 45°$
      $r = 15$ cm
   b) $\phi = 150°$
      $r = 13.5$ cm

2. Calculate the angle ϕ in each of the following sectors. The radius r and arc length a are given in each case.
   a) $r = 20$ mm
      $a = 95$ mm
   b) $r = 9$ cm
      $a = 9$ mm

3. Calculate the area of the sector shown below:

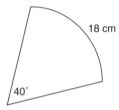

4. A sphere has a radius of 6.5 cm. Calculate to 1 d.p:
   a) its total surface area,
   b) its volume.

5. A pyramid with a base the shape of a regular hexagon is shown. If the length of each of its sloping edges is 24 cm, calculate:
   a) its total surface area,
   b) its volume.

24 cm

12 cm

Student Assessment 2

1. Calculate the arc length of the following sectors. The angle ϕ and radius r are given in each case.
   a) $\phi = 255°$
      $r = 40$ cm
   b) $\phi = 240°$
      $r = 16.3$ mm

2. Calculate the angle ϕ in each of the following sectors. The radius r and arc length a are given in each case.
   a) $r = 40$ cm
      $a = 100$ cm
   b) $r = 20$ cm
      $a = 10$ mm

3. Calculate the area of the sector shown below:

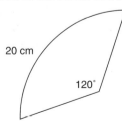

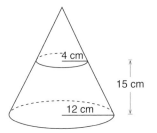

4. A hemisphere has a radius of 8 cm. Calculate to 1 d.p.:
a) its total surface area,
b) its volume.

5. A cone has its top cut as shown. Calculate:
a) the height of the large cone,
b) the volume of the small cone,
c) the volume of the truncated cone.

Student Assessment 3

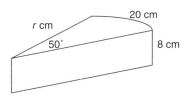

1. The prism here has a cross-sectional area in the shape of a sector.
Calculate:
a) the radius *r* cm,
b) the cross-sectional area of the prism,
c) the total surface area of the prism,
d) the volume of the prism.

2. The cone and sphere shown here have the same volume.

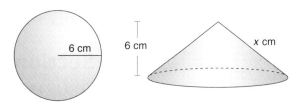

If the radius of the sphere and the height of the cone are both 6 cm, calculate:
a) the volume of the sphere,
b) the base radius of the cone,
c) the slant height *x* cm,
d) the surface area of the cone.

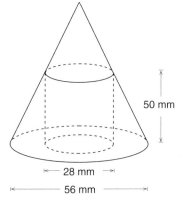

3. The top of a cone is cut off and a cylindrical hole is drilled out of the remaining truncated cone as shown.
Calculate:
a) the height of the original cone,
b) the volume of the original cone,
c) the volume of the solid truncated cone,
d) the volume of the cylindrical hole,
e) the volume of the remaining truncated cone.

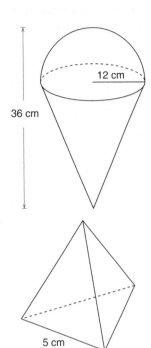

Student Assessment 4

1. A metal object is made from a hemisphere and a cone, both of base radius 12 cm. The height of the object when upright is 36 cm.
   Calculate:
   a) the volume of the hemisphere,
   b) the volume of the cone,
   c) the curved surface area of the hemisphere,
   d) the total surface area of the object.

2. A regular tetrahedron has edges of length 5 cm. Calculate:
   a) the surface area of the tetrahedron,
   b) the surface area of a tetrahedron with edge lengths of 10 cm.

3. A regular tetrahedron and a sphere have the same surface area. If the radius of the sphere is 10 cm, calculate:
   a) the area of one face of the tetrahedron,
   b) the length of each edge of the tetrahedron.
      (Hint: Use the trigonometric formula for the area of a triangle.)

ICT Section

In this chapter you will have seen that it is possible to construct a cone from a sector. The dimensions of the cone are dependent on the dimensions of the sector. In this activity you will be using a spreadsheet to investigate the maximum volume possible of a cone constructed from a sector of fixed radius.

Circles of radius 10 cm are cut from paper and used to construct cones. Different sized sectors are cut from the circles and then arranged to form a cone. e.g

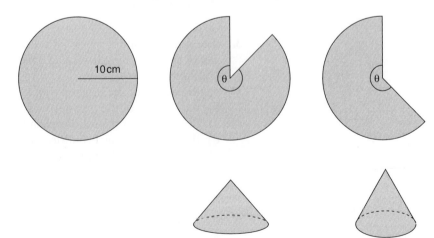

1. Using a spreadsheet similar to the one below calculate the maximum volume possible for a cone constructed from one of these circles:

	A	B	C	D	E	F
1	Angle of sector (θ)	Sector arc length (cm)	Base circumference of cone (cm)	Base radius of cone (cm)	Vertical height of cone (cm)	Volume of cone (cm³)
2	5	0.873	0.873	0.139	9.999	0.202
3	10	1.745	1.745	0.278	9.996	0.808
4	15	2.618	2.618	0.417	9.991	1.816
5	20					
6	25					
7	30					
8	Continue to 355⁰	Enter formulae here to calculate the results for each column				

2. Plot a graph to show how the volume changes as θ increases.

17 VECTORS

Core Section

> Describe a translation by using a column vector, $\overrightarrow{AB}$ or **a**; add vectors and multiply a vector by a scalar.

A **translation** (a sliding movement) can be described using column vectors. A column vector describes the movement of the object in both the x direction and the y direction.

Worked example

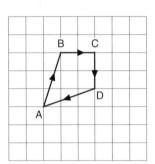

i) Describe the translation from A to B in the diagram (left) in terms of a column vector.

$$\overrightarrow{AB} = \begin{pmatrix} 1 \\ 3 \end{pmatrix}$$

i.e. 1 unit in the x direction, 3 units in the y direction

ii) Describe $\overrightarrow{BC}$ in terms of a column vector.

$$\overrightarrow{BC} = \begin{pmatrix} 2 \\ 0 \end{pmatrix}$$

iii) Describe $\overrightarrow{CD}$ in terms of a column vector.

$$\overrightarrow{CD} = \begin{pmatrix} 0 \\ -2 \end{pmatrix}$$

iv) Describe $\overrightarrow{DA}$ in terms of a column vector.

$$\overrightarrow{DA} = \begin{pmatrix} -3 \\ -1 \end{pmatrix}$$

Translations can also be named by a single letter. The direction of the arrow indicates the direction of the translation.

Worked example Define **a** and **b** in the diagram above using column vectors.

$$\mathbf{a} = \begin{pmatrix} 2 \\ 2 \end{pmatrix} \qquad\qquad \mathbf{b} = \begin{pmatrix} -2 \\ 1 \end{pmatrix}$$

Note: When you represent vectors by single letters, i.e. **a**, in handwritten work, you should write them as a̲.

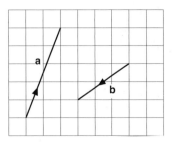

If $\mathbf{a} = \begin{pmatrix} 2 \\ 5 \end{pmatrix}$ and $\mathbf{b} = \begin{pmatrix} -3 \\ -2 \end{pmatrix}$, they can be represented diagrammatically as shown (left).

The diagrammatic representation of $-\mathbf{a}$ and $-\mathbf{b}$ is shown below.

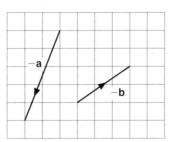

It can be seen from the diagram above that $-\mathbf{a} = \begin{pmatrix} -2 \\ -5 \end{pmatrix}$ and $-\mathbf{b} = \begin{pmatrix} 3 \\ 2 \end{pmatrix}$.

Exercise 17.1 In Q.1 and 2 describe each translation using a column vector.

1. a) $\overrightarrow{AB}$

b) $\overrightarrow{BC}$

c) $\overrightarrow{CD}$

d) $\overrightarrow{DE}$

e) $\overrightarrow{EA}$

f) $\overrightarrow{AE}$

g) $\overrightarrow{DA}$

h) $\overrightarrow{CA}$

i) $\overrightarrow{DB}$

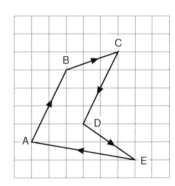

2. a) $\mathbf{a}$ b) $\mathbf{b}$ c) $\mathbf{c}$

d) $\mathbf{d}$ e) $\mathbf{e}$ f) $-\mathbf{b}$

g) $-\mathbf{c}$ h) $-\mathbf{d}$ i) $-\mathbf{a}$

3. Draw and label the following vectors on a square grid:

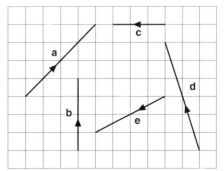

a) $\mathbf{a} = \begin{pmatrix} 2 \\ 4 \end{pmatrix}$ b) $\mathbf{b} = \begin{pmatrix} -3 \\ 6 \end{pmatrix}$ c) $\mathbf{c} = \begin{pmatrix} 3 \\ -5 \end{pmatrix}$

d) $\mathbf{d} = \begin{pmatrix} -4 \\ -3 \end{pmatrix}$ e) $\mathbf{e} = \begin{pmatrix} 0 \\ -6 \end{pmatrix}$ f) $\mathbf{f} = \begin{pmatrix} -5 \\ 0 \end{pmatrix}$

g) $-\mathbf{c}$ h) $-\mathbf{b}$ i) $-\mathbf{f}$

Addition and subtraction of vectors

Vectors can be added together and represented diagrammatically as shown (left).

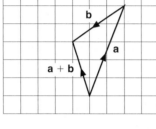

The translation represented by **a** followed by **b** can be written as a single transformation **a** + **b**:

i.e. $\begin{pmatrix} 2 \\ 5 \end{pmatrix} + \begin{pmatrix} -3 \\ -2 \end{pmatrix} = \begin{pmatrix} -1 \\ 3 \end{pmatrix}$

Worked example $\mathbf{a} = \begin{pmatrix} 2 \\ 5 \end{pmatrix}$ $\mathbf{b} = \begin{pmatrix} -3 \\ -2 \end{pmatrix}$

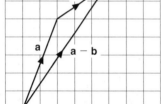

i) Draw a diagram to represent **a** − **b**, where

$$\mathbf{a} - \mathbf{b} = (\mathbf{a}) + (-\mathbf{b}).$$

ii) Calculate the vector represented by **a** − **b**.

$$\begin{pmatrix} 2 \\ 5 \end{pmatrix} - \begin{pmatrix} -3 \\ -2 \end{pmatrix} = \begin{pmatrix} 5 \\ 7 \end{pmatrix}$$

Exercise 17.2   In the following questions,

$$\mathbf{a} = \begin{pmatrix} 3 \\ 4 \end{pmatrix} \quad \mathbf{b} = \begin{pmatrix} -2 \\ 1 \end{pmatrix} \quad \mathbf{c} = \begin{pmatrix} -4 \\ -3 \end{pmatrix} \quad \mathbf{d} = \begin{pmatrix} 3 \\ -2 \end{pmatrix}.$$

1. Draw vector diagrams to represent the following:
   a) $\mathbf{a} + \mathbf{b}$ b) $\mathbf{b} + \mathbf{a}$ c) $\mathbf{a} + \mathbf{d}$
   d) $\mathbf{d} + \mathbf{a}$ e) $\mathbf{b} + \mathbf{c}$ f) $\mathbf{c} + \mathbf{b}$

2. What conclusions can you draw from your answers to Q.1 above?

3. Draw vector diagrams to represent the following:
   a) $\mathbf{b} - \mathbf{c}$ b) $\mathbf{d} - \mathbf{a}$ c) $-\mathbf{a} - \mathbf{c}$
   d) $\mathbf{a} + \mathbf{c} - \mathbf{b}$ e) $\mathbf{d} - \mathbf{c} - \mathbf{b}$ f) $-\mathbf{c} + \mathbf{b} + \mathbf{d}$

4. Represent each of the vectors in Q.3 by a single column vector.

Multiplying a vector by a scalar

Look at the two vectors in the diagram.

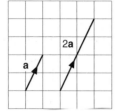

$$\mathbf{a} = \begin{pmatrix} 1 \\ 2 \end{pmatrix} \quad 2\mathbf{a} = 2\begin{pmatrix} 1 \\ 2 \end{pmatrix} = \begin{pmatrix} 2 \\ 4 \end{pmatrix}$$

Worked example If $\mathbf{a} = \begin{pmatrix} 2 \\ -4 \end{pmatrix}$ express the vectors **b**, **c**, **d** and **e** in terms of **a**.

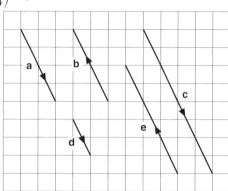

$$\mathbf{b} = -\mathbf{a} \qquad \mathbf{c} = 2\mathbf{a} \qquad \mathbf{d} = \tfrac{1}{2}\mathbf{a} \qquad \mathbf{e} = -\tfrac{3}{2}\mathbf{a}$$

Exercise 17.3

1. $\mathbf{a} = \begin{pmatrix} 1 \\ 4 \end{pmatrix} \qquad \mathbf{b} = \begin{pmatrix} -4 \\ -2 \end{pmatrix} \qquad \mathbf{c} = \begin{pmatrix} -4 \\ 6 \end{pmatrix}$

Express the following vectors in terms of either **a**, **b** or **c**.

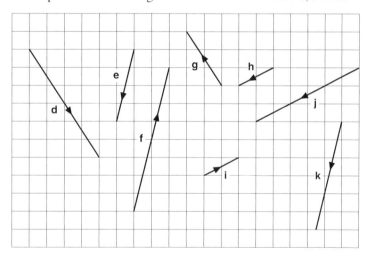

2. $\mathbf{a} = \begin{pmatrix} 2 \\ 3 \end{pmatrix} \qquad \mathbf{b} = \begin{pmatrix} -4 \\ -1 \end{pmatrix} \qquad \mathbf{c} = \begin{pmatrix} -2 \\ 4 \end{pmatrix}$

Represent each of the following as single column vectors:
a) **2a** b) **3b** c) **−c** d) **a + b** e) **b − c**
f) **3c − a** g) **2b − a** h) $\tfrac{1}{2}(\mathbf{a} - \mathbf{b})$ i) **2a − 3c**

3. $\mathbf{a} = \begin{pmatrix} -2 \\ 3 \end{pmatrix} \qquad \mathbf{b} = \begin{pmatrix} 0 \\ -3 \end{pmatrix} \qquad \mathbf{c} = \begin{pmatrix} 4 \\ -1 \end{pmatrix}$

Express each of the following vectors in terms of **a**, **b** and **c**:

a) $\begin{pmatrix} -4 \\ 6 \end{pmatrix}$ b) $\begin{pmatrix} 0 \\ 3 \end{pmatrix}$ c) $\begin{pmatrix} 4 \\ -4 \end{pmatrix}$

d) $\begin{pmatrix} -2 \\ 6 \end{pmatrix}$ e) $\begin{pmatrix} 8 \\ -2 \end{pmatrix}$ f) $\begin{pmatrix} 10 \\ -5 \end{pmatrix}$

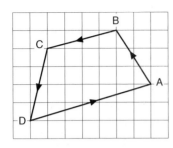

Core Section

Student Assessment 1

1. Using this diagram (left), describe the following translations using column vectors.
    a) $\overrightarrow{AB}$
    b) $\overrightarrow{DA}$
    c) $\overrightarrow{CA}$

2. Describe each of the translations shown below left using column vectors.

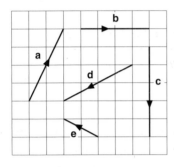

3. Using the vectors drawn in Q.2 above, draw diagrams to represent:
    a) $\mathbf{a} + \mathbf{b}$
    b) $\mathbf{e} - \mathbf{d}$
    c) $\mathbf{c} - \mathbf{e}$
    d) $2\mathbf{e} + \mathbf{b}$

4. In the following,
    $$\mathbf{a} = \begin{pmatrix} 2 \\ 6 \end{pmatrix} \quad \mathbf{b} = \begin{pmatrix} -3 \\ -1 \end{pmatrix} \quad \mathbf{c} = \begin{pmatrix} -2 \\ 4 \end{pmatrix}.$$
    Calculate:
    a) $\mathbf{a} + \mathbf{b}$
    b) $\mathbf{c} - \mathbf{b}$
    c) $2\mathbf{a} + \mathbf{b}$
    d) $3\mathbf{c} - 2\mathbf{b}$

Student Assessment 2

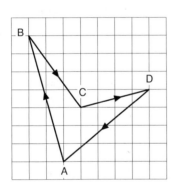

1. Using this diagram, describe the following translations with column vectors.
    a) $\overrightarrow{AB}$
    b) $\overrightarrow{DA}$
    c) $\overrightarrow{CA}$

2. Describe each of the translations below using column vectors.

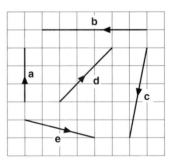

3. Using the vectors drawn in Q.2 above, draw diagrams to represent:
    a) $\mathbf{a} + \mathbf{e}$
    b) $\mathbf{c} - \mathbf{d}$
    c) $-\mathbf{c} - \mathbf{e}$
    d) $-\mathbf{b} + 2\mathbf{a}$

4. In the following,
    $$\mathbf{a} = \begin{pmatrix} 3 \\ -5 \end{pmatrix} \quad \mathbf{b} = \begin{pmatrix} 0 \\ 4 \end{pmatrix} \quad \mathbf{c} = \begin{pmatrix} -4 \\ 6 \end{pmatrix}$$
    Calculate:
    a) $\mathbf{a} - \mathbf{c}$
    b) $\mathbf{b} - \mathbf{a}$
    c) $2\mathbf{a} + \mathbf{b}$
    d) $3\mathbf{c} - 2\mathbf{a}$

VECTORS

Extended Section

Calculate the magnitude of a column vector. (Vectors will be printed as $\overrightarrow{AB}$ or **a** and their magnitudes denoted by modulus signs, e.g. $|\overrightarrow{AB}|$ or $|\mathbf{a}|$. In their answers to questions candidates are expected to indicate **a** in some definite way, e.g. by an arrow or by underlining, thus $\overrightarrow{AB}$ or a.)
Represent vectors by directed line segments; use the sum and difference of two vectors to express given vectors in terms of two coplanar vectors; use position vectors.

▨ **Magnitude**

The **magnitude** or size of a vector is represented by its length, i.e. the longer the length, the greater the magnitude. The magnitude of a vector **a** or $\overrightarrow{AB}$ is denoted by $|\mathbf{a}|$ or $|\overrightarrow{AB}|$ respectively and is calculated using Pythagoras' theorem.

Worked examples

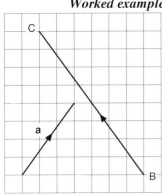

$$\mathbf{a} = \begin{pmatrix} 3 \\ 4 \end{pmatrix} \quad \overrightarrow{BC} = \begin{pmatrix} -6 \\ 8 \end{pmatrix}$$

a) Represent both of the above vectors diagramatically.

b) i) Calculate $|\mathbf{a}|$.

$$|\mathbf{a}| = \sqrt{(3^2 + 4^2)}$$
$$= \sqrt{25}$$
$$= 5$$

ii) Calculate $|\overrightarrow{BC}|$.

$$|\overrightarrow{BC}| = \sqrt{(-6)^2 + 8^2}$$
$$= \sqrt{100}$$
$$= 10$$

Exercise 17.4 **1.** Calculate the magnitude of the vectors shown below. Give your answers correct to 1 d.p.

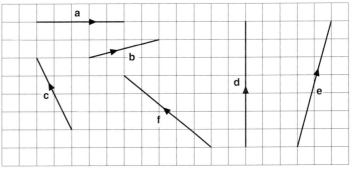

2. Calculate the magnitude of the following vectors, giving your answers to 1 d.p.

   a) $\overrightarrow{AB} = \begin{pmatrix} 0 \\ 4 \end{pmatrix}$ b) $\overrightarrow{BC} = \begin{pmatrix} 2 \\ 5 \end{pmatrix}$ c) $\overrightarrow{CD} = \begin{pmatrix} -4 \\ -6 \end{pmatrix}$

   d) $\overrightarrow{DE} = \begin{pmatrix} -5 \\ 12 \end{pmatrix}$ e) $2\overrightarrow{AB}$ f) $-\overrightarrow{CD}$

3. $\mathbf{a} = \begin{pmatrix} 4 \\ -3 \end{pmatrix}$ $\mathbf{b} = \begin{pmatrix} -5 \\ 7 \end{pmatrix}$ $\mathbf{c} = \begin{pmatrix} -1 \\ -8 \end{pmatrix}$

   Calculate the magnitude of the following, giving your answers to 1 d.p.

   a) $\mathbf{a} + \mathbf{b}$ b) $2\mathbf{a} - \mathbf{b}$ c) $\mathbf{b} - \mathbf{c}$
   d) $2\mathbf{c} + 3\mathbf{b}$ e) $2\mathbf{b} - 3\mathbf{a}$ f) $\mathbf{a} + 2\mathbf{b} - \mathbf{c}$

Position vectors

Sometimes a vector is fixed in position relative to a specific point. In the diagram (left), the position vector of A relative to

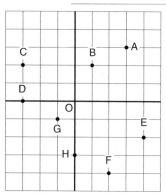

O is $\begin{pmatrix} 2 \\ 6 \end{pmatrix}$.

Exercise 17.5

1. Give the position vectors of A, B, C, D, E, F, G and H relative to O in the diagram (left).

Vector geometry

In general vectors are not fixed in position. If a vector **a** has a specific magnitude and direction, then any other vector with the same magnitude and direction as **a** can also be labelled **a**.

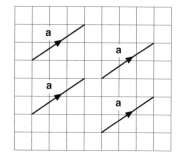

If $\mathbf{a} = \begin{pmatrix} 3 \\ 2 \end{pmatrix}$ then all the vectors

shown in this diagram can also be labelled **a**, as they all have the same magnitude and direction.

This property of vectors can be used to solve problems in vector geometry.

Worked example

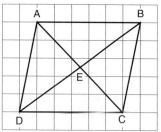

i) Name a vector equal to $\overrightarrow{AD}$.

$$\overrightarrow{BC} = \overrightarrow{AD}$$

ii) Write $\overrightarrow{BD}$ in terms of $\overrightarrow{BE}$.

$$\overrightarrow{BD} = 2\overrightarrow{BE}$$

iii) Express $\overrightarrow{CD}$ in terms of $\overrightarrow{AB}$.

$$\overrightarrow{CD} = \overrightarrow{BA} = -\overrightarrow{AB}$$

Exercise 17.6

1.

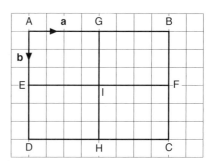

If $\overrightarrow{AG}$ = **a** and $\overrightarrow{AE}$ = **b**, express the following in terms of **a** and **b**:

a) $\overrightarrow{EI}$ b) $\overrightarrow{HC}$ c) $\overrightarrow{FC}$

d) $\overrightarrow{DE}$ e) $\overrightarrow{GH}$ f) $\overrightarrow{CD}$

g) $\overrightarrow{AI}$ h) $\overrightarrow{GE}$ i) $\overrightarrow{FD}$

2. If $\overrightarrow{LP}$ = **a** and $\overrightarrow{LR}$ = **b**, express the following in terms of **a** and **b**:

a) $\overrightarrow{LM}$ b) $\overrightarrow{PQ}$ c) $\overrightarrow{PR}$

d) $\overrightarrow{MQ}$ e) $\overrightarrow{MP}$ f) $\overrightarrow{NP}$

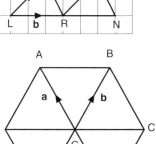

3. ABCDEF is a regular hexagon.

If $\overrightarrow{GA}$ = **a** and $\overrightarrow{GB}$ = **b**, express the following in terms of **a** and **b**:

a) $\overrightarrow{AD}$ b) $\overrightarrow{FE}$ c) $\overrightarrow{DC}$

d) $\overrightarrow{AB}$ e) $\overrightarrow{FC}$ f) $\overrightarrow{EC}$

g) $\overrightarrow{BF}$ h) $\overrightarrow{FD}$ i) $\overrightarrow{AE}$

4.

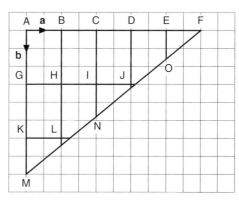

If $\overrightarrow{AB}$ = **a** and $\overrightarrow{AG}$ = **b**, express the following in terms of **a** and **b**:

a) $\overrightarrow{AF}$ b) $\overrightarrow{AM}$ c) $\overrightarrow{FM}$

d) $\overrightarrow{FO}$ e) $\overrightarrow{EI}$ f) $\overrightarrow{KF}$

g) $\overrightarrow{CN}$ h) $\overrightarrow{AN}$ i) $\overrightarrow{DN}$

Exercise 17.7

1. T is the midpoint of the line PS and R divides the line QS in the ratio 1 : 3.
   $$\overrightarrow{PT} = \mathbf{a} \text{ and } \overrightarrow{PQ} = \mathbf{b}.$$

   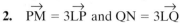

   a) Express each of the following in terms of **a** and **b**:

      i) $\overrightarrow{PS}$

      ii) $\overrightarrow{QS}$

      iii) $\overrightarrow{PR}$

   b) Show that $\overrightarrow{RT} = \frac{1}{4}(2\mathbf{a} - 3\mathbf{b})$.

2. $\overrightarrow{PM} = 3\overrightarrow{LP}$ and $\overrightarrow{QN} = 3\overrightarrow{LQ}$

   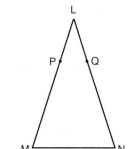

   Prove that:
   a) the line PQ is parallel to the line MN,
   b) the line MN is four times the length of the line PQ.

3. PQRS is a parallelogram. The point T divides the line PQ in the ratio 1 : 3, and U, V and W are the midpoints of SR, PS and QR respectively.

   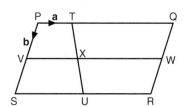

   $$\overrightarrow{PT} = \mathbf{a} \text{ and } \overrightarrow{PV} = \mathbf{b}.$$

   a) Express each of the following in terms of **a** and **b**:

      i) $\overrightarrow{PQ}$

      ii) $\overrightarrow{SU}$

      iii) $\overrightarrow{PU}$

      iv) $\overrightarrow{VX}$

   b) Show that $\overrightarrow{XR} = \frac{1}{2}(5\mathbf{a} + 2\mathbf{b})$.

4. ABC is an isosceles triangle. L is the midpoint of BC. M divides the line LA in the ratio 1 : 5, and N divides AC in the ratio 2 : 5.

   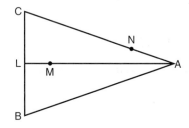

   a) $\overrightarrow{BC} = \mathbf{p}$ and $\overrightarrow{BA} = \mathbf{q}$. Express the following in terms of **p** and **q**:

      i) $\overrightarrow{LA}$

      ii) $\overrightarrow{AN}$

   b) Show that $\overrightarrow{MN} = \frac{1}{84}(46\mathbf{q} - 11\mathbf{p})$.

Extended Section

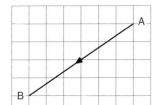

Student Assessment I

1. a) Calculate the magnitude of the vector $\overrightarrow{AB}$ shown in the diagram (left).
   b) Calculate the magnitude of the following vectors:

   $$\mathbf{a} = \begin{pmatrix} 2 \\ 9 \end{pmatrix} \quad \mathbf{b} = \begin{pmatrix} -7 \\ -4 \end{pmatrix} \quad \mathbf{c} = \begin{pmatrix} -5 \\ 12 \end{pmatrix}$$

2. $$\mathbf{p} = \begin{pmatrix} 3 \\ 2 \end{pmatrix} \quad \mathbf{q} = \begin{pmatrix} -4 \\ 1 \end{pmatrix} \quad \mathbf{r} = \begin{pmatrix} 3 \\ -4 \end{pmatrix}$$

   Calculate the magnitude of the following, giving your answers to 1 d.p.
   a) $3\mathbf{p} - 2\mathbf{q}$ b) $\frac{1}{2}\mathbf{r} + \mathbf{q}$

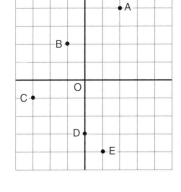

3. Give the position vectors of A, B, C, D and E relative to O for the diagram on the left.

4. a) Name another vector equal to $\overrightarrow{DE}$ in this diagram (right).
   b) Express $\overrightarrow{DF}$ in terms of $\overrightarrow{BC}$.
   c) Express $\overrightarrow{CF}$ in terms of $\overrightarrow{DE}$.

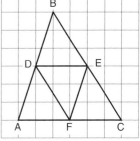

Student Assessment 2

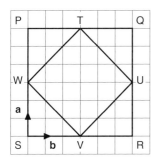

1. a) Calculate the magnitude of the vector $\overrightarrow{FG}$ in the diagram.
   b) Calculate the magnitude of each of the following vectors:

   $$\mathbf{a} = \begin{pmatrix} 1 \\ 6 \end{pmatrix} \quad \mathbf{b} = \begin{pmatrix} 12 \\ -3 \end{pmatrix} \quad \mathbf{c} = \begin{pmatrix} -10 \\ 10 \end{pmatrix}$$

2. $$\mathbf{p} = \begin{pmatrix} -3 \\ 5 \end{pmatrix} \quad \mathbf{q} = \begin{pmatrix} -4 \\ -4 \end{pmatrix} \quad \mathbf{r} = \begin{pmatrix} 8 \\ -2 \end{pmatrix}$$

   Calculate the magnitude of each of the following, giving your answers to 1 d.p.
   a) $4\mathbf{p} - \mathbf{r}$ b) $\frac{3}{2}\mathbf{q} - \mathbf{p}$

3. If $\overrightarrow{SW} = \mathbf{a}$ and $\overrightarrow{SV} = \mathbf{b}$ in the diagram (left), express each of the following in terms of $\mathbf{a}$ and $\mathbf{b}$:
   a) $\overrightarrow{SP}$ b) $\overrightarrow{QT}$ c) $\overrightarrow{TU}$

Student Assessment 3

1. In the triangle PQR, the point S divides the line PQ in the ratio 1 : 3, and T divides the line RQ in the ratio 3 : 2. $\overrightarrow{PR}$ = **a** and $\overrightarrow{PQ}$ = **b**.

   a) Express the following in terms of **a** and **b**:

   i) $\overrightarrow{PS}$ ii) $\overrightarrow{SR}$ iii) $\overrightarrow{TQ}$

   b) Show that $\overrightarrow{ST} = \frac{1}{20}(8\mathbf{a} + 7\mathbf{b})$.

2. In the triangle ABC, the point D divides the line AB in the ratio 1 : 3, and E divides the line AC also in the ratio 1 : 3. If $\overrightarrow{AD}$ = **a** and $\overrightarrow{AE}$ = **b** prove that:

   a) $\overrightarrow{BC} = 4\overrightarrow{DE}$,

   b) BCED is a trapezium.

3. The parallelogram ABCD shows the points P and Q dividing each of the lines AD and DC in the ratio 1 : 4.

   a) If $\overrightarrow{DA}$ = **a** and $\overrightarrow{DC}$ = **b** express the following in terms of **a** and **b**:

   i) $\overrightarrow{AC}$ ii) $\overrightarrow{CB}$ iii) $\overrightarrow{DB}$

   b) i) Find the ratio in which R divides DB.

   ii) Find the ratio in which R divides PQ.

Student Assessment 4

1. ABCDEFGH is a regular octagon. $\overrightarrow{AB}$ = **a** and $\overrightarrow{AH}$ = **b**. Express the following vectors in terms of **a** and **b**:

   a) $\overrightarrow{FE}$ b) $\overrightarrow{ED}$ c) $\overrightarrow{BG}$

2. In the triangle ABC, $\overrightarrow{AB}$ = **a** and $\overrightarrow{AD}$ = **b**. D divides the side AC in the ratio 1 : 4 and E is the midpoint of BC. Express the following in terms of **a** and **b**:

   a) $\overrightarrow{AC}$ b) $\overrightarrow{BC}$ c) $\overrightarrow{DE}$

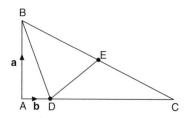

3. In the square PQRS, T is the midpoint of the side PQ and U is the midpoint of the side SR. $\overrightarrow{PQ}$ = **a** and $\overrightarrow{PS}$ = **b**.

   a) Express the following in terms of **a** and **b**:

   i) $\overrightarrow{PT}$ ii) $\overrightarrow{QS}$

   b) Calculate the ratio $\overrightarrow{PV} : \overrightarrow{PU}$.

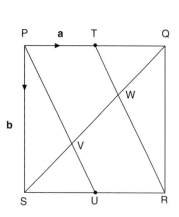

ICT Section

Using Autograph or another appropriate software package, prepare a help sheet for your revision that demonstrates the addition, subtraction and multiplication of vectors. An example is shown below:

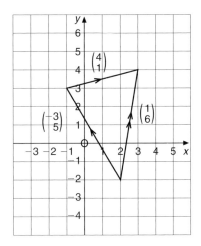

Vector addition:

$$\begin{pmatrix} -3 \\ 5 \end{pmatrix} + \begin{pmatrix} 4 \\ 1 \end{pmatrix} = \begin{pmatrix} 1 \\ 6 \end{pmatrix}$$

18 MATRICES

Display information in the form of a matrix of any order; calculate the sum and product (where appropriate) of two matrices; calculate the product of a matrix and a scalar quantity; use the algebra of 2 × 2 matrices including the zero and identity 2 × 2 matrices; calculate the determinant and inverse $\mathbf{A}^{-1}$ of a non-singular matrix $\mathbf{A}$.

A matrix represents another way of writing information. Here the information is written as a rectangular array. For example, two pupils Lea and Pablo sit a maths exam, a science exam and an English exam. Lea scores 73%, 67% and 81% respectively, whilst Pablo scores 64%, 82% and 48% respectively. This can be written as

$$\begin{pmatrix} 73 & 67 & 81 \\ 64 & 82 & 48 \end{pmatrix}$$

A matrix can take any size.

$$\mathbf{A} = \begin{pmatrix} 4 & 1 & 3 \\ 2 & 6 & 4 \\ 9 & 1 & 0 \end{pmatrix} \qquad \mathbf{B} = \begin{pmatrix} 7 & 1 & 4 & 5 \\ 6 & 8 & 2 & 1 \end{pmatrix}$$

$$\mathbf{C} = (9 \quad 6 \quad 2) \qquad \mathbf{D} = \begin{pmatrix} 4 \\ 0 \\ 6 \end{pmatrix} \qquad \mathbf{E} = \begin{pmatrix} 1 & 0 \\ 0 & 1 \end{pmatrix}$$

A size of a matrix is known as its **order** and is denoted by the number of rows times the number of columns. Therefore the order of matrix $\mathbf{A}$ is 3 × 3, whilst the order of matrix $\mathbf{B}$ is 2 × 4. Each of the numbers in the matrix is called an **element**. A 2 × 4 matrix consists of 8 elements.

Matrix $\mathbf{A}$ and matrix $\mathbf{E}$ above are called **square matrices** as they have the same number of rows and columns. Matrix $\mathbf{C}$ is called a **row matrix** as it consists of only one row, and matrix $\mathbf{D}$ is called a **column matrix** as it consists of only one column.

Therefore, for any matrix of order $m \times n$:

- m is the number of rows,
- n is the number of columns,
- if $m = n$, it is a square matrix,
- if $m = 1$, it is a row matrix,
- if $n = 1$, it is a column matrix.

Exercise 18.1

1. Give the order of the following matrices:

a) $\mathbf{P} = \begin{pmatrix} 9 & 0 & 3 \\ 4 & 6 & 2 \end{pmatrix}$

b) $\mathbf{Q} = \begin{pmatrix} 8 & 6 & 5 & -6 \\ 7 & 2 & 4 & 0 \end{pmatrix}$

c) $\mathbf{R} = \begin{pmatrix} 1 & 12 \\ 4 & 0 \\ 9 & 10 \\ 2 & -6 \end{pmatrix}$

d) $\mathbf{S} = \begin{pmatrix} 8 & 2 & 1 & 4 & -3 \\ 6 & 7 & 9 & 3 & 12 \\ 8 & 5 & 1 & 6 & 1 \\ 7 & 3 & 2 & 8 & 9 \end{pmatrix}$

e)
$$T = \begin{pmatrix} 4 \\ 0 \\ -9 \\ 8 \\ 7 \end{pmatrix}$$

f) $\mathbf{F} = (6 \quad 6 \quad 8 \quad 4 \quad 2)$

2. Write matrices of the following orders:
   a) 3×2 b) 2×3 c) 4×1
   d) 1×4 e) 4×4 f) 2×2

3. A small factory produces televisions and videos. In 1995 it manufactured 6500 televisions and 900 videos. In 1996 it made 7200 televisions and 1100 videos, and in 1997 it made 7300 televisions and 1040 videos. Write this information in a 3×2 matrix.

4. A shop selling beds records the number and type of beds it sells over a three-week period. In the first week it sells three cots, four single beds, two double beds and one king-size bed. In the second week it sells only six single beds and two double beds. In the third week it sells one cot, three single beds and two king-size beds. Write this information as a 3×4 matrix.

5. A shoe shop sells shoes for girls, boys, ladies and gentlemen. One Saturday it sells eight pairs of girls' shoes, six pairs of boys' shoes, nine pairs of ladies' shoes and three pairs of men's shoes. Write this information as a row matrix.

6. Four students sit two tests. Carlos achieved 37% in the first test and 49% in the second. Cristina achieved 74% in the first test and 58% in the second. Ali got 76% in the first test and 62% in the second. Helena got 89% in the first test and 56% in the second. Write this information in a 4×2 matrix.

7. The pie charts below show the nationalities of students at three different schools A, B and C.

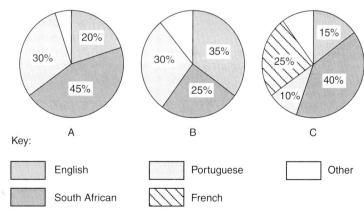

Key:

English   Portuguese   Other

South African   French

Write this information as a matrix.

8. The graph below shows the number of units sold by a computer manufacturer for each quarter of the years 1995, 1996 and 1997.

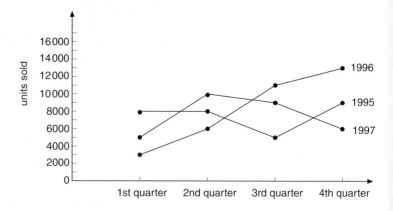

Represent this information as a matrix.

9. The percentage of pupils achieving each of the pass grades in a maths exam for the years 1995, 1996 and 1997 is shown below.

1995

A	B	C	D	E	F	G
6%	12%	43%	18%	6%	9%	6%

1996

A	B	C	D	E	F	G
9%	15%	28%	18%	12%	12%	6%

1997

A	B	C	D	E	F	G
12%	19%	30%	12%	9%	9%	9%

Represent this information as a matrix.

10. Collect some data of your own, either from newspapers, books or from your own surveys, and write it as a matrix.

Addition and subtraction of matrices

Worked examples **a)** A music store sells music in three formats: records, cassettes and CDs. It also sells the following types of music: classical, rock/pop and jazz. It records the number of each type and format sold on two consecutive Saturdays. The results of this are presented in the matrices below:

1st Saturday

	Record	Cassette	CD
Classical	28	14	46
Rock/pop	56	91	15
Jazz	17	5	7

2nd Saturday

	Record	Cassette	CD
Classical	24	10	51
Rock/pop	35	82	24
Jazz	15	8	6

i) Calculate the total number of records sold on the 1st Saturday.

$$\text{Records} = 28 + 56 + 17$$
$$= 101$$

ii) Calculate the total sales of jazz on the 2nd Saturday.

$$\text{Jazz} = 15 + 8 + 6$$
$$= 29$$

iii) Calculate the total sales of each type of music and format. Express your answer as a matrix.

$$\begin{pmatrix} 28 & 14 & 46 \\ 56 & 91 & 15 \\ 17 & 5 & 7 \end{pmatrix} + \begin{pmatrix} 24 & 10 & 51 \\ 35 & 82 & 24 \\ 15 & 8 & 6 \end{pmatrix} = \begin{pmatrix} 52 & 24 & 97 \\ 91 & 173 & 39 \\ 32 & 13 & 13 \end{pmatrix}$$

To add matrices together they must be of the same order. Corresponding elements are then added together.

b) A shop sells shoes for both adults and children. Matrix **A** below shows how many of each type it has in stock. Matrix **B** shows the number of each type it sells in a particular week.

A

	Male	Female
Child	65	42
Adult	111	154

B

	Male	Female
Child	15	21
Adult	19	28

i) Calculate how many shoes the shop has in stock at the start of the week.

$$\text{Total stock} = 65 + 42 + 111 + 154$$
$$= 372$$

ii) Calculate the number of shoes the shop sells over the week to females.

$$\text{Total female sales} = 21 + 28$$
$$= 49$$

iii) Calculate the number of each type of shoe still in stock at the end of the week. Give your answer as a matrix.

$$\begin{pmatrix} 65 & 42 \\ 111 & 154 \end{pmatrix} - \begin{pmatrix} 15 & 21 \\ 19 & 28 \end{pmatrix} = \begin{pmatrix} 50 & 21 \\ 92 & 126 \end{pmatrix}$$

To subtract matrices from each other, they also need to be of the same order. Corresponding elements are then subtracted from each other.

Exercise 18.2

1. Add the following matrices:

a) $\begin{pmatrix} 8 & 2 \\ 7 & 9 \end{pmatrix} + \begin{pmatrix} 6 & 6 \\ 0 & 7 \end{pmatrix}$

b) $\begin{pmatrix} 0 & 1 & 3 \\ 12 & 15 & 9 \end{pmatrix} + \begin{pmatrix} 9 & 2 & 21 \\ 7 & 20 & -4 \end{pmatrix}$

c) $\begin{pmatrix} 9 \\ 3 \\ 12 \end{pmatrix} + \begin{pmatrix} 15 \\ 0 \\ 6 \end{pmatrix}$

d) $\begin{pmatrix} -14 & 7 & 8 \\ 0 & -5 & 12 \\ 1 & 3 & 6 \end{pmatrix} + \begin{pmatrix} 4 & 6 & 9 \\ 2 & -8 & 6 \\ 4 & 7 & 5 \end{pmatrix}$

e) $(-1 \quad -1 \quad 1) + (-2 \quad 2 \quad -2)$

f) $\begin{pmatrix} 6 & -1 \\ 9 & -8 \\ 4 & 6 \end{pmatrix} + \begin{pmatrix} 8 & -8 \\ 6 & 3 \\ -9 & 4 \end{pmatrix}$

2. Subtract the following matrices:

a) $\begin{pmatrix} 8 & 6 \\ 7 & 5 \end{pmatrix} - \begin{pmatrix} 3 & 2 \\ 6 & 5 \end{pmatrix}$

b) $\begin{pmatrix} 9 & 12 & 8 \\ 15 & 7 & 2 \end{pmatrix} - \begin{pmatrix} 7 & 10 & 6 \\ 13 & 5 & 0 \end{pmatrix}$

c) $\begin{pmatrix} 15 \\ 6 \\ 8 \end{pmatrix} - \begin{pmatrix} 20 \\ 6 \\ 4 \end{pmatrix}$

d) $\begin{pmatrix} 8 & -9 & 7 \\ 6 & 12 & 10 \\ -5 & 6 & 4 \end{pmatrix} - \begin{pmatrix} 4 & 5 & 8 \\ 4 & 12 & 6 \\ 4 & 7 & 9 \end{pmatrix}$

e) $\begin{pmatrix} -6 & 14 \\ 9 & -12 \end{pmatrix} - \begin{pmatrix} -3 & -4 \\ 10 & 0 \end{pmatrix}$

f) $\begin{pmatrix} 6 & -8 \\ -15 & 4 \\ 9 & -6 \end{pmatrix} - \begin{pmatrix} 9 & -4 \\ 6 & -9 \\ 12 & -8 \end{pmatrix}$

3. Four matrices are given below:

$$\mathbf{A} = \begin{pmatrix} 2 & 7 \\ 6 & 3 \\ 1 & 4 \end{pmatrix} \quad \mathbf{B} = \begin{pmatrix} 7 & 4 \\ 9 & 2 \\ 8 & 4 \end{pmatrix}$$

$$\mathbf{C} = \begin{pmatrix} -3 & -2 \\ 0 & 3 \\ 6 & -9 \end{pmatrix} \quad \mathbf{D} = \begin{pmatrix} 8 & -4 \\ -9 & 11 \\ 2 & 0 \end{pmatrix}$$

Calculate the following:
a) $\mathbf{A} + \mathbf{B}$ b) $\mathbf{B} + \mathbf{A}$ c) $\mathbf{A} + \mathbf{C}$
d) $\mathbf{D} + \mathbf{A}$ e) $\mathbf{B} - \mathbf{C}$ f) $\mathbf{D} - \mathbf{C}$
g) $\mathbf{A} - \mathbf{C}$ h) $\mathbf{A} + \mathbf{D} - \mathbf{B}$ i) $\mathbf{B} + \mathbf{D} - \mathbf{A}$

4. Three teams compete in a two-day athletics competition. On day 1, team A won two gold medals, three silver medals and one bronze medal. Team B won three gold, one silver and four bronze medals. Team C won three gold, four silver medal and three bronze medals. On day 2, team A won five gold medals, one silver medal and one bronze medal. Team B won one gold, four silver and three bronze medals and team C won one gold, two silver and three bronze medals.
   a) Write this information down in two matrices.
   b) What was the total number of races over the two days?
   c) Write down a matrix to represent the total number of each medal won by each team for the whole competition.

5. A shop selling clothes keeps a record of its stock. Matrix **A** below shows the number and type of shirts and blouses it has in stock at the start of the week. Matrix **B** shows the number and type of shirts and blouses it has at the end of the week.

	A				**B**	
	Shirts	Blouses			Shirts	Blouses
Child	265	312		Child	189	204
Adult	140	132		Adult	121	68

   a) Calculate the total number of blouses at the start of the week.
   b) Write the matrix which shows the number of each type sold over the week.
   c) What is the total number of shirts and blouses sold over the whole week?

Multiplying matrices by a scalar quantity

Worked example   Two children A and B record the number of hours of television and video they watch over the period of one week. This information is represented in the matrix below.

$$\begin{array}{cc} & \text{A} \quad \text{B} \\ \begin{array}{c} \text{TV} \\ \text{Video} \end{array} & \begin{pmatrix} 8 & 6 \\ 4 & 2 \end{pmatrix} \end{array}$$

i)   The following week, both children watch twice as much TV and video as the first week. Write a second matrix to show the amount of TV and video they watch in the second week.

$$2\begin{pmatrix} 8 & 6 \\ 4 & 2 \end{pmatrix} = \begin{pmatrix} 16 & 12 \\ 8 & 4 \end{pmatrix}$$

ii)   In the third week they watch half as much TV and video as they did in the first week. Write another matrix to show the amount of TV and video they watched in the third week.

$$\frac{1}{2}\begin{pmatrix} 8 & 6 \\ 4 & 2 \end{pmatrix} = \begin{pmatrix} 4 & 3 \\ 2 & 1 \end{pmatrix}$$

When multiplying a matrix by a scalar quantity, each element in the matrix is multiplied by that quantity.

Exercise 18.3

Evaluate the following:

1. a) $2\begin{pmatrix} 4 & 6 \\ 7 & 3 \end{pmatrix}$ b) $3\begin{pmatrix} 7 & 2 \\ 1 & 0 \end{pmatrix}$

c) $1\begin{pmatrix} 9 & 3 \\ 0 & 6 \end{pmatrix}$ d) $5\begin{pmatrix} 2 & 1 \\ 0 & 4 \end{pmatrix}$

e) $4\begin{pmatrix} 7 & 5 \\ 4 & 8 \end{pmatrix}$ f) $3.5\begin{pmatrix} 6 & 12 \\ 4 & 2 \end{pmatrix}$

2. a) $\frac{1}{2}\begin{pmatrix} 8 & 2 \\ 0 & 4 \end{pmatrix}$ b) $\frac{1}{3}\begin{pmatrix} 12 & 9 \\ 6 & 3 \end{pmatrix}$

c) $\frac{3}{4}\begin{pmatrix} 4 & 8 \\ 1 & 2 \end{pmatrix}$ d) $\frac{2}{5}\begin{pmatrix} 15 & 10 \\ 2 & 5 \end{pmatrix}$

e) $\frac{5}{8}\begin{pmatrix} 4 & 12 \\ 12 & 16 \end{pmatrix}$ f) $\frac{3}{7}\begin{pmatrix} 28 & 7 \\ 7 & 14 \end{pmatrix}$

Multiplying a matrix by another matrix

Worked examples   **a)**   Paula and Gregori have a choice of shopping at one of two supermarkets, X and Y. The first matrix below shows the type and quantity of certain foods they both wish to buy and the second matrix shows the cost of the items (in cents) at each of the supermarkets.

$$\begin{array}{c}\text{Cereal packets}\\\text{Loaves of bread}\\\text{Potatoes (kg)}\end{array}$$

$$\begin{array}{cc} & \text{X} \quad \text{Y}\\\begin{array}{c}\text{Paula}\\\text{Gregori}\end{array}\begin{pmatrix} 2 & 4 & 5\\ 1 & 7 & 3\end{pmatrix} & \begin{pmatrix} 120 & 110\\ 55 & 60\\ 35 & 30\end{pmatrix}\begin{array}{l}\text{Cereal packets}\\\text{Loaves of bread}\\\text{Potatoes (kg)}\end{array}\end{array}$$

Calculate their shopping bill for these items at each supermarket and decide where they should buy their food.

The shopping bill for each can be calculated by multiplying the two matrices together.

Paula at X $= (2 \times 120) + (4 \times 55) + (5 \times 35) = 635$
Paula at Y $= (2 \times 110) + (4 \times 60) + (5 \times 30) = 610$
Gregori at X $= (1 \times 120) + (7 \times 55) + (3 \times 35) = 610$
Gregori at Y $= (1 \times 110) + (7 \times 60) + (3 \times 30) = 620$

This can also be written as a matrix:

$$\begin{array}{cc} & \text{X} \qquad \text{Y}\\\begin{array}{c}\text{Paula}\\\text{Gregori}\end{array}\begin{pmatrix} 635 & 610\\ 610 & 620\end{pmatrix}\end{array}$$

Paula should therefore shop at Y and Gregori at X.

Multiplying matrices together involves multiplying the elements in the rows of the first matrix by the elements in the columns of the second matrix.

b) $\begin{pmatrix} 6 & 3\\ 2 & 4\\ 0 & 1\end{pmatrix} \times \begin{pmatrix} 2 & 4\\ 1 & 3\end{pmatrix}$

$$\begin{pmatrix} (6 \times 2) + (3 \times 1) & (6 \times 4) + (3 \times 3)\\ (2 \times 2) + (4 \times 1) & (2 \times 4) + (4 \times 3)\\ (0 \times 2) + (1 \times 1) & (0 \times 4) + (1 \times 3)\end{pmatrix} = \begin{pmatrix} 15 & 33\\ 8 & 20\\ 1 & 3\end{pmatrix}$$

Note: Not all matrices can be multiplied together. For matrices to be multiplied together, the number of columns in the first must be equal to the number of rows in the second. This can be seen clearly if their orders are considered.

e.g.

$$\begin{pmatrix} 4 & 3 & 2\\ 1 & 2 & 7\end{pmatrix}\begin{pmatrix} 4 & 2 & 8\\ 6 & 1 & 7\\ 7 & 2 & 0\end{pmatrix} = \begin{pmatrix} 48 & 15 & 44\\ 65 & 18 & 16\end{pmatrix}$$

Order: $\quad 2 \times \mathbf{3} \qquad \mathbf{3} \times 3 \quad = \quad 2 \times 3$

For multiplication to be possible, the two middle numbers (in bold) must be the same. The result will be a matrix of the order of the outer two numbers (i.e. 2×3).

Exercise 18.4 Multiply the following pairs of matrices (remember row into column):

1. a) $\begin{pmatrix} 2 & 8 \\ 6 & 4 \end{pmatrix}\begin{pmatrix} 1 & 3 \\ 2 & 9 \end{pmatrix}$ b) $\begin{pmatrix} 3 & 6 \\ 4 & 0 \end{pmatrix}\begin{pmatrix} 0 & 6 \\ 2 & 9 \end{pmatrix}$

2. a) $\begin{pmatrix} 1 & 6 \\ 3 & 4 \end{pmatrix}\begin{pmatrix} 4 & 6 & 0 \\ 8 & 1 & 3 \end{pmatrix}$ b) $\begin{pmatrix} 1 & 4 \\ 2 & 1 \\ 3 & 0 \end{pmatrix}\begin{pmatrix} 2 & 2 \\ 4 & 4 \end{pmatrix}$

3. a) $\begin{pmatrix} 4 & -1 & 2 \\ 3 & 6 & -4 \end{pmatrix}\begin{pmatrix} 3 & 1 & 0 \\ -2 & 4 & 6 \\ 7 & 2 & -1 \end{pmatrix}$

b) $(8 \ -2 \ -3 \ 0)\begin{pmatrix} 2 & 2 \\ -1 & 6 \\ -3 & 7 \\ 2 & -4 \end{pmatrix}$

4. a) $\begin{pmatrix} -2 \\ 1 \\ 4 \\ -6 \end{pmatrix}(-3 \ -2 \ 1)$

b) $\begin{pmatrix} 2 & -6 \\ 8 & 1 \\ -1 & -3 \\ -4 & -9 \end{pmatrix}\begin{pmatrix} 3 & 2 & 1 \\ -4 & -3 & -6 \end{pmatrix}$

Exercise 18.5 In the following, calculate $\mathbf{V} \times \mathbf{W}$ and where possible $\mathbf{W} \times \mathbf{V}$.

1. $\mathbf{V} = \begin{pmatrix} -3 & 2 \\ 0 & 4 \end{pmatrix}$ $\mathbf{W} = \begin{pmatrix} 4 & -1 \\ 6 & 2 \end{pmatrix}$

2. $\mathbf{V} = \begin{pmatrix} 4 & 6 & 1 \\ 3 & 2 & -3 \end{pmatrix}$ $\mathbf{W} = \begin{pmatrix} 2 & 1 \\ -4 & -3 \\ 6 & 5 \end{pmatrix}$

3. $\mathbf{V} = (2 \ -5 \ 9 \ 2)$ $\mathbf{W} = \begin{pmatrix} 2 \\ 0 \\ -3 \\ 6 \end{pmatrix}$

4. $\mathbf{V} = \begin{pmatrix} 3 & 2 \\ -4 & -2 \end{pmatrix}$ $\mathbf{W} = \begin{pmatrix} -1 & -3 & 5 \\ 5 & 8 & -3 \end{pmatrix}$

5. $\mathbf{V} = \begin{pmatrix} -3 & -2 & 5 & 8 \\ -1 & 4 & -3 & 6 \end{pmatrix}$ $\mathbf{W} = \begin{pmatrix} 3 & -3 & 6 \\ 2 & -2 & 1 \\ -4 & 4 & 2 \\ 0 & 1 & 0 \end{pmatrix}$

In general if matrix **A** is multiplied by matrix **B** then this is not the same as matrix **B** multiplied by matrix **A**.

i.e. $\mathbf{A} \times \mathbf{B} \neq \mathbf{B} \times \mathbf{A}$

However there are some exceptions to this.

The identity matrix

The matrix $\begin{pmatrix} 1 & 0 \\ 0 & 1 \end{pmatrix}$ is known as the **identity matrix** of order 2.
The identity matrix is always represented by **I**.

Exercise 18.6 In the following calculate, where possible, a) **AI**, b) **IA**.

1. $\mathbf{A} = \begin{pmatrix} 2 & 1 \\ 3 & 2 \end{pmatrix}$

2. $\mathbf{A} = \begin{pmatrix} -2 & -4 \\ 3 & 6 \end{pmatrix}$

3. $\mathbf{A} = \begin{pmatrix} 4 & 8 \\ -2 & 4 \end{pmatrix}$

4. $\mathbf{A} = \begin{pmatrix} 3 & 2 \\ 1 & 6 \\ -2 & 5 \end{pmatrix}$

5. $\mathbf{A} = (-5 \quad -6)$

6. $\mathbf{A} = \begin{pmatrix} 4 & -3 \\ 5 & -6 \\ 3 & 2 \\ 1 & 4 \end{pmatrix}$

7. What conclusions can you make about **AI** in each case?

8. What conclusions can you make about **AI** and **IA** where **A** is a 2×2 matrix?

The zero matrix

A matrix in which all the elements are zero is called a zero matrix. Multiplying a matrix by a zero matrix gives a zero matrix.

i.e. $\begin{pmatrix} 4 & 2 \\ -3 & 0 \end{pmatrix} \begin{pmatrix} 0 & 0 \\ 0 & 0 \end{pmatrix} = \begin{pmatrix} 0 & 0 \\ 0 & 0 \end{pmatrix}$

The determinant of a 2×2 matrix

Worked examples **a)** Find the determinant |**R**| of matrix **R** if $\mathbf{R} = \begin{pmatrix} 4 & 2 \\ 3 & 5 \end{pmatrix}$.

The product of the elements in the leading diagonal $= 4 \times 5 = 20$.
The product of the elements in the secondary diagonal $= 2 \times 3 = 6$.

$|\mathbf{R}| = 20 - 6$
$\quad\;\; = 14$

b) If $\mathbf{S} = \begin{pmatrix} 3 & -4 \\ 2 & 1 \end{pmatrix}$, calculate |**S**|.

The product of the elements in the leading diagonal $= 3 \times 1 = 3$.
The product of the elements in the secondary diagonal $= -4 \times 2 = -8$.

$|\mathbf{S}| = 3 - (-8)$
$\quad\;\; = 11$

Exercise 18.7 Calculate the determinant of each of the matrices in Q.1–3.

1. a) $\begin{pmatrix} 6 & 3 \\ 5 & 3 \end{pmatrix}$ b) $\begin{pmatrix} 5 & 7 \\ 6 & 9 \end{pmatrix}$

   c) $\begin{pmatrix} 6 & 10 \\ 5 & 9 \end{pmatrix}$ d) $\begin{pmatrix} 7 & 9 \\ 6 & 8 \end{pmatrix}$

2. a) $\begin{pmatrix} 4 & 9 \\ 4 & 8 \end{pmatrix}$ b) $\begin{pmatrix} 3 & 7 \\ 4 & 6 \end{pmatrix}$

   c) $\begin{pmatrix} 2 & 8 \\ 3 & 7 \end{pmatrix}$ d) $\begin{pmatrix} 5 & 4 \\ 9 & 7 \end{pmatrix}$

3. a) $\begin{pmatrix} -3 & -5 \\ 6 & 4 \end{pmatrix}$ b) $\begin{pmatrix} 5 & 6 \\ -4 & -2 \end{pmatrix}$

   c) $\begin{pmatrix} 8 & 6 \\ -3 & -9 \end{pmatrix}$ d) $\begin{pmatrix} 1 & 4 \\ -1 & 0 \end{pmatrix}$

4. Write two matrices with a determinant of 5.

5. Write two matrices with a determinant of 0.

6. Write two matrices with a determinant of -7.

7. $\mathbf{A} = \begin{pmatrix} 4 & 2 \\ 3 & 8 \end{pmatrix}$ $\mathbf{B} = \begin{pmatrix} -3 & 2 \\ -4 & 4 \end{pmatrix}$ $\mathbf{C} = \begin{pmatrix} -5 & 3 \\ -2 & -6 \end{pmatrix}$

   Calculate:
   a) $|\mathbf{A} + \mathbf{B}|$ b) $|\mathbf{A} - \mathbf{C}|$ c) $|\mathbf{C} - \mathbf{B}|$
   d) $|\mathbf{AC}|$ e) $|\mathbf{BA}|$ f) $|2\mathbf{BC}|$
   g) $|3\mathbf{A} - 2\mathbf{B}|$ h) $|2\mathbf{CB}|$ i) $|\mathbf{B} + \mathbf{C} - \mathbf{A}|$

The inverse of a matrix

Consider the two matrices $\begin{pmatrix} 2 & 3 \\ 3 & 5 \end{pmatrix}$ and $\begin{pmatrix} 5 & -3 \\ -3 & 2 \end{pmatrix}$.

$$\begin{pmatrix} 2 & 3 \\ 3 & 5 \end{pmatrix}\begin{pmatrix} 5 & -3 \\ -3 & 2 \end{pmatrix} = \begin{pmatrix} 1 & 0 \\ 0 & 1 \end{pmatrix}$$

The product of these two matrices gives the identity matrix.

If $\mathbf{A} = \begin{pmatrix} 2 & 3 \\ 3 & 5 \end{pmatrix}$ then $\begin{pmatrix} 5 & -3 \\ -3 & 2 \end{pmatrix}$ is known as the **inverse** of $\mathbf{A}$

and is written as $\mathbf{A}^{-1}$.

Finding the inverse of a matrix can be done in two ways: by simultaneous equations, and by use of a formula.

Simultaneous equations

Worked example Find the inverse of $\begin{pmatrix} 6 & 8 \\ 2 & 3 \end{pmatrix}$.

$$\begin{pmatrix} 6 & 8 \\ 2 & 3 \end{pmatrix}\begin{pmatrix} w & y \\ x & z \end{pmatrix} = \begin{pmatrix} 1 & 0 \\ 0 & 1 \end{pmatrix}$$

$$6w + 8x = 1 \qquad (1)$$
$$2w + 3x = 0 \qquad (2)$$

Multiplying eq.(2) by 3 and subtracting it from eq.(1) gives:

$$6w + 8x = 1$$
$$6w + 9x = 0$$
$$\overline{\qquad -x = 1} \qquad \text{therefore } x = -1$$

Substituting $x = -1$ into eq.(1) gives:

$$6w + 8(-1) = 1$$
$$6w - 8 = 1$$
$$6w = 9 \qquad \text{therefore } w = 1.5$$

$$6y + 8z = 0 \qquad (3)$$
$$2y + 3z = 1 \qquad (4)$$

Multiplying eq.(4) by 3 and subtracting it from eq.(3) gives:

$$6y + 8z = 0$$
$$6y + 9z = 3$$
$$\overline{\qquad -z = -3} \qquad \text{therefore } z = 3$$

Substituting $z = 3$ into eq.(4) gives:

$$2y + 9 = 1$$
$$2y = -8 \qquad \text{therefore } y = -4$$

The inverse of $\begin{pmatrix} 6 & 8 \\ 2 & 3 \end{pmatrix} = \begin{pmatrix} 1.5 & -4 \\ -1 & 3 \end{pmatrix}$.

Use of a formula

If $\mathbf{A} = \begin{pmatrix} w & y \\ x & z \end{pmatrix}$, $\mathbf{A}^{-1} = \dfrac{1}{wz - xy}\begin{pmatrix} z & -y \\ -x & w \end{pmatrix}$

Therefore if $\mathbf{A} = \begin{pmatrix} 6 & 8 \\ 2 & 3 \end{pmatrix}$

$$\mathbf{A}^{-1} = \frac{1}{18 - 16}\begin{pmatrix} 3 & -8 \\ -2 & 6 \end{pmatrix}$$
$$= \frac{1}{2}\begin{pmatrix} 3 & -8 \\ -2 & 6 \end{pmatrix}$$
$$= \begin{pmatrix} 1.5 & -4 \\ -1 & 3 \end{pmatrix}$$

Exercise 18.8

1. Using simultaneous equations find the inverse of each of the following matrices:

   a) $\begin{pmatrix} 9 & 5 \\ 7 & 4 \end{pmatrix}$

   b) $\begin{pmatrix} 10 & 7 \\ 7 & 5 \end{pmatrix}$

   c) $\begin{pmatrix} 5 & 5 \\ 4 & 5 \end{pmatrix}$

   d) $\begin{pmatrix} 6 & -9 \\ 3 & -4 \end{pmatrix}$

   e) $\begin{pmatrix} -5 & -4 \\ 10 & 9 \end{pmatrix}$

   f) $\begin{pmatrix} -3 & 2 \\ 6 & -4 \end{pmatrix}$

2. Using the formula find the inverse of the matrices in Q.1.

3. Explain why (f) in the questions above has no inverse.

4. Which of the following four matrices have no inverse?

   $$\mathbf{A} = \begin{pmatrix} 6 & 8 \\ 3 & 4 \end{pmatrix} \qquad \mathbf{B} = \begin{pmatrix} 6 & 9 \\ 4 & 6 \end{pmatrix}$$

   $$\mathbf{C} = \begin{pmatrix} 6 & 15 \\ -2 & 5 \end{pmatrix} \qquad \mathbf{D} = \begin{pmatrix} -4 & -2 \\ 8 & 4 \end{pmatrix}$$

5. $\mathbf{L} = \begin{pmatrix} 3 & 4 \\ 7 & 9 \end{pmatrix} \qquad \mathbf{M} = \begin{pmatrix} 2 & 3 \\ 3 & 6 \end{pmatrix}$

   $$\mathbf{N} = \begin{pmatrix} -4 & -3 \\ 8 & 5 \end{pmatrix} \qquad \mathbf{O} = \begin{pmatrix} 4 & 3 \\ -8 & -5 \end{pmatrix}$$

   Calculate the following:

   a) $\mathbf{L}^{-1}$

   b) $\mathbf{O}^{-1}$

   c) $(\mathbf{LO})^{-1}$

   d) $(\mathbf{M} + \mathbf{N})^{-1}$

   e) $(\mathbf{NM})^{-1}$

   f) $(\mathbf{MO} + \mathbf{LN})^{-1}$

Student Assessment I

1. Calculate the following:

a) $\begin{pmatrix} 2 & 4 \\ -3 & 5 \end{pmatrix} + \begin{pmatrix} 2 & -5 \\ -4 & 3 \end{pmatrix}$ b) $\begin{pmatrix} 6 & 8 \\ 4 & -6 \\ 9 & -1 \end{pmatrix} + \begin{pmatrix} 0 & 3 \\ -5 & -6 \\ 1 & -4 \end{pmatrix}$

c) $\begin{pmatrix} -5 & -3 \\ 7 & 2 \end{pmatrix} - \begin{pmatrix} 4 & 3 \\ 8 & 1 \end{pmatrix}$ d) $\begin{pmatrix} 6 & -9 \\ 4 & -3 \\ 0 & 7 \end{pmatrix} - \begin{pmatrix} 1 & -5 \\ 3 & -8 \\ -1 & 2 \end{pmatrix}$

e) $2\begin{pmatrix} 3 & 8 & -4 \\ 1 & -6 & 7 \end{pmatrix}$ f) $\frac{1}{4}\begin{pmatrix} 4 & -8 \\ 0 & -6 \end{pmatrix}$

2. Multiply the following matrices:

a) $\begin{pmatrix} 8 & -1 \\ 3 & 6 \\ -4 & 7 \end{pmatrix}\begin{pmatrix} 3 & -2 \\ 6 & 1 \end{pmatrix}$ b) $(1 \quad -4 \quad 6)\begin{pmatrix} 3 & 4 \\ 8 & 0 \\ -9 & -3 \end{pmatrix}$

3. $\mathbf{A} = \begin{pmatrix} 4 & 9 \\ 3 & 7 \end{pmatrix}$ $\mathbf{B} = \begin{pmatrix} -6 & 7 \\ -5 & 6 \end{pmatrix}$

Calculate: a) $|\mathbf{A}|$ b) $|\mathbf{A} + \mathbf{B}|$ c) $|\mathbf{B} - \mathbf{A}|$ d) $|2\mathbf{BA}|$

4. $\mathbf{X} = \begin{pmatrix} 7 & 4 \\ 3 & 2 \end{pmatrix}$ $\mathbf{Y} = \begin{pmatrix} 8 & 7 \\ 9 & 8 \end{pmatrix}$

Calculate: a) $\mathbf{X}^{-1}$ b) $\mathbf{Y}^{-1}$ c) $(\mathbf{X} + \mathbf{Y})^{-1}$ d) $(\mathbf{Y} - \mathbf{X})^{-1}$

Student Assessment 2

1. Calculate the following:

a) $\begin{pmatrix} 3 & 6 \\ -2 & 0 \end{pmatrix} + \begin{pmatrix} 0 & -7 \\ -6 & 2 \end{pmatrix}$ b) $\begin{pmatrix} 7 & 1 \\ 8 & -9 \\ 4 & -2 \end{pmatrix} + \begin{pmatrix} 4 & 6 \\ -2 & -3 \\ 4 & -3 \end{pmatrix}$

c) $\begin{pmatrix} -7 & -2 \\ 2 & 1 \end{pmatrix} - \begin{pmatrix} 3 & 5 \\ 6 & 0 \end{pmatrix}$ d) $\begin{pmatrix} 5 & -6 \\ 1 & -2 \\ 4 & 3 \end{pmatrix} - \begin{pmatrix} 2 & -7 \\ 5 & -3 \\ -2 & 6 \end{pmatrix}$

e) $3\begin{pmatrix} 1 & 4 & -1 \\ 2 & -4 & 5 \end{pmatrix}$ f) $\frac{1}{3}\begin{pmatrix} 3 & -6 \\ 0 & -12 \end{pmatrix}$

2. Multiply the following matrices:

a) $\begin{pmatrix} 1 & -2 \\ 4 & 2 \\ -5 & 3 \end{pmatrix}\begin{pmatrix} 9 & -1 \\ 3 & 7 \end{pmatrix}$ b) $(2 \quad -3 \quad 5)\begin{pmatrix} 6 & 3 \\ 2 & 1 \\ -5 & -7 \end{pmatrix}$

3. $\mathbf{A} = \begin{pmatrix} 2 & 5 \\ 3 & 8 \end{pmatrix}$ $\mathbf{B} = \begin{pmatrix} -7 & 10 \\ -3 & 4 \end{pmatrix}$

Calculate: a) $|\mathbf{A}|$ b) $|\mathbf{A} - \mathbf{B}|$ c) $|\mathbf{B} - \mathbf{A}|$ d) $|3\mathbf{AB}|$

4. $\mathbf{X} = \begin{pmatrix} 9 & 8 \\ 10 & 9 \end{pmatrix}$ $\mathbf{Y} = \begin{pmatrix} 8 & 6 \\ 9 & 7 \end{pmatrix}$

Calculate: a) $\mathbf{X}^{-1}$ b) $\mathbf{Y}^{-1}$ c) $(\mathbf{X} - \mathbf{Y})^{-1}$ d) $(\mathbf{YX})^{-1}$

ICT Section

Set up a spreadsheet in such a way that it will be able to carry out the following types of matrix calculation:

1. Addition of two 2×2 matrices.

2. Multiplication of a 3×3 matrix by a scalar quantity.

3. Multiplication of two matrices of order 2×3 and 3×3.

4. The determinant of a 2×2 matrix.

5. The inverse of a 2×2 matrix.

An example is shown below:

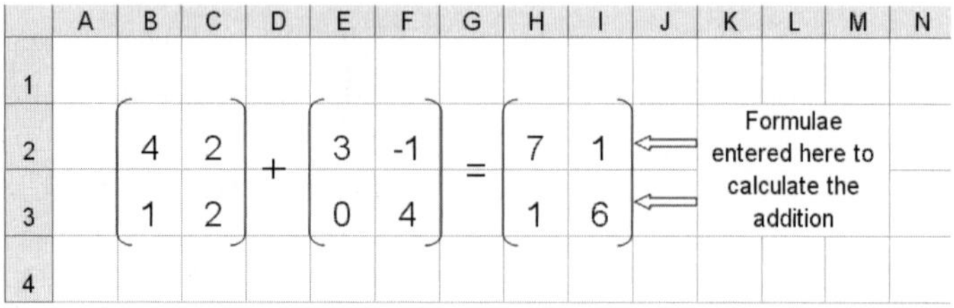

	A	B	C	D	E	F	G	H	I	J	K	L	M	N
1														
2		4	2	+	3	-1	=	7	1	⇐		Formulae entered here to		
3		1	2		0	4		1	6	⇐		calculate the addition		
4														

19 SYMMETRY

Recognise symmetry properties of the prism (including cylinder) and the pyramid (including cone); use the following symmetry properties of circles:

- equal chords are equidistant from the centre
- the perpendicular bisector of a chord passes through the centre
- tangents from an external point are equal in length.

NB: All diagrams are not drawn to scale.

A **line of symmetry** divides a two-dimensional (flat) shape into two congruent (identical) shapes.

e.g.

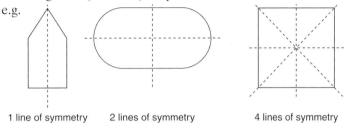

1 line of symmetry   2 lines of symmetry   4 lines of symmetry

A **plane of symmetry** divides a three-dimensional (solid) shape into two congruent solid shapes.

e.g.

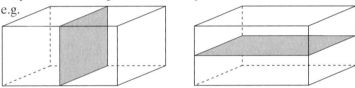

A cuboid has at least three planes of symmetry, two of which are shown above.

A shape has **reflective symmetry** if it has one or more lines or planes of symmetry.

A two-dimensional shape has **rotational symmetry** if, when rotated about a central point, it fits its outline. The number of times it fits its outline during a complete revolution is called the **order of rotational symmetry**.

e.g.

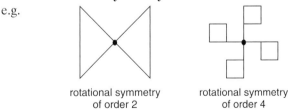

rotational symmetry   rotational symmetry
of order 2   of order 4

A three-dimensional shape has **rotational symmetry** if, when rotated about a central axis, it looks the same at certain intervals.

e.g.

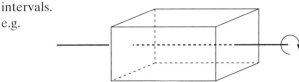

This cuboid has rotational symmetry of order 2 about the axis shown.

Exercise 19.1 **1.** Draw each of the solid shapes below twice, then:
  i) on each drawing of the shape, draw a different plane of symmetry,
  ii) state how many planes of symmetry the shape has in total.

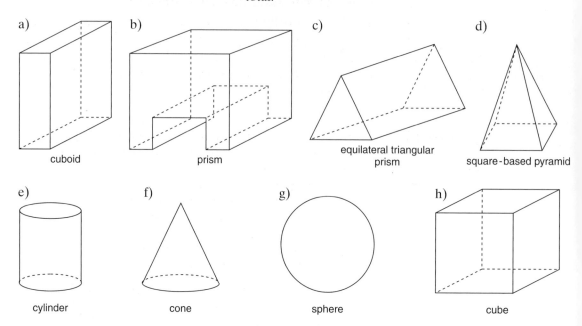

a) cuboid

b) prism

c) equilateral triangular prism

d) square-based pyramid

e) cylinder

f) cone

g) sphere

h) cube

2. For each of the solid shapes shown below determine the order of rotational symmetry about the axis shown.

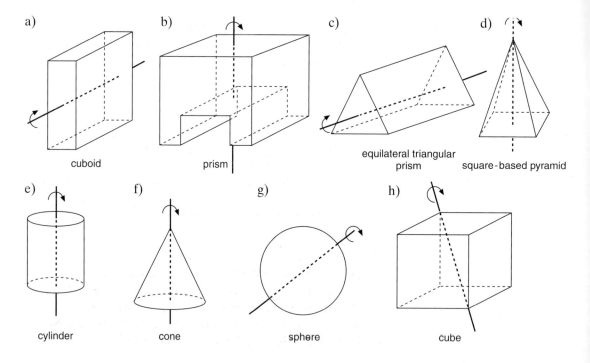

a) cuboid

b) prism

c) equilateral triangular prism

d) square-based pyramid

e) cylinder

f) cone

g) sphere

h) cube

Circle properties

Equal chords and perpendicular bisectors

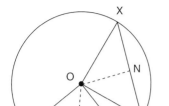

If chords AB and XY are of equal length, then, since OA, OB, OX and OY are radii, the triangles OAB and OXY are congruent isosceles triangles. It follows that:

- the section of a line of symmetry OM through △ OAB is the same length as the section of a line of symmetry ON through △ OXY,
- OM and ON are perpendicular bisectors of AB and XY respectively.

Exercise 19.2

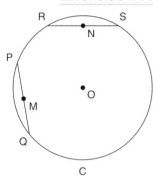

1. In the diagram (left) O is the centre of the circle, PQ and RS are chords of equal length and M and N are their respective midpoints.
   a) What kind of triangle is △ POQ?
   b) Describe the line ON in relation to RS.
   c) If ∠POQ is 80°, calculate ∠OQP.
   d) Calculate ∠ORS.
   e) If PQ is 6 cm calculate the length OM.
   f) Calculate the diameter of the circle.

2. In the diagram (left) O is the centre of the circle. AB and CD are equal chords and the points R and S are their midpoints respectively.
   State whether the statements below are true or false, giving reasons for your answers.
   a) ∠COD = 2 × ∠AOR
   b) OR = OS
   c) If ∠ROB is 60° then △ AOB is equilateral.
   d) OR and OS are perpendicular bisectors of AB and CD respectively.

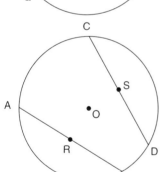

3. Using the diagram (left) state whether the following statements are true or false, giving reasons for your answer.
   a) If △ VOW and △ TOU are isosceles triangles, then T, U, V and W would all lie on the circumference of a circle with its centre at O.
   b) If △ VOW and △ TOU are congruent isosceles triangles, then T, U, V and W would all lie on the circumference of a circle with its centre at O.

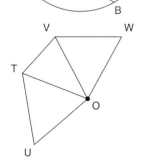

Tangents from an external point

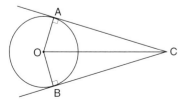

Triangles OAC and OBC are congruent since ∠OAC and ∠OBC are right angles, OA = OB because they are both radii, and OC is common to both triangles. Hence AC = BC.

In general, therefore, tangents being drawn to the same circle from an external point are equal in length.

Exercise 19.3

1. Copy each of the diagrams below and calculate the size of the angle marked $x°$ in each case. Assume that the lines drawn from points on the circumference are tangents.

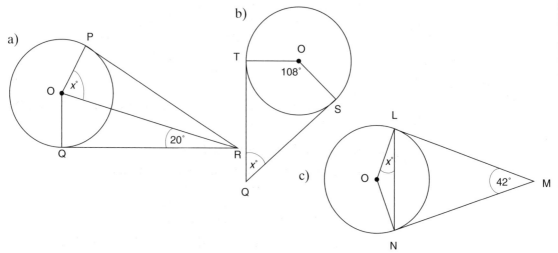

a)

b)

c)

2. Copy each of the diagrams below and calculate the length of the side marked y cm in each case. Assume that the lines drawn from points on the circumference are tangents.

a)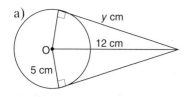

b)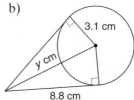

c)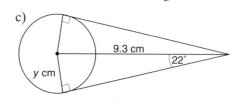

Student Assessment 1

1. Draw a shape with exactly:
   a) one line of reflective symmetry,
   b) two lines of reflective symmetry,
   c) three lines of reflective symmetry.

2. Draw and name a shape with:
   a) two planes of symmetry,
   b) four planes of symmetry.

3. If O is the centre of the circle and the lengths AB and XY are equal, prove that $\triangle$ AOB and $\triangle$ XOY are congruent.

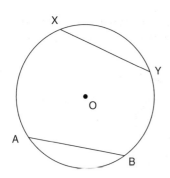

4. Given that PQ and QR are both tangents to the circle, calculate the size of the angle marked $x°$.

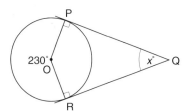

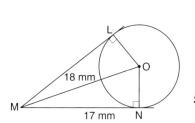

5. Calculate the diameter of the circle given that LM and MN are both tangents to the circle and O is its centre.

Student Assessment 2

1. Draw a two-dimensional shape with exactly:
   a) rotational symmetry of order 2,
   b) rotational symmetry of order 4,
   c) rotational symmetry of order 6.

2. Draw and name a three-dimensional shape with the following orders of rotational symmetry. Mark the position of the axis of symmetry clearly.
   a) Order 2
   b) Order 3
   c) Order 8

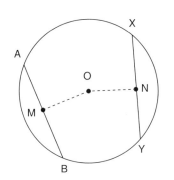

3. In the diagram (left), OM and ON are perpendicular bisectors of AB and XY respectively. OM = ON.
   Prove that AB and XY are chords of equal length.

4. In the diagram, XY and YZ are both tangents to the circle with centre O.
   a) Calculate ∠OZX.
   b) Calculate the length XZ.

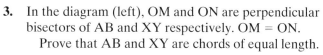

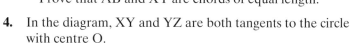

5. In the diagram, LN and MN are both tangents to the circle centre O. If ∠LNO is 35°, calculate the circumference of the circle.

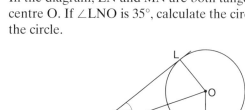

ICT Section

Using Cabri or other appropriate software, demonstrate that:

1. The perpendicular bisector of a chord passes through the centre of the circle.

2. Tangents from an external point to a circle are of equal length.

20 TRANSFORMATIONS

Core Section

Recognise and describe reflections, rotations, translations and enlargements.

An object undergoing a transformation changes either in position or shape. In its simplest form this change can occur as a result of either a **reflection**, **rotation**, **translation** or **enlargement**. If an object undergoes a transformation, then its new position or shape is known as the **image**.

Reflection

If an object is reflected it undergoes a 'flip' movement about a dashed (broken) line known as the **mirror line**, as shown in the diagram.

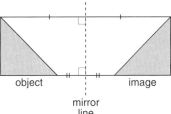

A point on the object and its equivalent point on the image are equidistant from the mirror line. This distance is measured at right angles to the mirror line. The line joining the point to its image is perpendicular to the mirror line.

Exercise 20.1   In each of the following, copy the diagram and draw in the position of the object under reflection in the dashed line(s).

1.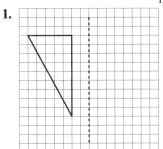

2.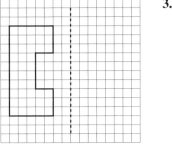

3.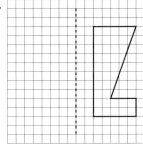

4.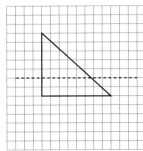

5.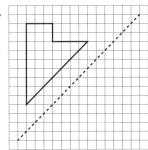

6.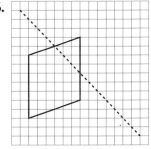

7.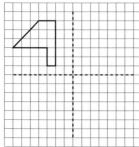

8.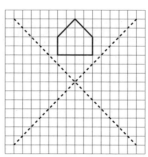

Exercise 20.2 Copy the following objects and images and in each case draw in the position of the mirror line(s).

1.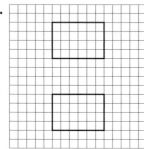

2.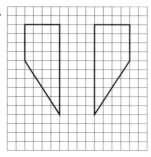

3.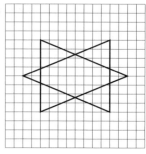

4.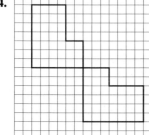

5.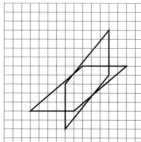

6.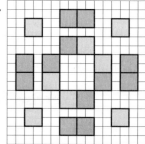

▥ Rotation

If an object is rotated it undergoes a 'turning' movement about a specific point known as the **centre of rotation**. When describing a rotation it is necessary to identify not only the position of the centre of rotation, but also the angle and direction of the turn, as shown in the diagram.

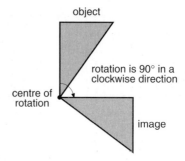

object

rotation is 90° in a clockwise direction

centre of rotation

image

Exercise 20.3 In the following, the object and centre of rotation have both been given. Copy each diagram and draw the object's image under the stated rotation about the marked point.

1.

rotation 180°

2.

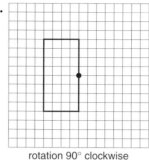

rotation 90° clockwise

3.

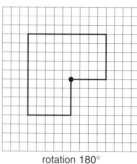

rotation 180°

4.

rotation 90° clockwise

5.

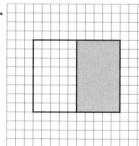

rotation 90° anti-clockwise

6.

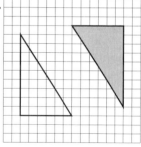

rotation 90° clockwise

Exercise 20.4 In the following, the object (unshaded) and image (shaded) have been drawn. Copy each diagram.
a) Mark the centre of rotation.
b) Calculate the angle and direction of rotation.

1.

2.

3.

4.

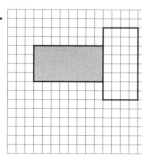

5.

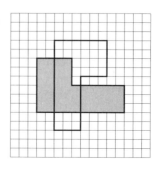

6.

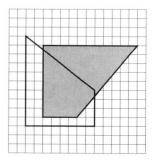

Translation

If an object is translated, it undergoes a 'straight sliding' movement. When describing a translation it is necessary to give the translation vector. As no rotation is involved, each point on the object moves in the same way to its corresponding point on the image, e.g.

1.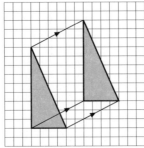

2.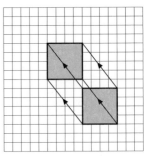

$$\text{Vector} = \begin{pmatrix} 6 \\ 3 \end{pmatrix}$$

$$\text{Vector} = \begin{pmatrix} -4 \\ 5 \end{pmatrix}$$

Exercise 20.5 In the following diagrams, object A has been translated to each of images B and C. Give the translation vectors in each case.

1.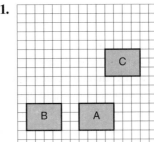

2.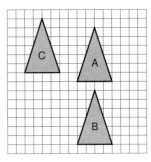

3.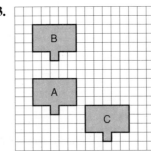

4.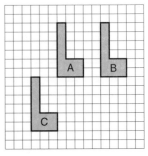

Exercise 20.6 Copy each of the following diagrams and draw the object. Translate the object by the vector given in each case and draw the image in its position. (Note that a bigger grid than the one shown may be needed.)

1.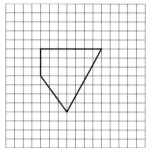

Vector $= \begin{pmatrix} 3 \\ 5 \end{pmatrix}$

2.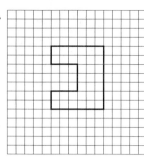

Vector $= \begin{pmatrix} 5 \\ -4 \end{pmatrix}$

3.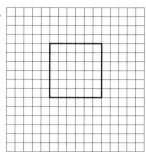

Vector $= \begin{pmatrix} -4 \\ 6 \end{pmatrix}$

4.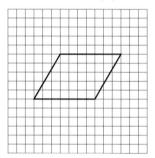

Vector $= \begin{pmatrix} -2 \\ -5 \end{pmatrix}$

5.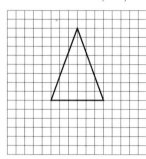

Vector $= \begin{pmatrix} -6 \\ 0 \end{pmatrix}$

6.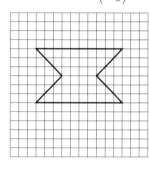

Vector $= \begin{pmatrix} 0 \\ -1 \end{pmatrix}$

▓ Enlargement

If an object is enlarged, the result is an image which is mathematically similar to the object but of a different size. The image can be either larger or smaller than the original object. When describing an enlargement two pieces of information need to be given, the position of the **centre of enlargement** and the **scale factor of enlargement**.

Worked examples **a)** In the diagram below, triangle ABC is enlarged to form triangle A'B'C'.

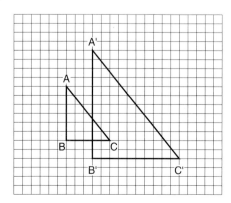

i)   Find the centre of enlargement.

The centre of enlargement is found by joining corresponding points on the object and image with a straight line. These lines are then extended until they meet. The point at which they meet is the centre of enlargement O.

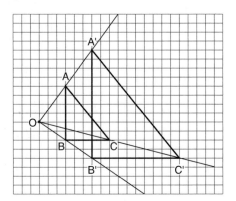

ii) Calculate the scale factor of enlargement.

The scale factor of enlargement can be calculated in one of two ways. From the diagram above it can be seen that the distance OA' is twice the distance OA. Similarly OC' and OB' are both twice OC and OB respectively, hence the scale factor of enlargement is 2.

Alternatively the scale factor can be found by considering the ratio of the length of a side on the image to the length of the corresponding side on the object. i.e.

$$\frac{A'B'}{AB} = \frac{12}{6} = 2$$

Hence the scale factor of enlargement is 2.

b) In the diagram below, the rectangle ABCD undergoes a transformation to form rectangle A'B'C'D'.

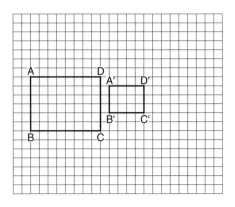

i)   Find the centre of enlargement.

By joining corresponding points on both the object and the image the centre of enlargement is found at O.

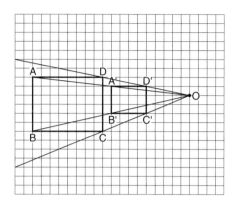

ii)   Calculate the scale factor of enlargement.

$$\text{The scale factor of enlargement} = \frac{A'B'}{AB} = \frac{3}{6} = \frac{1}{2}$$

Note: If the scale factor of enlargement is greater than 1, then the image is larger than the object. If the scale factor lies between 0 and 1, then the resulting image is smaller than the object. In these cases, although the image is smaller than the object, the transformation is still known as an enlargement.

Exercise 20.7   Copy the following diagrams and find:
a)   the centre of enlargement,
b)   the scale factor of enlargement.

1.

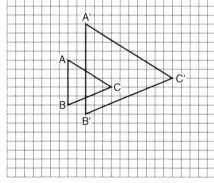

2.

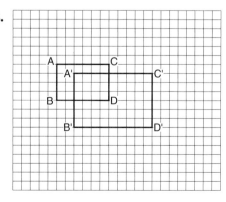

3.

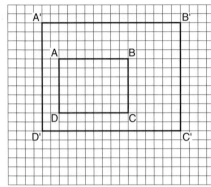

4.

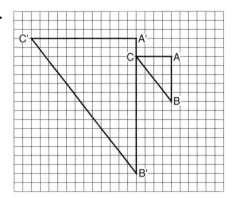

5.

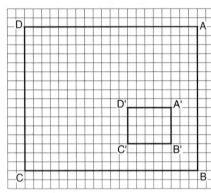

Exercise 20.8 Copy the following diagrams and enlarge the objects by the scale factor given and from the centre of enlargement shown. Grids larger than those shown may be needed.

1.

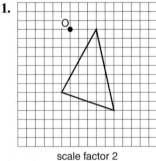

scale factor 2

2.

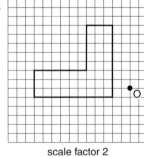

scale factor 2

3.

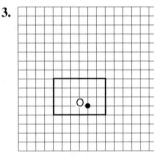

scale factor 3

4.

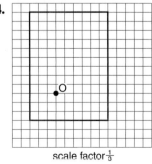

scale factor $\frac{1}{3}$

The diagram below shows an example of **negative enlargement**.

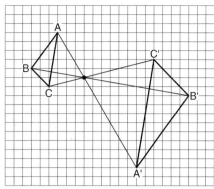

scale factor of enlargement is −2

With negative enlargement each point and its image are on opposite sides of the centre of enlargement. The scale factor of enlargement is calculated in the same way, remembering, however, to write a '−' sign before the number.

Exercise 20.9 **1.** Copy the following diagram and then calculate the scale factor of enlargement and show the position of the centre of enlargement.

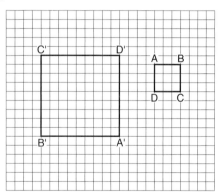

2. The scale factor of enlargement and centre of enlargement are both given. Copy and complete the diagram.

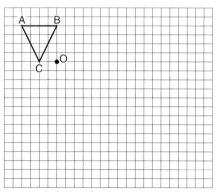

scale factor of enlargement is −2.5

3. The scale factor of enlargement and centre of enlargement are both given. Copy and complete the diagram.

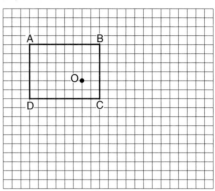

scale factor of enlargement is −2

4. Copy the following diagram and then calculate the scale factor of enlargement and show the position of the centre of enlargement.

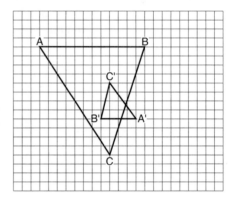

5. An object and part of its image under enlargement are given in the diagram below. Copy the diagram and complete the image. Also find the centre of enlargement and calculate the scale factor of enlargement.

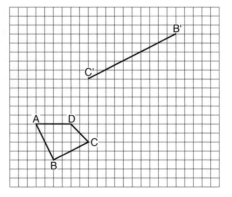

6. In the diagram below, part of an object in the shape of a quadrilateral and its image under enlargement are drawn. Copy and complete the diagram. Also find the centre of enlargement and calculate the scale factor of enlargement.

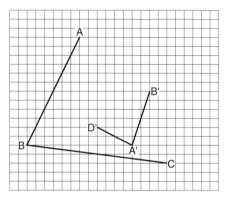

Core Section

Student Assessment 1

1. Reflect the object below in the mirror line shown.

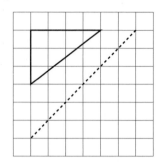

2. Rotate the object below 90° anti-clockwise about the centre of rotation O.

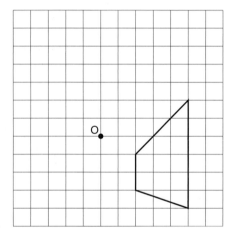

3. Write down the column vector of the translation which maps:
   a) rectangle A to rectangle B,
   b) rectangle B to rectangle C.

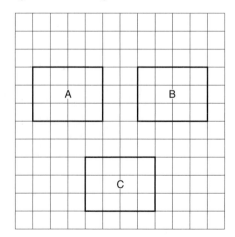

4. Enlarge the triangle below by a scale factor 2 and from the centre of enlargement O.

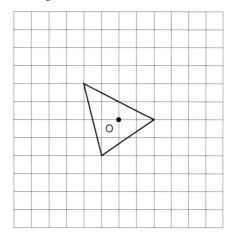

5. An object ABCD and its image A'B'C'D' are shown below.
   a) Find the position of the centre of enlargement.
   b) Calculate the scale factor of enlargement.

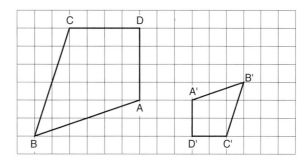

Student Assessment 2

1. Reflect the object below in the mirror line shown.

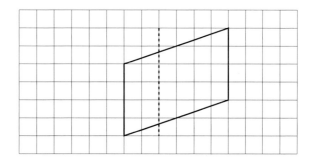

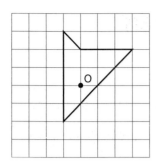

2. Rotate the object (left) 180° about the centre of rotation O.

3. Write down the column vector of the translation which maps:
   a) triangle A to triangle B,
   b) triangle B to triangle C.

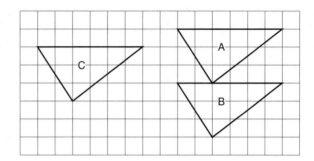

4. Enlarge the rectangle below by a scale factor 1.5 and from the centre of enlargement O.

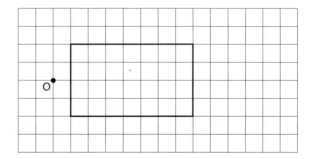

5. An object WXYZ and its image W'X'Y'Z' are shown below.
   a) Find the position of the centre of enlargement.
   b) Calculate the scale factor of enlargement.

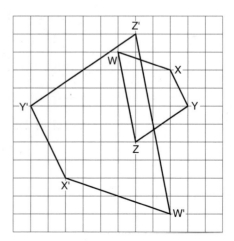

TRANSFORMATIONS

Extended Section

> Use the following transformations of the plane: reflection, rotation, translation, enlargement, shear, stretching and their combinations.
> Identify and give precise descriptions of transformations connecting given figures; describe transformations using coordinates and matrices (singular matrices are excluded).

The core section of this chapter dealt with the elementary aspects of transformation. However, as with most branches of mathematics, a basic principle can be extended.

▓ Reflection

The position of the mirror line is essential when describing a reflection. At times its equation as well as its position will be required.

Worked examples **a)** Find the equation of the mirror line in the reflection given in the diagram (left).

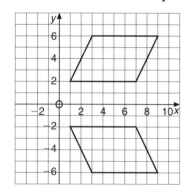

Here the mirror line is the *x*-axis. The equation of the mirror line is therefore $y = 0$.

b) A reflection is shown below.
 i) Draw the position of the mirror line.

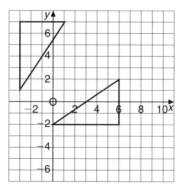
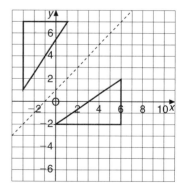

 ii) Give the equation of the mirror line.

Equation of mirror line: $y = x + 1$.

Exercise 20.10 Copy each of the following diagrams, then:
a) draw the position of the mirror line(s),
b) give the equation of the mirror line(s).

1.

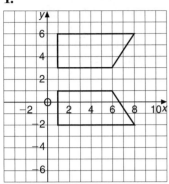

2.

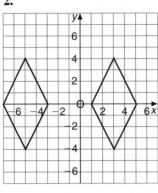

3.

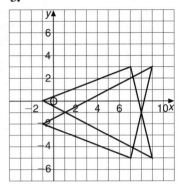

4.

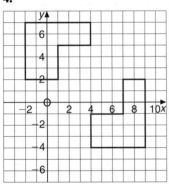

5.

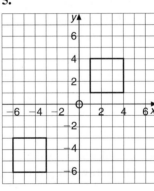

6.

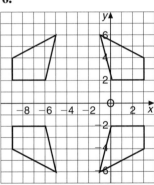

7.

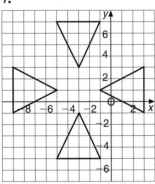

8.

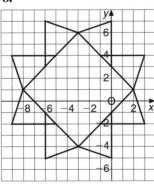

Exercise 20.11

In Q.1 and 2, copy each diagram four times and reflect the object in each of the lines given.

1.
a) $x = 2$
b) $y = 0$
c) $y = x$
d) $y = -x$

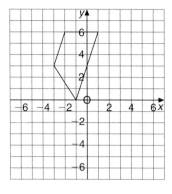

2.
a) $x = -1$
b) $y = -x - 1$
c) $y = x + 2$
d) $x = 0$

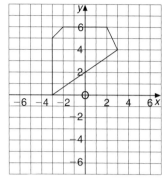

3. Copy the diagram (right), and reflect the triangles in the following lines:

$x = 1$ and
$y = -3$.

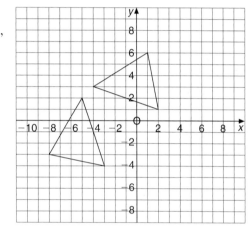

Rotation

As discussed earlier, to describe a rotation, three pieces of information need to be given. These are the centre of rotation, the angle of rotation and the direction of rotation.

Finding the centre and angle of rotation

Worked example Consider the triangle ABC and its new position A'B'C' after being rotated.

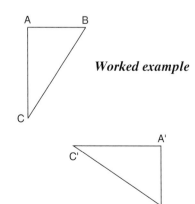

i) Find the centre of rotation.

The centre of rotation is found in the following way:

- Join a point on the object to its corresponding point on the image, e.g. AA'.
- Find the perpendicular bisector of this line.
- Repeat this for another pair of points, e.g. BB'.

Where the two perpendicular bisectors meet gives the centre of rotation O.

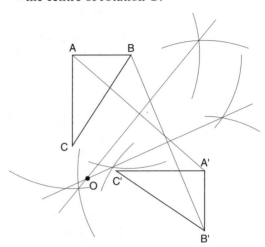

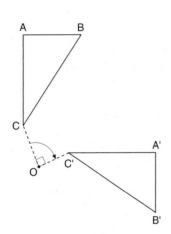

ii) Find the angle and direction of rotation.

Angle of rotation = 90° in a clockwise direction

Exercise 20.12 For each of the following, draw two identical shapes in approximately the same positions as shown. For each pair, assuming the left or upper shape is the initial object, find:
a) the centre of rotation,
b) the angle and direction of rotation.

Check the accuracy of your results using tracing paper.

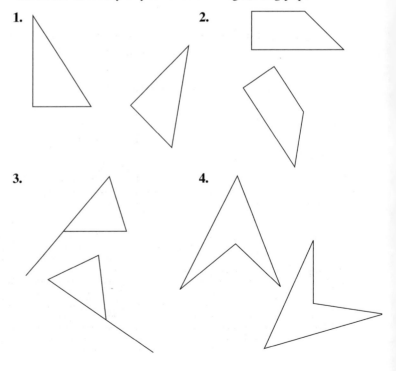

1.

2.

3.

4.

5. Draw a shape of your choice then draw its image after undergoing a rotation of 60° in a clockwise direction. Mark the centre of rotation.

6. Draw a shape of your choice then draw its image after undergoing a rotation of 240°. Mark on the centre of rotation.

Shear

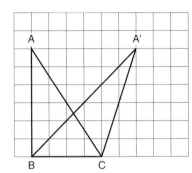

The transformation known as **shear** refers to a sliding of different layers in an object in relation to an **invariant line**. When describing a shear, the invariant line needs to be given as well as the **shear factor**.

BC is the invariant line as the position of every point on it is the same before and after the shear. The perpendicular distance of A from the invariant line is 6 units. The effect of the shear on A is to move it 6 units to the right.

$$\text{Shear factor} = \frac{\text{distance a point moves due to shear}}{\text{perpendicular distance of point from the invariant line}}$$

$$= \frac{6}{6}$$

$$= 1$$

Note that the shear factor will be the same calculated from any point on the object with the exception of those on the invariant line.

Exercise 20.13

Calculate the shear factor for each of the following shears. The object is in **bold** and XY is the invariant line in each case.

1.

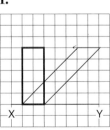

2.

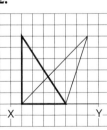

3.

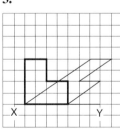

4.

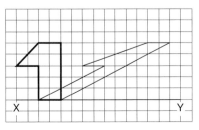

5.

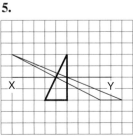

6.

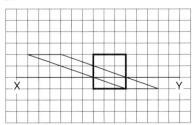

Exercise 20.14 Copy each of the diagrams below. Shear each of the objects to the left by the shear factor shown. The line XY is the invariant in each case.

1.

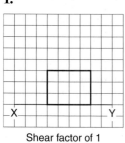

Shear factor of 1

2.

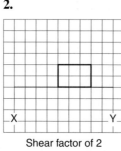

Shear factor of 2

3.

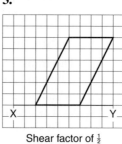

Shear factor of ½

4.

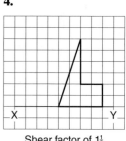

Shear factor of 1½

Copy each of the diagrams shown below. A shear maps point A on the object to point A' on the image. In each case complete the image and calculate the shear factor.

5.

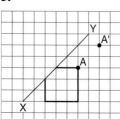

6.

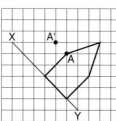

7.

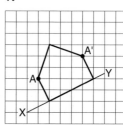

8.

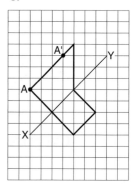

■ Stretch

If an object undergoes a **stretch**, the effect is a lengthening in one direction only. When describing a stretch, two pieces of information need to be given, namely the scale factor and the invariant line, as shown in the diagram.

XY is the invariant line as the position of every point on it remains fixed. The perpendicular distance of A from the invariant line is 3 units. A' is the image of A after being stretched. Its distance from the invariant line is 9 units.

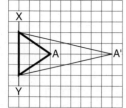

$$\text{Scale factor} = \frac{\text{perpendicular distance of A' from XY}}{\text{perpendicular distance of A from XY}}$$

$$= \frac{9}{3}$$

$$= 3$$

Exercise 20.15
In the following, the object is outlined in **bold**. XY is the invariant line. Calculate the scale factor for each of the stretches shown.

1.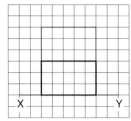

2.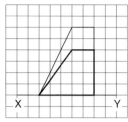

3.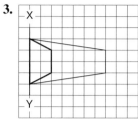

4.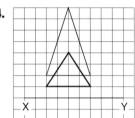

5.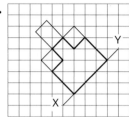

6.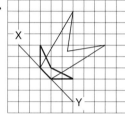

Exercise 20.16
In each of the following, both the object (in **bold**) and the image have been drawn. Determine the position of the invariant line and calculate the stretch scale factor in each case.

1.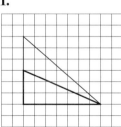

2.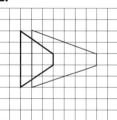

3.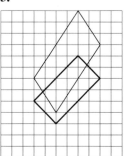

4.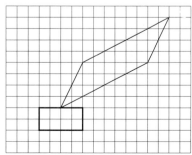

■ Combinations of transformations

An object need not be subjected to just one type of transformation. It can undergo a succession of different transformations.

Worked examples

a) A rectangle ABCD maps onto A'B'C'D' after a stretch of scale factor 1.5, keeping the line $y = -2$ as invariant. A'B'C'D' maps onto A"B"C"D" after undergoing a rotation of 180° about the point $(0, 6)$.
   i) Draw and label the image A'B'C'D'.
   ii) Draw and label the image A"B"C"D".

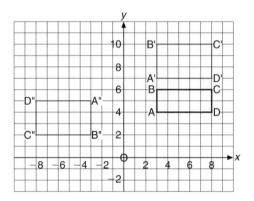

b) A triangle ABC maps onto A'B'C' after an enlargement of scale factor 3 from the centre of enlargement $(0, 7)$. A'B'C' is then mapped onto A"B"C" by a reflection in the line $x = 1$.
   i) Draw and label the image A'B'C'.
   ii) Draw and label the image A"B"C".

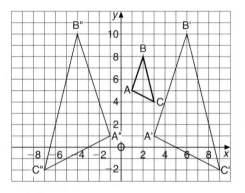

Exercise 20.17 In each of the following questions, copy the diagram. After each transformation, draw the image on the same grid and label it clearly.

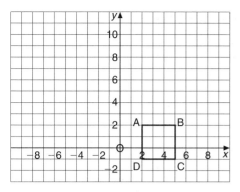

1. The square ABCD is mapped onto A'B'C'D' by a reflection in the line $y = 3$. A'B'C'D' then maps onto A"B"C"D" as a result of a 90° rotation in a clockwise direction about the point $(-2, 5)$.

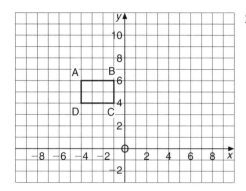

2. The rectangle ABCD is mapped onto A'B'C'D' by an enlargement of scale factor -2 with its centre at $(0, 5)$. A'B'C'D' then maps onto A"B"C"D" as a result of a reflection in the line $y = -x + 7$.

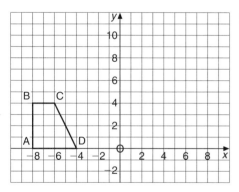

3. The trapezium ABCD is mapped onto A'B'C'D' by a stretch of scale factor 2 with $y = 0$ as the invariant line. A'B'C'D' then maps onto A"B"C"D" as a result of an enlargement of scale factor $-\frac{1}{2}$ with its centre at $(2, 4)$.

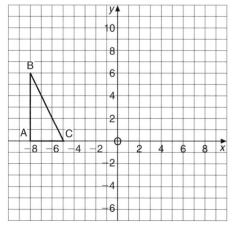

4. The triangle ABC is mapped onto A'B'C' by a shear to the right, of shear factor 2 with $y = 0$ as the invariant line. A'B'C' then maps onto A"B"C" as a result of a reflection in the line $x = -1$. A"B"C" is mapped onto A‴B‴C‴ by a 90° clockwise rotation about the origin.

▥ **Transformations and matrices**

A transformation can be represented by a matrix.

Worked example

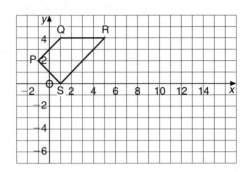

i) Express the vertices of the trapezium PQRS in the form of a matrix.

$$\begin{array}{cccc} \text{P} & \text{Q} & \text{R} & \text{S} \end{array}$$
$$\begin{pmatrix} -1 & 1 & 5 & 1 \\ 2 & 4 & 4 & 0 \end{pmatrix}$$

ii) Find the coordinates of the vertices of its image if PQRS undergoes a transformation by the matrix $\begin{pmatrix} 2 & 1 \\ 0 & -1 \end{pmatrix}$.

$$\begin{pmatrix} 2 & 1 \\ 0 & -1 \end{pmatrix}\begin{pmatrix} -1 & 1 & 5 & 1 \\ 2 & 4 & 4 & 0 \end{pmatrix} = \begin{matrix} \text{P'} & \text{Q'} & \text{R'} & \text{S'} \\ \begin{pmatrix} 0 & 6 & 14 & 2 \\ -2 & -4 & -4 & 0 \end{pmatrix} \end{matrix}$$

iii) Plot the object PQRS and its image P'Q'R'S'.

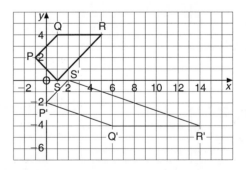

Exercise 20.18 In Q.1–6, transform the object shown in the diagram below by the matrix given. Draw a diagram for each transformation, and plot both the object and its image on the same grid.

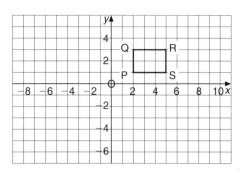

1. Transformation matrix is $\begin{pmatrix} 0 & 1 \\ 1 & 0 \end{pmatrix}$.

2. Transformation matrix is $\begin{pmatrix} 0 & -1 \\ -1 & 0 \end{pmatrix}$.

3. Transformation matrix is $\begin{pmatrix} 0 & 1 \\ -1 & 0 \end{pmatrix}$.

4. Transformation matrix is $\begin{pmatrix} 0 & -1 \\ 1 & 0 \end{pmatrix}$.

5. Transformation matrix is $\begin{pmatrix} 2 & 0 \\ 0 & 2 \end{pmatrix}$.

6. Transformation matrix is $\begin{pmatrix} -2 & 0 \\ 0 & -2 \end{pmatrix}$.

7. Describe in geometrical terms each of the transformations in Q.1–6.

8. a) Draw the image of triangle XYZ under the transformation

   matrix $\begin{pmatrix} 1 & 1 \\ -2 & -1 \end{pmatrix}$

   Label it X'Y'Z'.
   b) Calculate the area of triangle XYZ.
   c) Calculate the area of triangle X'Y'Z'.
   d) Calculate the area scale factor.
   e) Calculate the determinant of the transformation matrix.

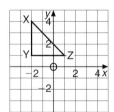

9. a) Draw the image of rectangle PQRS under the transformation matrix $\begin{pmatrix} 2 & 2 \\ 2 & 3 \end{pmatrix}$.

   Label it P'Q'R'S'.

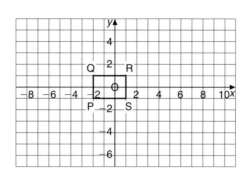

b) Calculate the area of PQRS.
c) Calculate the area of P'Q'R'S'.
d) Calculate the area scale factor.
e) Calculate the determinant of the transformation matrix.

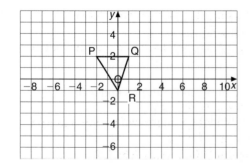

10. a) Draw the image of ABC under the transformation matrix $\begin{pmatrix} -3 & 0 \\ 0 & -1 \end{pmatrix}$.

Label it A'B'C'.
b) Calculate the area of ABC.
c) Calculate the area of A'B'C'.
d) Calculate the area scale factor.
e) Calculate the determinant of the transformation matrix.

11. From your answers in Q.8–10, what can you say about the relationship between the area scale factor and the determinant of a transformation matrix?

Transformations and inverse matrices

If an object is transformed by a matrix **A**, then the inverse matrix $\mathbf{A}^{-1}$ gives the inverse transformation, i.e. it maps the image back onto the object.

Worked example i) The matrix $\begin{pmatrix} 2 & -2 \\ -1 & 2 \end{pmatrix}$ maps $\triangle$ PQR onto $\triangle$ P'Q'R'.

Draw the image of $\triangle$ P'Q'R'.

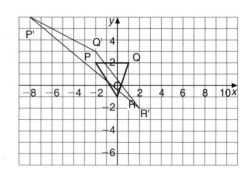

ii) Calculate the area of Δ P'Q'R'.

Area of Δ PQR = 4.5 units²
The determinant of the transformation matrix is 2. As the determinant is numerically equal to the area factor the area factor is also 2.

Area of Δ P'Q'R' = 2 × 4.5
= 9 units²

iii) Calculate the matrix which maps Δ P'Q'R' back onto Δ PQR.

This is calculated by finding the inverse of the transformation matrix.

$$\begin{pmatrix} 2 & -2 \\ -1 & 2 \end{pmatrix}^{-1} = \frac{1}{2 \times 2 - (-1)(-2)} \begin{pmatrix} 2 & 2 \\ 1 & 2 \end{pmatrix}$$

$$= \frac{1}{2} \begin{pmatrix} 2 & 2 \\ 1 & 2 \end{pmatrix}$$

$$= \begin{pmatrix} 1 & 1 \\ 0.5 & 1 \end{pmatrix}$$

Exercise 20.19

1. a) On a grid, draw a quadrilateral of your choice and label its vertices P, Q, R and S.
   b) Draw its image under the transformation of the matrix
   $$\begin{pmatrix} 0 & -1 \\ 2 & 1 \end{pmatrix}$$
   and label the vertices P'Q'R'S'.
   c) By using the determinant of the transformation matrix, calculate the area of P'Q'R'S'.
   d) What matrix maps P'Q'R'S' onto PQRS?

2. a) On a grid, draw a triangle of your choice and label its vertices A, B and C.
   b) Draw its image under the transformation of the matrix
   $$\begin{pmatrix} -3 & -2 \\ 2 & 1 \end{pmatrix}$$
   and label the vertices A'B'C'.
   c) By using the determinant of the transformation matrix, calculate the area of Δ A'B'C'.
   d) What matrix maps Δ A'B'C' onto Δ ABC?

3. a) On a grid, draw a kite of your choice and label its vertices A, B, C and D.
   b) Draw its image under the transformation of the matrix
   $$\begin{pmatrix} -1 & 1 \\ 1 & -2.5 \end{pmatrix}$$
   and label the vertices A'B'C'D'.
   c) By using the determinant of the transformation matrix, calculate the area of A'B'C'D'.
   d) What matrix maps A'B'C'D' onto ABCD?

4. a) On a grid, draw a square of your choice and label its vertices W, X, Y and Z.

b) Draw its image under the transformation of the matrix

$$\begin{pmatrix} 0 & -2 \\ -1.5 & 0 \end{pmatrix}$$

and label the vertices W'X'Y'Z'.

c) By using the determinant of the transformation matrix, calculate the area of W'X'Y'Z'.

d) What matrix maps W'X'Y'Z' onto WXYZ?

5. a) On a grid, draw an isosceles triangle of your choice and label its vertices L, M and N.

b) Draw its image under the transformation of the matrix

$$\begin{pmatrix} 2 & 0 \\ -2 & 1.5 \end{pmatrix}$$

and label the vertices L'M'N'.

c) By using the determinant of the transformation matrix, calculate the area of $\triangle$L'M'N'.

d) What matrix maps $\triangle$L'M'N' onto $\triangle$LMN?

6. a) On a grid, draw a square of your choice and label its vertices A, B, C and D.

b) Draw its image under the transformation of the matrix

$$\begin{pmatrix} -1 & 0 \\ 1 & -2 \end{pmatrix}$$

and label its vertices A'B'C'D'.

c) A'B'C'D' is then mapped onto A"B"C"D" under the transformation of the matrix $\begin{pmatrix} 1 & 1 \\ -2 & -1 \end{pmatrix}$.

On the same grid, plot A"B"C"D".

d) Calculate the area of A'B'C'D'.

e) Calculate the area of A"B"C"D".

f) Find the matrix that maps A"B"C"D" onto A'B'C'D'.

g) Find the matrix that maps A'B'C'D' onto ABCD.

h) Find the matrix that maps ABCD onto A"B"C"D".

i) Find the matrix that maps A"B"C"D" onto ABCD.

Combinations of transformations

An object can be transformed by a series of transformation matrices. These matrices can be replaced by a single matrix which maps the original object onto the final image.

Worked example $\mathbf{A} = \begin{pmatrix} 2 & 1 \\ -3 & 1 \end{pmatrix}$ $\mathbf{B} = \begin{pmatrix} -1 & 1 \\ 0 & 1 \end{pmatrix}$

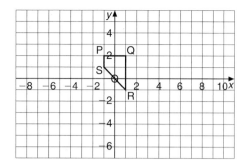

The trapezium PQRS is mapped onto P"Q"R"S" under the transformation of matrix **A**.

i)   Draw and label the position of P"Q"R"S".

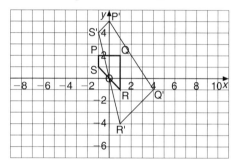

P'Q'R'S' is mapped onto P"Q"R"S" under the transformation of matrix **B**.

ii) Draw and label the position of P"Q"R"S".

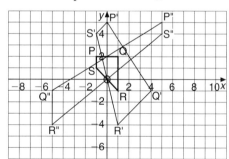

iii) Find the matrix which maps PQRS onto P"Q"R"S".

PQRS undergoes a transformation firstly by matrix **A** and then by matrix **B**. The matrix which maps PQRS directly onto P"Q"R"S" is given by **BA**, i.e.

$$\begin{pmatrix} -1 & 1 \\ 0 & 1 \end{pmatrix}\begin{pmatrix} 2 & 1 \\ -3 & 1 \end{pmatrix} = \begin{pmatrix} -5 & 0 \\ -3 & 1 \end{pmatrix}$$

iv) Find the matrix which maps P"Q"R"S" onto PQRS.

This is the inverse of the matrix which maps PQRS onto P"Q"R"S" , i.e.

$$\begin{pmatrix} -5 & 0 \\ -3 & 1 \end{pmatrix}^{-1} = \begin{pmatrix} -0.2 & 0 \\ -0.6 & 1 \end{pmatrix}$$

Exercise 20.20

1. $A = \begin{pmatrix} -1 & 1 \\ -1 & 3 \end{pmatrix}$ $B = \begin{pmatrix} 0 & -1 \\ 1 & 1 \end{pmatrix}$

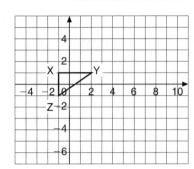

Triangle XYZ is mapped onto Z'Y'Z' under the transformation of matrix **A**. $\triangle$ X'Y'Z' is subsequently mapped onto $\triangle$ X"Y"Z" under the transformation of matrix **B**.

a) Copy the grid (left) and plot the position of $\triangle$ X'Y'Z'.
b) Plot the position of $\triangle$ X"Y"Z".
c) Calculate the area of $\triangle$ X'Y'Z'.
d) Calculate the area of $\triangle$ X"Y"Z".
e) Find the matrix which maps $\triangle$ XYZ onto $\triangle$ X"Y"Z".
f) Find the matrix which maps $\triangle$ X"Y"Z" onto $\triangle$ XYZ.

2. $P = \begin{pmatrix} 0.5 & -1 \\ -1 & 3 \end{pmatrix}$ $Q = \begin{pmatrix} 0 & 1 \\ -3 & 1 \end{pmatrix}$

a) Draw a square on a grid and label the vertices ABCD.
b) Draw the image of ABCD under the transformation with matrix **P** and label its vertices A'B'C'D'.
c) Draw the image of A'B'C'D' under the transformation with matrix **Q** and label its vertices A"B"C"D".
d) Calculate the area of A'B'C'D'.
e) Calculate the area of A"B"C"D".
f) Find the matrix which maps ABCD directly onto A"B"C"D".
g) Find the matrix which maps A"B"C"D" directly onto ABCD.

3. The vertices of a triangle ABC are given by the coordinates A(−3, 2), B(1, 1), C(−3, −1). $\triangle$ ABC is mapped onto $\triangle$ A'B'C', the coordinates of its vertices being: A'(−3, −5), B'(1, 0), C'(−3, −2).
a) Find the matrix which maps $\triangle$ ABC onto $\triangle$ A'B'C'.
b) Find the matrix which maps $\triangle$ A'B'C' onto $\triangle$ ABC.

4. The coordinates of the vertices of a square ABCD after two transformations are as follows:

A(−2, 2)	B(2, 2)	C(2, −2)	D(−2, −2)
A'(−2, 6)	B'(6, 2)	C'(2, −6)	D'(−6, −2)
A"(−6, 8)	B"(−2, −4)	C"(6, −8)	D"(2, 4)

a) Find the matrix which maps ABCD onto A'B'C'D'.
b) Find the matrix which maps A'B'C'D' onto A"B"C"D".
c) Find the matrix which maps ABCD onto A"B"C"D".
d) Find the matrix which maps A"B"C"D" onto ABCD.
e) Calculate the area of ABCD.
f) Calculate the area of A'B'C'D'.
g) Calculate the area of A"B"C"D".

Extended Section

Student Assessment I

1. Copy the diagram below, which shows an object and its reflected image.

    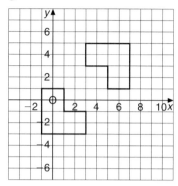

    a) Draw on your diagram the position of the mirror line.
    b) Find the equation of the mirror line.

2. The triangle ABC is mapped onto triangle A'B'C' by a rotation (left).

    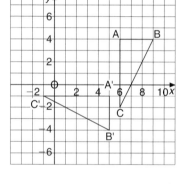

    a) Find the coordinates of the centre of rotation.
    b) Give the angle and direction of rotation.

3. The objects below (in **bold**) have been sheared. In each case calculate the shear factor if the line AB is the invariant line.

    a)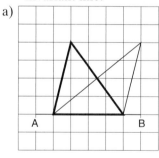

    b)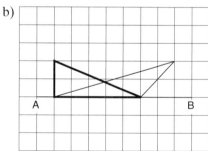

4. The objects below (in **bold**) have been stretched. In each case calculate the stretch factor if the line XY is the invariant line.

    a)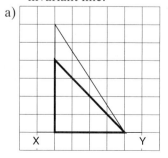

    b)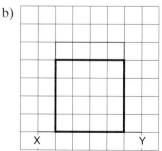

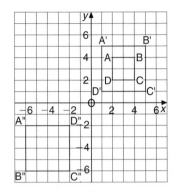

5. The square ABCD is mapped onto A'B'C'D'. A'B'C'D' is subsequently mapped onto A"B"C"D".
   a) Describe in full the transformation which maps ABCD onto A'B'C'D'.
   b) Describe in full the transformation which maps A'B'C'D' onto A"B"C"D".

Student Assessment 2

1. Copy the diagram (below left).
   a) Draw in the mirror line with equation $y = x - 1$.
   b) Reflect the object in the mirror line.

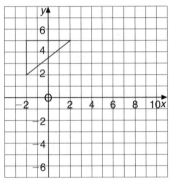

2. The objects below are to be sheared to the right by the shear factor shown. Copy each diagram and draw the position of the image if AB is the invariant line.

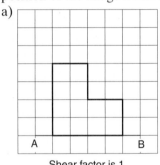
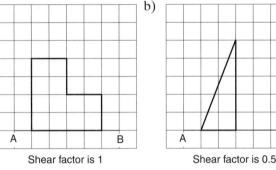

Shear factor is 1   Shear factor is 0.5

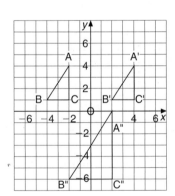

3. The triangle ABC is mapped onto A'B'C'. A'B'C' is subsequently mapped onto A"B"C".
   a) Describe in full the transformation which maps ABC onto A'B'C'.
   b) Describe in full the transformation which maps A'B'C' onto A"B"C".

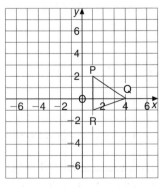

4. The triangle PQR undergoes a transformation by the matrix $\begin{pmatrix} -1 & 2 \\ 0 & -2 \end{pmatrix}$.
   a) Draw the image of $\triangle$ PQR after the transformation and label its vertices P'Q'R'.
   b) Calculate the area scale factor from $\triangle$ PQR to $\triangle$ P'Q'R'.

5. Calculate the inverse of each of the following matrices:
   a) $\begin{pmatrix} 4 & -2 \\ -1 & 1 \end{pmatrix}$
   b) $\begin{pmatrix} 1 & 0 \\ 1 & -1 \end{pmatrix}$

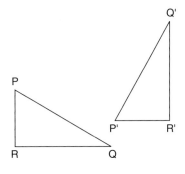

Student Assessment 3

1. Draw a triangle PQR and a triangle P'Q'R' in positions similar to those shown (left).
   a) Find, by construction, the centre of the rotation which maps Δ PQR onto Δ P'Q'R'.
   b) Calculate the angle and direction of the rotation.

2. The objects below (in **bold**) have been stretched. If XY is the invariant line, calculate the stretch factor in each case.

a)

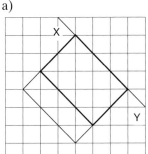

b)

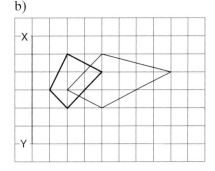

3. Square ABCD is mapped onto square A'B'C'D'. Square A'B'C'D' is subsequently mapped onto square A"B"C"D".

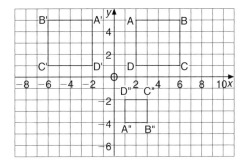

   a) Describe fully the transformation which maps ABCD onto A'B'C'D'.
   b) Describe fully the transformation which maps A'B'C'D' onto A"B"C"D".

4. The triangle JKL (left) is mapped onto triangle J'K'L' by the matrix $\begin{pmatrix} 0 & -1 \\ -1 & 0 \end{pmatrix}$.

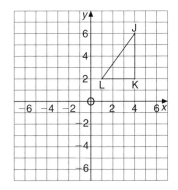

Δ J'K'L' is subsequently mapped onto Δ J"K"L" by the matrix $\begin{pmatrix} 0 & 1 \\ -1 & 0 \end{pmatrix}$.

   a) Copy the grid (left), and draw and label the position of J'K'L'.
   b) On the same axes, draw and label the position of J"K"L".
   c) Calculate the matrix which maps JKL directly onto J"K"L".

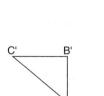

Student Assessment 4

1. Draw a triangle ABC and a triangle A'B'C' in positions similar to those shown (left).
    a) Find, by construction, the centre of the rotation which maps Δ ABC onto Δ A'B'C'.
    b) Calculate the angle and direction of the rotation.

2. Δ LMN below is mapped onto Δ L'M'N' by a stretching of scale factor 2 with $y = x + 3$ as the invariant line. Δ L'M'N' is subsequently mapped onto Δ L"M"N" by a rotation of 180° about the point (0, 1).
    a) Copy the grid below, and plot and label the position of Δ L'M'N'.
    b) On the same axes plot and label the position of Δ L"M"N".

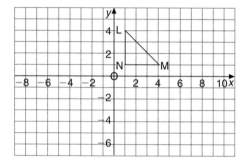

3. The rhombus PQRS is mapped onto P'Q'R'S' by the matrix $\begin{pmatrix} 1 & 2 \\ 1 & -3 \end{pmatrix}$.

    P'Q'R'S' is mapped onto P"Q"R"S" by the matrix $\begin{pmatrix} -2 & -2 \\ -1 & 0 \end{pmatrix}$.

    a) Copy the grid (left), and plot and label the position of P'Q'R'S'.
    b) On the same axes, plot and label the position of P"Q"R"S".
    c) Calculate the area of PQRS.
    d) Calculate the area of P'Q'R'S'.
    e) Calculate the area of P"Q"R"S".
    f) Find the matrix which maps P"Q"R"S" onto P'Q'R'S'.
    g) Find the matrix which would map PQRS directly onto P"Q"R"S".
    h) Find the matrix which would map P"Q"R"S" directly onto PQRS.

ICT Section

Using Autograph or other appropriate software, investigate the effect of different transformation matrices on an object.

1. Draw an object such as a rectangle or other basic shape.

2. Apply a 2×2 matrix to the object in the form $\begin{pmatrix} a & b \\ c & d \end{pmatrix}$

3. By changing the values of a, b, c, and d to $\begin{pmatrix} 0 & 1 \\ 1 & 0 \end{pmatrix}$ describe the transformation on the original object.

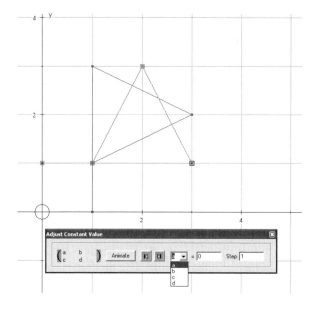

4. Apply each of the following transformation matrices in turn. Describe in geometrical terms each of the transformations:

a) $\begin{pmatrix} 0 & -1 \\ -1 & 0 \end{pmatrix}$

b) $\begin{pmatrix} 0 & 1 \\ -1 & 0 \end{pmatrix}$

c) $\begin{pmatrix} 0 & -1 \\ 1 & 0 \end{pmatrix}$

d) $\begin{pmatrix} 2 & 0 \\ 0 & 2 \end{pmatrix}$

e) $\begin{pmatrix} -2 & 0 \\ 0 & -2 \end{pmatrix}$

5. Prepare a small display of your findings in Q.3 and 4 above.

STATISTICS AND PROBABILITY

21 *HANDLING DATA*

Understand and use mean, median, mode and correlation; construct and read histograms with equal and unequal intervals (areas proportional to frequencies and vertical axis labelled 'frequency density'); construct and use cumulative frequency diagrams; estimate the median, percentiles, quartiles and inter-quartile range; interpret the median; calculate an estimate of the mean for grouped and continuous data; identify the modal class from a grouped frequency distribution.

■ Average

'Average' is a word which in general use is taken to mean somewhere in the middle. For example, a woman may describe herself as being of average height. A student may think he or she is of average ability in maths. Mathematics is more exact and uses three principal methods to measure average.

■ The **mode** is the value occurring the most often.
■ The **median** is the middle value when all the data is arranged in order of size.
■ The **mean** is found by adding together all the values of the data and then dividing that total by the number of data values.

Worked examples **a)** Find the mean, median and mode of the data listed below.
$$1, 0, 2, 4, 1, 2, 1, 1, 2, 5, 5, 0, 1, 2, 3$$

$$\text{Mean} = \frac{1 + 0 + 2 + 4 + 1 + 2 + 1 + 1 + 2 + 5 + 5 + 0 + 1 + 2 + 3}{15}$$

$$= 2$$

Arranging all the data in order and then picking out the middle number gives the median:

$$0, 0, 1, 1, 1, 1, 1, \boxed{2,} \; 2, 2, 2, 3, 4, 5, 5$$

The mode is the number which appeared most often.

Therefore the mode is 1.

b) The frequency chart (left) shows the score out of 10 achieved by a class in a maths test.
Calculate the mean, median and mode for this data.

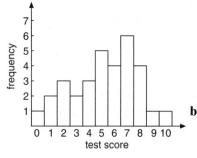

Transferring the results to a frequency table gives:

Test score	0	1	2	3	4	5	6	7	8	9	10	Total
Frequency		1	2	3	2	3	5	4	6	4	1	1	32
Frequency × score	0	2	6	6	12	25	24	42	32	9	10	168

By looking at the total column we can ascertain the number of students taking the test, i.e. 32, and also the total number of marks obtained by all the students, i.e. 168.

Therefore mean score $= \dfrac{168}{32} = 5.25$

Arranging all the scores in order gives:

0, 1, 1, 2, 2, 2, 3, 3, 4, 4, 4, 5, 5, 5, 5, (5, 6,) 6, 6, 6, 7, 7, 7, 7, 7, 7, 8, 8, 8, 8, 9, 10

Because there is an even number of students there isn't one middle number. There is a middle pair. The median is $\dfrac{(5 + 6)}{2} = 5.5$.

The mode is 7 as it is the score which occurs most often.

Exercise 21.1

In Q.1–5, find the mean, median and mode for each set of data.

1. A hockey team plays 15 matches. Below is a list of the number of goals scored in each match.

   1, 0, 2, 4, 0, 1, 1, 1, 2, 5, 3, 0, 1, 2, 2

2. The total scores when two dice are thrown 20 times are:

   7, 4, 5, 7, 3, 2, 8, 6, 8, 7, 6, 5, 11, 9, 7, 3, 8, 7, 6, 5

3. The ages of a group of girls are:

   14 years 3 months, 14 years 5 months, 13 years 11 months, 14 years 3 months, 14 years 7 months, 14 years 3 months, 14 years 1 month

4. The number of students present in a class over a three-week period is:

   28, 24, 25, 28, 23, 28, 27, 26, 27, 25, 28, 28, 28, 26, 25

5. An athlete keeps a record in seconds of her training times for the 100 m race:

   14.0, 14.3, 14.1, 14.3, 14.2, 14.0, 13.9, 13.8, 13.9, 13.8, 13.8, 13.7, 13.8, 13.8, 13.8

6. The mean mass of the 11 players in a football team is 80.3 kg. The mean mass of the team plus a substitute is 81.2 kg. Calculate the mass of the substitute.

7. After eight matches a basketball player had scored a mean of 27 points. After three more matches his mean was 29. Calculate the total number of points he scored in the last three games.

Exercise 21.2

1. An ordinary dice was rolled 60 times. The results are shown in the table below. Calculate the mean, median and mode of the scores.

Score	1	2	3	4	5	6
Frequency	12	11	8	12	7	10

2. Two dice were thrown 100 times. Each time their combined score was recorded. Below is a table of the results. Calculate the mean score.

Score	2	3	4	5	6	7	8	9	10	11	12
Frequency	5	6	7	9	14	16	13	11	9	7	3

3. Sixty flowering bushes are planted. At their flowering peak, the number of flowers per bush is counted and recorded. The results are shown in the table below.

Flowers per bush	0	1	2	3	4	5	6	7	8
Frequency	0	0	0	6	4	6	10	16	18

a) Calculate the mean, median and mode of the number of flowers per bush.
b) Which of the mean, median and mode would be most useful when advertising the bush to potential buyers?

▓ The mean for grouped data

The mean for grouped data can only be an estimate as the position of the data within a group is not known. An estimate is made by calculating the mid-interval value for a group and then assigning all of the data within the group that mid-interval value.

Worked example

The history test scores for a group of 40 students are shown in the grouped frequency table below.

Score	Frequency	Mid-interval value	Frequency × mid-interval value
$0 \leqslant S < 20$	2	9.5	19
$20 \leqslant S < 40$	4	29.5	118
$40 \leqslant S < 60$	14	49.5	693
$60 \leqslant S < 80$	16	69.5	1112
$80 \leqslant S < 100$	4	89.5	358

i) Calculate an estimate for the mean test result.

$$\text{Mean} = \frac{19 + 118 + 693 + 1112 + 358}{40} = 57.5$$

ii) What is the modal class?

This refers to the class with the greatest frequency density. Therefore the modal class is $60 \leqslant S < 80$.

Exercise 21.3

1. The heights of 50 basketball players attending a tournament are recorded in the grouped frequency table. Note: 1.8– means $1.8 \leqslant H < 1.9$.

    a) Copy the table and complete it to include the necessary data with which to calculate the mean height of the players.

    b) Estimate the mean height of the players.

Height (m)	Frequency
1.8–	2
1.9–	5
2.0–	10
2.1–	22
2.2–	7
2.3–2.4	4

    c) What is the modal class height of the players?

Hours of overtime	Frequency
0–9	12
10–19	18
20–29	22
30–39	64
40–49	32
50–59	20

2. The number of hours of overtime worked by employees at a factory over a period of a month is given in the table (left).

    a) Calculate an estimate for the mean number of hours of overtime worked by the employees that month.

    b) What is the modal class?

3. The length of the index finger of 30 students in a class is measured. The results were recorded and are shown in the grouped frequency table.

Length (cm)	Frequency
5.0–	3
5.5–	8
6.0–	10
6.5–	7
7.0–7.5	2

    a) Calculate an estimate for the mean index finger length of the students.

    b) What is the modal class?

Scatter graphs

Scatter graphs are particularly useful if we wish to see if there is a **correlation** (relationship) between two sets of data. The two values of data collected represent the coordinates of each point plotted. How the points lie when plotted indicates the type of relationship between the two sets of data.

Worked example The heights and weights (masses) of 20 children under the age of five were recorded. The heights were recorded in centimetres and the weights in kilograms. The data is shown below.

Height	32	34	45	46	52
Weight	5.834	3.792	9.037	4.225	10.149
Height	59	63	64	71	73
Weight	6.188	9.891	16.010	15.806	9.929
Height	86	87	95	96	96
Weight	11.132	16.443	20.895	16.181	14.000
Height	101	108	109	117	121
Weight	19.459	15.928	12.047	19.423	14.331

a) Plot a scatter graph of the above data.
b) Comment on any relationship you see.
c) If another child was measured as having a height of 80 cm, approximately what weight would you expect him or her to be?

a)

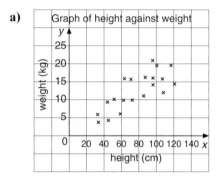

b) The points tend to lie in a diagonal direction from bottom left to top right. This suggests that as height increases then, in general, weight increases too. Therefore there is a **positive correlation** between height and weight.

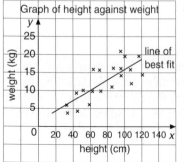

c) We assume that this child will follow the trend set by the other 20 children. To deduce an approximate value for the weight, we draw a **line of best fit**. This is a solid straight line which passes through the points as closely as possible, as shown (left).

The line of best fit can now be used to give an approximate solution to the question. If a child has a height of 80 cm, you would expect his or her weight to be in the region of 13 kg.

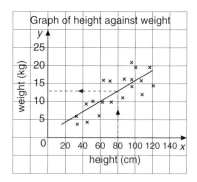

Types of correlation

There are several types of correlation, depending on the arrangement of the points plotted on the scatter graph. These are described below.

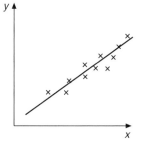

A **strong positive correlation** between the variables x and y. The points lie very close to the line of best fit.
As x increases, so does y.

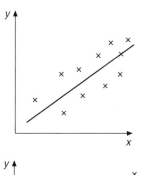

A **weak positive correlation**. Although there is direction to the way the points are lying, they are not tightly packed around the line of best fit.
As x increases, y tends to increase too.

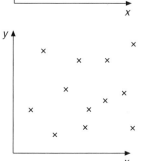

No correlation. As there is no pattern to the way in which the points are lying, there is no correlation between the variables x and y. As a result there can be no line of best fit.

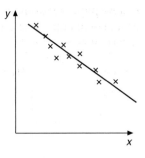

A **strong negative correlation**. The points lie close around the line of best fit.

As x increases, y decreases.

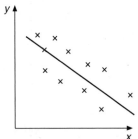

A **weak negative correlation**. The points are not tightly packed around the line of best fit.

As x increases, y tends to decrease.

Exercise 21.4   **1.**   State what type of correlation you might expect, if any, if the following data was collected and plotted on a scatter graph. Give reasons for your answer.

a) A student's score in a maths exam and their score in a science exam.

b) A student's hair colour and the distance they have to travel to school.

c) The outdoor temperature and the number of cold drinks sold by a shop.

d) The age of a motorcycle and its second-hand selling price.

e) The number of people living in a house and the number of rooms the house has.

f) The number of goals your opponents score and the number of times you win.

g) A child's height and the child's age.

h) A car's engine size and its fuel consumption.

2.   A newspaper gives daily readings for the number of hours of sunshine and the amount of rainfall in millimetres for several places in the UK. The table below is a summary.

Place	Hours of sunshine	Rainfall (mm)
Aberdeen	9.8	0.3
Aviemore	2.8	0
Belfast	4.1	6.6
Birmingham	6.4	0.8
Bournemouth	9.3	4.3
Bristol	10.2	1.8
Cardiff	10.8	2.8
Folkestone	10.5	0
Hastings	7.2	0.3
Isle of Man	9.3	2.5
Isle of Wight	10.7	0.3
London	10.1	1.5
Manchester	4.6	2.0
Margate	10.9	0
Newcastle	8.2	0.5
Newquay	4.1	2.3
Oxford	10.1	3.0
Scarborough	10.6	1.8
Skegness	12.0	3.3
Southport	8.2	7.4
Torquay	8.8	1.5

a)   Plot a scatter graph of hours of sunshine against amount of rainfall. Use a spreadsheet if possible.

b)   What type of correlation, if any, is there between the two variables? Comment on whether this is what you would expect.

3. The United Nations keeps an up-to-date database of statistical information on its member countries.
The table below shows some of the information available.

Country	Life expectancy at birth (years, 1990–99)		Adult illiteracy rate (%, 1995)	Infant mortality rate (per 1000 births, 1990–99)
	Female	Male		
Australia	81	76	0	6
Barbados	79	74	2.6	12
Brazil	71	63	16.8	42
Chad	49	46	51.9	112
China	72	68	18.5	41
Colombia	74	67	9.6	30
Congo	51	46	25.6	90
Cuba	78	74	4.4	9
Egypt	68	65	48.9	51
France	82	74	0	6
Germany	80	74	0	5
India	63	62	48	72
Iraq	64	61	42	95
Israel	80	76	4.9	8
Japan	83	77	0	4
Kenya	53	51	22.7	66
Mexico	76	70	10.5	31
Nepal	57	58	64.1	83
Portugal	79	72	10	9
Russian Federation	73	61	0.9	18
Saudi Arabia	73	70	27.8	23
United Kingdom	80	75	0	7
United States of America	80	73	0	7

a) By plotting a scatter graph, decide if there is a correlation between the adult illiteracy rate and the infant mortality rate.

b) Are your findings in part a) what you expected? Explain your answer.

c) Without plotting a graph, decide if you think there is likely to be a correlation between male and female life expectancy at birth. Explain your reasons.

d) Plot a scatter graph to test if your predictions for part c) were correct.

Histograms

A histogram displays the frequency of either continuous or grouped discrete data in the form of bars. There are several important features of a histogram which distinguish it from a bar chart.

■ The bars are joined together.
■ The bars can be of varying width.
■ The frequency of the data is represented by the area of the bar and not the height (though in the case of bars of equal width, the area is directly proportional to the height of the bar and so the height is usually used as the measure of frequency).

Worked example

Test marks	Frequency
1–10	0
11–20	0
21–30	1
31–40	2
41–50	5
51–60	8
61–70	7
71–80	6
81–90	2
91–100	1

The table (left) shows the marks out of 100 in a maths test for a class of 32 students. Draw a histogram representing this data.

All the class intervals are the same. As a result the bars of the histogram will all be of equal width, and the frequency can be plotted on the vertical axis. The histogram is shown below.

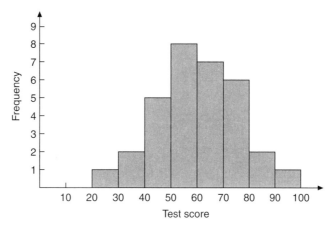

Exercise 21.5

1. The table (right) shows the distances travelled to school by a class of 30 students. Represent this information on a histogram.

Distance (km)	Frequency
$0 \leqslant d < 1$	8
$1 \leqslant d < 2$	5
$2 \leqslant d < 3$	6
$3 \leqslant d < 4$	3
$4 \leqslant d < 5$	4
$5 \leqslant d < 6$	2
$6 \leqslant d < 7$	1
$7 \leqslant d < 8$	1

Height (cm)	Frequency
145–	1
150–	2
155–	4
160–	7
165–	6
170–	3
175–	2
180–185	1

2. The heights of students in a class were measured. The results are shown in the table (left). Draw a histogram to represent this data.

Note that both questions in Exercise 21.5 deal with **continuous data**. In these questions equal class intervals are represented in different ways. However, they mean the same thing. In Q.2, 145– means the students whose heights fall in the range $145 \leqslant h < 150$.

So far the work on histograms has only dealt with problems in which the class intervals are of the same width. This, however, need not be the case.

Worked example   The heights of 25 sunflowers were measured and the results recorded in the table (below left).

If a histogram were drawn with frequency plotted on the vertical axis, then it could look like the one drawn below.

Height (m)	Frequency
$0 \quad \leqslant h < 1.0$	6
$1.0 \quad \leqslant h < 1.5$	3
$1.5 \quad \leqslant h < 2.0$	4
$2.0 \quad \leqslant h < 2.25$	3
$2.25 \leqslant h < 2.50$	5

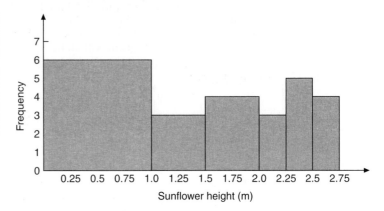

This graph is misleading because it leads people to the conclusion that most of the sunflowers were under 1 m, simply because the area of the bar is so great. In actual fact only approximately one quarter of the sunflowers were under 1 m.

When class intervals are different it is the area of the bar which represents the frequency not the height. Instead of frequency being plotted on the vertical axis, **frequency density** is plotted.

$$\text{Frequency density} = \frac{\text{frequency}}{\text{class width}}$$

The results of the sunflower measurements in the example above can therefore be written as:

Height (m)	Frequency	Frequency density
$0 \quad \leqslant h < 1.0$	6	$6 \div 1 \quad = \ 6$
$1.0 \leqslant h < 1.5$	3	$3 \div 0.5 \ = \ 6$
$1.5 \leqslant h < 2.0$	4	$4 \div 0.5 \ = \ 8$
$2.0 \leqslant h < 2.25$	3	$3 \div 0.25 = 12$
$2.25 \leqslant h < 2.50$	5	$5 \div 0.25 = 20$
$2.50 \leqslant h < 2.75$	4	$4 \div 0.25 = 16$

The histogram can therefore be redrawn as shown below giving a more accurate representation of the data.

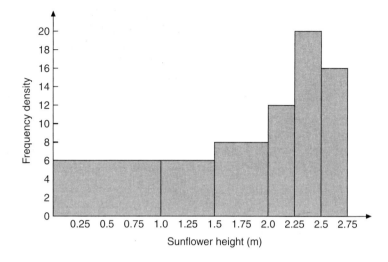

1. The table below shows the time taken, in minutes, by 40
students to travel to school.

Time (min)	0–	10–	15–	20–	25–	30–	40–60
Frequency	6	3	13	7	3	4	4
Frequency density							

   a) Copy the table and complete it by calculating the
      frequency density.
   b) Represent the information on a histogram.

2. On Sundays Maria helps her father feed their chickens.
   Over a period of one year she kept a record of how long it
   took. Her results are shown in the table below.

Time (min)	Frequency	Frequency density
$0 \leqslant t < 30$	8	
$30 \leqslant t < 45$	5	
$45 \leqslant t < 60$	8	
$60 \leqslant t < 75$	9	
$75 \leqslant t < 90$	10	
$90 \leqslant t < 120$	12	

   a) Copy the table and complete it by calculating the
      frequency density. Give the answers correct to
      1 d.p.
   b) Represent the information on a histogram.

3. Frances and Ali did a survey of the ages of the people living in their village. Part of their results are set out in the table below.

Age (years)	0–	1–	5–	10–	20–	40–	60–90
Frequency	35			180	260		150
Frequency density		12	28			14	

a) Copy the table and complete it by calculating either the frequency or the frequency density.
b) Represent the information on a histogram.

4. The table below shows the ages of 150 people, chosen randomly, taking the 6 a.m. train into a city.

Age (years)	0–	15–	20–	25–	30–	40–	50–80
Frequency	3	25	20	30	32	30	10

The histogram below shows the results obtained when the same survey was carried out on the 11 a.m. train.

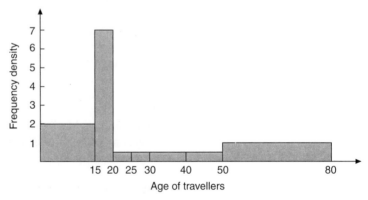

a) Draw a histogram for the 6 a.m. train.
b) Compare the two sets of data and give two possible reasons for the differences.

Cumulative frequency

Calculating the **cumulative frequency** is done by adding up the frequencies as we go along. A cumulative frequency graph is particularly useful when trying to calculate the median of a large set of data, grouped or continuous data, or when trying to establish how consistent a set of results are.

Worked example   The duration of two different brands of battery, A and B, is tested. 50 batteries of each type are randomly selected and tested in the same way. The duration of each battery is then recorded. The results of the tests are shown in the table below.

Type A		
Duration (h)	**Frequency**	**Cumulative frequency**
$0 \leqslant t < 5$	3	3
$5 \leqslant t < 10$	5	8
$10 \leqslant t < 15$	8	16
$15 \leqslant t < 20$	10	26
$20 \leqslant t < 25$	12	38
$25 \leqslant t < 30$	7	45
$30 \leqslant t < 35$	5	50

Type B		
Duration (h)	**Frequency**	**Cumulative frequency**
$0 \leqslant t < 5$	1	1
$5 \leqslant t < 10$	1	2
$10 \leqslant t < 15$	10	12
$15 \leqslant t < 20$	23	35
$20 \leqslant t < 25$	9	44
$25 \leqslant t < 30$	4	48
$30 \leqslant t < 35$	2	50

i)   Plot a cumulative frequency curve for each brand of battery.

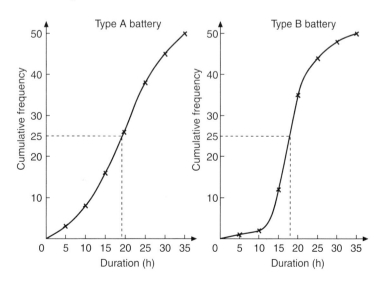

Both cumulative frequency curves are plotted above.
Notice how the points are plotted at the upper boundary of each class interval and *not* at the middle of the interval.

ii) Calculate the median duration for each brand.

The median value is the value which occurs half-way up the cumulative frequency axis. Therefore:

Median for type A batteries ≈ 19 h
Median for type B batteries ≈ 18 h

This tells us that the same number of batteries are still working as have stopped working after 19 h for A and 18 h for B.

Exercise 21.7

1. Sixty athletes enter a cross-country race. Their finishing times are recorded and are shown in the table below:

Finishing time (h)	0–	0.5–	1.0–	1.5–	2.0–	2.5–	3.0–3.5
Frequency	0	0	6	34	16	3	1
Cumulative freq.							

a) Copy the table and calculate the values for the cumulative frequency.
b) Draw a cumulative frequency curve of the results.
c) Show how your graph could be used to find the approximate median finishing time.
d) What does the median value tell us?

2. Three mathematics classes take the same test in preparation for their final exam. Their raw scores are shown in the table below:

Class A	12, 21, 24, 30, 33, 36, 42, 45, 53, 53, 57, 59, 61, 62, 74, 88, 92, 93
Class B	48, 53, 54, 59, 61, 62, 67, 78, 85, 96, 98, 99
Class C	10, 22, 36, 42, 44, 68, 72, 74, 75, 83, 86, 89, 93, 96, 97, 99, 99

a) Using the class intervals $0 \leqslant x < 20$, $20 \leqslant x < 40$ etc. draw up a grouped frequency and cumulative frequency table for each class.
b) Draw a cumulative frequency curve for each class.
c) Show how your graph could be used to find the median score for each class.
d) What does the median value tell us?

3. The table below shows the heights of students in a class over a three-year period.

Height (cm)	Frequency 1996	Frequency 1997	Frequency 1998
150–	6	2	2
155–	8	9	6
160–	11	10	9
165–	4	4	8
170–	1	3	2
175–	0	2	2
180–185	0	0	1

a) Construct a cumulative frequency table for each year.
b) Draw the cumulative frequency curve for each year.
c) Show how your graph could be used to find the median height for each year.
d) What does the median value tell us?

▨ Quartiles and the inter-quartile range

The cumulative frequency axis can also be represented in terms of **percentiles**. A percentile scale divides the cumulative frequency scale into hundredths. The maximum value of cumulative frequency is found at the 100th percentile. Similarly the median, being the middle value, is called the 50th percentile. The 25th percentile is known as the **lower quartile**, and the 75th percentile is called the **upper quartile**.

The **range** of a distribution is found by subtracting the lowest value from the highest value. Sometimes this will give a useful result, but often it will not. A better measure of spread is given by looking at the spread of the middle half of the results, i.e. the difference between the upper and lower quartiles. This result is known as the **inter-quartile range**.

The graph shows the terms mentioned above.

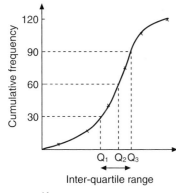

Key:
Q_1 Lower quartile
Q_2 Median
Q_3 Upper quartile

Worked example Consider again the two types of batteries A and B discussed earlier (page 359)

i) Using the graphs, estimate the upper and lower quartiles for each battery.

Lower quartile of type A ≈ 13 h
Upper quartile of type A ≈ 25 h
Lower quartile of type B ≈ 15 h
Upper quartile of type B ≈ 21 h

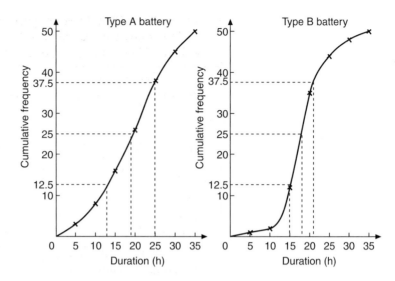

ii)   Calculate the inter-quartile range for each type of battery.

Inter-quartile range of type A ≈ 12 h
Inter-quartile range of type B ≈ 6 h

iii) Based on these results, how might the manufacturers advertise the two types of battery?

Type A: on 'average' the longer-lasting battery
Type B: the more reliable battery

Exercise 21.8   **1.**   Using the results obtained from Q.2 in Exercise 21.7:
a) find the inter-quartile range of each of the classes taking the mathematics test,
b) analyse your results and write a brief summary comparing the three classes.

2.   Using the results obtained from Q.3 in Exercise 21.7:
a) find the inter-quartile range of the students' heights each year,
b) analyse your results and write a brief summary comparing the three years.

3.   Forty boys enter for a school javelin competition. The distances thrown are recorded below:

Distance thrown (m)	0–	20–	40–	60–	80–100
Frequency	4	9	I5	10	2

a) Construct a cumulative frequency table for the above results.
b) Draw a cumulative frequency curve.
c) If the top 20% of boys are considered for the final, estimate (using the graph) the qualifying distance.
d) Calculate the inter-quartile range of the throws.
e) Calculate the median distance thrown.

4. The masses of two different types of oranges are compared. Eighty oranges are randomly selected from each type and weighed. The results are shown below.

Type A		Type B	
Mass (g)	Frequency	Mass (g)	Frequency
75–	4	75–	0
100–	7	100–	16
125–	15	125–	43
150–	32	150–	10
175–	14	175–	7
200–	6	200–	4
225–250	2	225–250	0

a) Construct a cumulative frequency table for each type of orange.
b) Draw a cumulative frequency graph for each type of orange.
c) Calculate the median mass for each type of orange.
d) Using your graphs estimate:
    i) the lower quartile,
    ii) the upper quartile,
    iii) the inter-quartile range
    for each type of orange.
e) Write a brief report comparing the two types of orange.

5. Two competing brands of battery are compared. A hundred batteries of each brand are tested and the duration of each is recorded. The results of the tests are shown in the cumulative frequency graphs below.

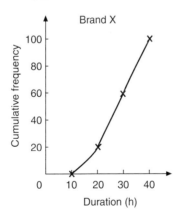
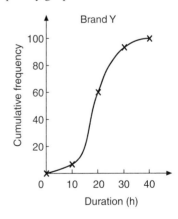

a) The manufacturers of brand X claim that on average their batteries will last at least 40% longer than those of brand Y. Showing your method clearly, decide whether this claim is true.
b) The manufacturers of brand X also claim that their batteries are more reliable than those of brand Y. Is this claim true? Show your working clearly.

Student Assessment 1

1. A rugby team scores the following number of points in 12 matches:

    21, 18, 3, 12, 15, 18, 42, 18, 24, 6, 12, 3

    Calculate for the 12 matches:
    a) the mean score,
    b) the median score,
    c) the mode.

2. The bar chart (left) shows the marks out of 10 for an English test taken by a class of students.
    a) Calculate the number of students who took the test.
    b) Calculate for the class:
       i) the mean test result,
       ii) the median test result,
       iii) the mode test result.

Mass (kg)	Frequency
$15 \leqslant M < 16$	0
$16 \leqslant M < 17$	3
$17 \leqslant M < 18$	6
$18 \leqslant M < 19$	14
$19 \leqslant M < 20$	18
$20 \leqslant M < 21$	8
$21 \leqslant M < 22$	1

3. Fifty sacks of grain are weighed as they are unloaded from a truck. The mass of each is recorded in the grouped frequency table (left).
    a) Calculate the mean mass of the 50 sacks.
    b) State the modal class.

4. Athletico football team keep a record of attendance at their matches over a 36-game season. The attendances for the first 30 games are listed below:

18 418	16 161	15 988	13 417	12 004
11 932	8 461	10 841	19 000	19 214
16 645	14 782	17 935	8 874	19 023
19 875	16 472	14 840	18 450	16 875
13 012	17 858	19 098	6 972	8 452
7 882	11 972	16 461	11 311	19 458

    a) Copy and complete the table below.

Attendance	Tally	Frequency
0–3999		
4000–7999		
8000–11 999		
12 000–15 999		
16 000–19 999		

    b) Illustrate the above table using a frequency chart.
    c) Calculate an estimate for the mean attendance.

    The club then sign on a world class player. The mean attendance for the final six games of the season is calculated as being 28 540.
    d) Calculate the mean attendance for the club over the whole season.

5. Thirty students sit a maths exam. Their marks are given as percentages and are shown in the table below.

Mark	20–	30–	40–	50–	60–	70–	80–	90–100
Frequency	2	3	5	7	6	4	2	1

    a) Construct a cumulative frequency table of the above results.
    b) Draw a cumulative frequency curve of the results.
    c) Using the graph, estimate a value for:
       i) the median,
       ii) the upper and lower quartiles,
       iii) the inter-quartile range.

6. The table below gives the average time taken for 30 pupils in a class to get to school each morning, and the distance they live from the school.

Distance (km)	2	10	18	15	3	4	6	2	25	23	3	5	7	8	2
Time (min)	5	17	32	38	8	14	15	7	31	37	5	18	13	15	8
Distance (km)	19	15	11	9	2	3	4	3	14	14	4	12	12	7	1
Time (min)	27	40	23	30	10	10	8	9	15	23	9	20	27	18	4

    a) Plot a scatter graph of distance travelled against time taken.
    b) Describe the correlation between the two variables.
    c) Explain why some pupils who live further away may get to school more quickly than some of those who live nearer.
    d) Draw a line of best fit on your scatter graph.
    e) A new pupil joins the class. Use your line of best fit to estimate how far away from school she might live if she takes, on average, 19 minutes to get to school each morning.

7. A golf club has four classes of member based on age. The membership numbers for each class are shown below.

	Age	Number	Frequency density
Juniors	0–	32	
Intermediates	16–	80	
Full members	21–	273	
Seniors	60–80	70	

    a) Calculate the frequency density for each class of member.
    b) Illustrate your data on a histogram.

Student Assessment 2

1. A javelin thrower keeps a record of her best throws over ten competitions. These are shown in the table below.

Competition	1	2	3	4	5	6	7	8	9	10
Distance (m)	77	75	78	86	92	93	93	93	92	89

Find the mean, median and mode of her throws.

2. The bar chart shows the marks out of 10 for a maths test taken by a class of students.
    a) Calculate the number of pupils who took the test.
    b) Calculate for the class:
        i) the mean test result,
        ii) the median test result,
        iii) the mode test result.

3. A hundred sacks of coffee with a nominal mass of 10 kg are unloaded from a train. The mass of each sack is checked and the results are presented in the table.
    a) Calculate an estimate for the mean mass.
    b) What is the modal class?

Mass (kg)	Frequency
$9.8 \leqslant M < 9.9$	14
$9.9 \leqslant M < 10.0$	22
$10.0 \leqslant M < 10.1$	36
$10.1 \leqslant M < 10.2$	20
$10.2 \leqslant M < 10.3$	8

4. 400 students sit an IGCSE exam. Their marks (as percentages) are shown in the table below.

Mark (%)	Frequency	Cumulative frequency
31–40	21	
41–50	55	
51–60	125	
61–70	74	
71–80	52	
81–90	45	
91–100	28	

a) Copy and complete the above table by calculating the cumulative frequency.
b) Draw a cumulative frequency curve of the results.
c) Using the graph, estimate a value for:
    i) the median result,
    ii) the upper and lower quartiles,
    iii) the inter-quartile range.

5. Eight hundred students sit an exam. Their marks (as percentages) are shown in the table below.

Mark (%)	Frequency	Cumulative frequency
1–10	10	
11–20	30	
21–30	40	
31–40	50	
41–50	70	
51–60	100	
61–70	240	
71–80	160	
81–90	70	
91–100	30	

a) Copy and complete the above table by calculating the cumulative frequency.
b) Draw a cumulative frequency curve of the results.
c) An 'A' grade is awarded to a student at or above the 80th percentile. What mark is the minimum requirement for an 'A' grade?
d) A 'C' grade is awarded to any student between and including the 55th and the 70th percentile. What marks form the lower and upper boundaries of a 'C' grade?
e) Calculate the inter-quartile range for this exam.

6. A department store decides to investigate whether there is a correlation between the number of pairs of gloves it sells and the outside temperature. Over a one-year period the store records, every two weeks, how many pairs of gloves are sold and the mean daytime temperature during the same period. The results are given in the table below.

Mean temperature (°C)	3	6	8	10	10	11	12	14	16	16	17	18	18
Number of pairs of gloves	61	52	49	54	52	48	44	40	51	39	31	43	35
Mean temperature (°C)	19	19	20	21	22	22	24	25	25	26	26	27	28
Number of pairs of gloves	26	17	36	26	46	40	30	25	11	7	3	2	0

a) Plot a scatter graph of mean temperature against number of pairs of gloves.
b) What type of correlation is there between the two variables?
c) How might this information be useful for the department store in the future?

7. The grouped frequency table below shows the number of points scored by a school basketball player.

Points	0–4	5–9	10–14	15–24	25–34	35–49
Number of games	2	3	8	9	12	3
Frequency density						

a) Copy and complete the table by calculating the frequency densities. Give your answers to 1 d.p.

b) Draw a histogram to illustrate the data.

ICT Section

In this activity you will be collecting the height data of all the pupils in your class and plotting a cumulative frequency graph of the results.

1. Measure the heights of all the pupils in your class.

2. Arrange the data into appropriate sized groups.

3. Enter your data into graphing software such as Autograph.

4. Produce a cumulative frequency graph of the results.

5. From your graph deduce:
   a) the median height of the class,
   b) the inter-quartile range of the heights.

6. Compare the cumulative frequency graph from your class with one produced from data collected from another class in a different year group. Comment on any differences/ similarities between the two.

22 *PROBABILITY*

Calculate the probability of simple combined events, using possibility diagrams and tree diagrams where appropriate (in possibility diagrams outcomes will be represented by points on a grid and in tree diagrams outcomes will be written at the end of branches and probabilities by the side of branches).

Probability is the study of chance, or the likelihood of an event happening. In this chapter we will be looking at theoretical probability. But, because probability is based on chance, what theory predicts does not necessarily happen in practice.

A favourable outcome refers to the event in question actually happening. The total number of possible outcomes refers to all the different types of outcome one can get in a particular situation. In general:

Probability of an event =

$$\frac{\text{number of favourable outcomes}}{\text{total number of equally likely outcomes}}$$

Therefore

if the probability = 0 it implies the event is impossible
if the probability = 1 it implies the event is certain to happen

Worked examples   **a)**   An ordinary, fair dice is rolled. Calculate the probability of getting a 6.

Number of favourable outcomes = 1   (i.e. getting a 6)
Total number of possible outcomes = 6
(i.e. getting a 1, 2, 3, 4, 5 or 6)
Probability of getting a six $= \frac{1}{6}$

b)   An ordinary, fair dice is rolled. Calculate the probability of getting an even number.

Number of favourable outcomes = 3
(i.e. getting a 2, 4 or 6)
Total number of possible outcomes = 6
(i.e. getting a 1, 2, 3, 4, 5 or 6)
Probability of getting an even number $= \frac{3}{6} = \frac{1}{2}$

Exercise 22.1   **1.**   Calculate the theoretical probability, when rolling an ordinary, fair dice, of getting each of the following:
a) a score of 1,   b) a score of 5,
c) an odd number,   d) a score less than 6,
e) a score of 7,   f) a score less than 7.

2.   a) Calculate the probability of:
i) being born on a Wednesday,
ii) not being born on a Wednesday.
b) Explain the result of adding the answers to a) i) and ii) together.

3. 250 tickets are sold for a raffle. What is the probability of winning if you buy:
   a) 1 ticket,   b) 5 tickets,
   c) 250 tickets,   d) 0 tickets?

4. In a class there are 25 girls and 15 boys. The teacher takes in all of their books in a random order. Calculate the probability that the teacher will
   a) mark a book belonging to a girl first,
   b) mark a book belonging to a boy first.

5. Tiles, each lettered with one different letter of the alphabet, are put into a bag. If one file is drawn out at random, calculate the probability that it is:
   a) an A or P,   b) a vowel,
   c) a consonant,   d) an X, Y or Z,
   e) a letter in your first name.

6. A boy was late for school 5 times in the previous 30 school days. If tomorrow is a school day, calculate the probability that he will arrive late.

7. a) Three red, 10 white, 5 blue and 2 green counters are put into a bag. If one is picked at random calculate the probability that it is:
      i) a green counter,   ii) a blue counter.
   b) If the first counter taken out is green and it is not put back into the bag, calculate the probability that the second counter picked is:
      i) a green counter,   ii) a red counter.

8. A roulette wheel has the numbers 0 to 36 equally spaced around its edge. Assuming that it is unbiased, calculate the probability on spinning it of getting:
   a) the number 5,   b) an even number,
   c) an odd number,   d) zero,
   e) a number greater than 15, f) a multiple of 3,
   g) a multiple of 3 or 5,   h) a prime number.

9. The letters R, C and A can be combined in several different ways.
   a) Write the letters in as many different combinations as possible.
   If a computer writes these three letters at random, calculate the probability that:
   b) the letters will be written in alphabetical order,
   c) that the letter R is written before both the letters A and C,
   d) that the letter C is written after the letter A,
   e) the computer will spell the word CART if the letter T is added.

10. A normal pack of playing cards contains 52 cards. These are made up of four suits (hearts, diamonds, clubs and spades). Each suit consists of 13 cards. These are labelled Ace, 2, 3, 4, 5, 6, 7, 8, 9, 10, Jack, Queen and King. The hearts and diamonds are red; the clubs and spades are black.

If a card is picked at random from a normal pack of cards calculate the probability of picking:

a) a heart,
b) a black card,
c) a four,
d) a red King,
e) a Jack, Queen or King,
f) the ace of spades,
g) an even numbered card,
h) a seven or a club.

Combined events

Combined events look at the probability of two or more events.

Worked example

i) Two coins are tossed. Show in a two-way table all the possible outcomes.

ii) Calculate the probability of getting two heads.

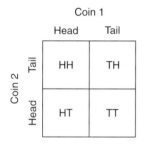

All four outcomes are equally likely: therefore, the probability of getting HH is $\frac{1}{4}$.

iii) Calculate the probability of getting a head and a tail in any order.

The probability of getting a head and a tail in any order, i.e. HT or TH, is $\frac{2}{4}=\frac{1}{2}$.

Exercise 22.2

1. a) Two fair tetrahedral dice are rolled. If each is numbered 1–4, draw a two-way table to show all the possible outcomes.

b) What is the probability that both dice show the same number?

c) What is the probability that the number on one dice is double the number on the other?

d) What is the probability that the sum of both numbers is prime?

2. Two fair dice are rolled. Copy and complete the diagram (left) to show all the possible combinations.

What is the probability of getting:

a) a double 3,
b) any double,
c) a total score of 11,
d) a total score of 7,
e) an even number on both dice,
f) an even number on at least one dice,
g) a 6 or a double,
h) scores which differ by 3,
i) a total which is either a multiple of 2 or 5?

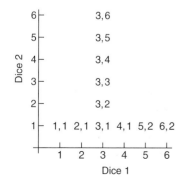

▨ Tree diagrams

When more than two combined events are being considered then two-way tables cannot be used and therefore another method of representing information diagrammatically is needed. Tree diagrams are a good way of doing this.

Worked example
i) If a coin is tossed three times, show all the possible outcomes on a tree diagram, writing each of the probabilities at the side of the branches.

ii) What is the probability of getting three heads?

There are eight equally likely outcomes, therefore the probability of getting HHH is $\frac{1}{8}$.

iii) What is the probability of getting two heads and one tail in any order?

The successful outcomes are HHT, HTH, THH. Therefore the probability is $\frac{3}{8}$.

iv) What is the probability of getting at least one head?

This refers to any outcome with either one, two or three heads, i.e. all of them *except* TTT. Therefore the probability is $\frac{7}{8}$.

v) What is the probability of getting no heads?

The only successful outcome for this event is TTT. Therefore the probability is $\frac{1}{8}$.

Exercise 22.3
1. a) A computer uses the numbers 1, 2 or 3 at random to make three-digit numbers. Assuming that a number can be repeated, show on a tree diagram all the possible combinations that the computer can print.

b) Calculate the probability of getting:
   i) the number 131,　ii) an even number,
   iii) a multiple of 11,　iv) a multiple of 3,
   v) a multiple of 2 or 3,　vi) a palindromic number.

2. a) A family has four children. Draw a tree diagram to show all the possible combinations of boys and girls. [assume P (girl) = P (boy)]
   b) Calculate the probability of getting:
      i) all girls,   ii) two girls and two boys,
      iii) at least one girl,   iv) more girls than boys.

3. a) A netball team plays three matches. In each match the team is equally likely to win, lose or draw. Draw a tree diagram to show all the possible outcomes over the three matches.
   b) Calculate the probability that the team:
      i) wins all three matches,
      ii) wins more times than loses,
      iii) loses at least one match,
      iv) either draws or loses all three matches.
   c) Explain why it is not very realistic to assume that the outcomes are equally likely in this case.

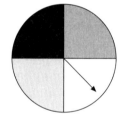

4. A spinner is split into quarters.
   a) If it is spun twice, draw a probability tree showing all the possible outcomes.
   b) Calculate the probability of getting:
      i) two dark greys,
      ii) two greys of either shade,
      iii) a black and a white in any order.

In each of the cases considered so far, all of the outcomes have been assumed to be equally likely. However, this need not be the case.

Worked example   In winter, the probability that it rains on any one day is $\frac{5}{7}$.

i)   Using a tree diagram show all the possible combinations for two consecutive days.
ii) Write each of the probabilities by the sides of the branches.

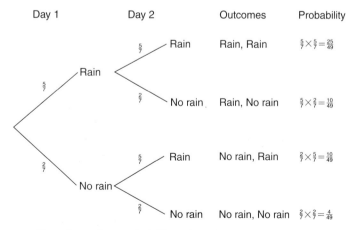

Note how the probability of each outcome is arrived at by multiplying the probabilities of the branches.

iii) Calculate the probability that it will rain on both days.

$$P(R, R) = \tfrac{5}{7} \times \tfrac{5}{7} = \tfrac{25}{49}$$

iv) Calculate the probability that it will rain on the first but not the second day.

$$P(R, NR) = \tfrac{5}{7} \times \tfrac{2}{7} = \tfrac{10}{49}$$

v) Calculate the probability that it will rain on at least one day.

The outcomes which satisfy this event are (R, R) (R, NR) and (NR, R).
Therefore the probability is $\tfrac{25}{49} + \tfrac{10}{49} + \tfrac{10}{49} = \tfrac{45}{49}$.

Exercise 22.4

1. A particular board game involves players rolling a dice. However, before a player can start, he or she needs to roll a 6.
   a) Copy and complete the tree diagram below showing all the possible combinations for the first two rolls of the dice.

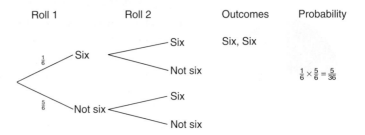

   b) Calculate the probability of the following:
      i) getting a six on the first throw,
      ii) starting within the first two throws,
      iii) starting on the second throw,
      iv) not starting within the first three throws,
      v) starting within the first three throws.
   c) If you add the answers to b) iv) and v) what do you notice? Explain.

2. In Italy $\tfrac{3}{5}$ of the cars are foreign made. By drawing a tree diagram and writing the probabilities next to each of the branches, calculate the following probabilities:
   a) the next two cars to pass a particular spot are both Italian,
   b) two of the next three cars are foreign,
   c) at least one of the next three cars is Italian.

3. The probability that a morning bus arrives on time is 65%.
   a) Draw a tree diagram showing all the possible outcomes for three consecutive mornings.
   b) Label your tree diagram and use it to calculate the probability that:
      i) the bus is on time on all three mornings,
      ii) the bus is late the first two mornings,
      iii) the bus is on time two out of the three mornings,
      iv) the bus is on time at least twice.

4. A normal pack of 52 cards is shuffled and three cards are picked at random. Draw a tree diagram to help calculate the probability of picking:
   a) two clubs first,
   b) three clubs,
   c) no clubs,
   d) at least one club.

5. A bowl of fruit contains one apple, one banana, two oranges and two pears. Two pieces of fruit are chosen at random and eaten.
   a) Draw a probability tree showing all the possible combinations of the two pieces of fruit.
   b) Use your tree diagram to calculate the probability that:
      i) both the pieces of fruit eaten are oranges,
      ii) an apple and a banana are eaten,
      iii) at least one pear is eaten.

6. Light bulbs are packaged in cartons of three. 10% of the bulbs are found to be faulty. Calculate the probability of finding two faulty bulbs in a single carton.

7. A volleyball team has a 0.25 chance of losing a game. Calculate the probability of the team achieving:
   a) two consecutive wins,
   b) three consecutive wins,
   c) 10 consecutive wins.

Student Assessment I

1. Calculate the theoretical probability of:
   a) being born on a Saturday,
   b) being born on the 5th of a month in a non-leap year,
   c) being born on the 20th June in a non-leap year,
   d) being born on the 29th February.

2. When throwing an ordinary, fair dice, calculate the theoretical probability of getting:
   a) a 2,   b) an even number,
   c) a 3 or more,   d) less than 1.

3. A bag contains 12 white counters, 7 black counters and 1 red counter.
   a) If, when a counter is taken out, it is not replaced, calculate the probability that:
      i) the first counter is white,
      ii) the second counter removed is red, given that the first was black.
   b) If, when a counter is picked, it is then put back in the bag, how many attempts will be needed before it is mathematically certain that a red counter will have been picked out?

4. A coin is tossed and an ordinary, fair dice is rolled.
   a) Draw a two-way table showing all the possible combinations.
   b) Calculate the probability of getting:
      i) a head and a six,
      ii) a tail and an odd number,
      iii) a head and a prime number.

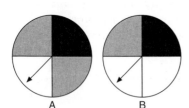

A   B

5. Two spinners A and B are split into quarters and coloured as shown. Both spinners are spun.
   a) Draw a fully labelled tree diagram showing all the possible combinations on the two spinners. Write beside each branch the probability of each outcome.
   b) Use your tree diagram to calculate the probability of getting:
      i) two blacks,
      ii) two greys,
      iii) a grey on spinner A and a white on spinner B.

6. A coin is tossed three times.
   a) Draw a tree diagram to show all the possible outcomes.
   b) Use your tree diagram to calculate the probability of getting:
      i) three tails,
      ii) two heads,
      iii) no tails,
      iv) at least one tail.

7. A goalkeeper expects to save one penalty out of every three. Calculate the probability that he:
a) saves one penalty out of the next three,
b) fails to save any of the next three penalties,
c) saves two out of the next three penalties.

8. A board game uses a fair dice in the shape of a tetrahedron. The sides of the dice are numbered 1, 2, 3 and 4. Calculate the probability of:
a) not throwing a 4 in two throws,
b) throwing two consecutive 1s,
c) throwing a total of 5 in two throws.

9. A normal pack of 52 cards is shuffled and three cards picked at random. Calculate the probability that all three cards are picture cards.

Student Assessment 2

1. If a card is picked from a complete pack of 52 playing cards, calculate the probability of getting:
a) a nine,   b) a heart,
c) the seven of clubs,   d) a black Jack, Queen or King.

2. Two normal and fair dice are rolled and their scores added together.
a) Using a two-way table, show all the possible scores that can be achieved.
b) Using your two-way table, calculate the probability of getting:
i) a score of 12,   ii) a score of 7,
iii) a score less than 4,   iv) a score of 7 or more.
c) Two dice are rolled 180 times. In theory, how many times would you expect to get a total score of 6?

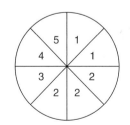

3. A spinner is numbered as shown.
a) If it is spun once, calculate the probability of getting:
i) a 1,
ii) a 2.
b) If it is spun twice, calculate the probability of getting:
i) a 2 followed by a 4,
ii) a 2 and a 4 in any order,
iii) at least one 1,
iv) at least one 2.

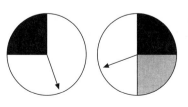

4. Two spinners are coloured as shown.
a) They are both spun. Draw and label a tree diagram showing all the possible outcomes.
b) Using your tree diagram calculate the probability of getting:
i) two blacks,   ii) two whites,
iii) a white and a grey,   iv) at least one white.

5. Two spinners are labelled as shown:

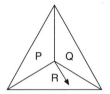

Calculate the probability of getting:
a) A and P,
b) A or B and R,
c) C but not Q.

6. A vending machine accepts €1 and €2 coins. The probability of a €2 coin being rejected is 0.2. The probability of a €1 coin being rejected is 0.1.
A sandwich costing €3 is bought. Calculate the probability of getting a sandwich first time if:
a) one of each coin is used,
b) three €1 coins are used.

7. A biased coin is tossed three times. On each occasion the probability of getting a head is 0.6.
a) Draw a tree diagram to show all the possible outcomes after three tosses. Label each branch clearly with the probability of each outcome.
b) Using your tree diagram calculate the probability of getting:
i) three heads,
ii) three tails,
iii) at least two heads.

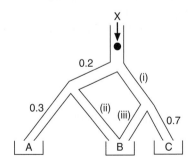

8. A ball enters a chute at X.
a) What are the probabilities of the ball going down each of the chutes labelled (i), (ii) and (iii)?
b) Calculate the probability of the ball landing in:
i) tray A,
ii) tray C,
iii) tray B.

ICT Section

1. Investigate how to generate random numbers in Excel.

2. Adapt your findings in Q.1 to generate random integers between 1 and 6.

3. Simulate the rolling of a dice 100 times by getting your Excel spreadsheet to randomly generate 100 integers between 1 and 6.

4. Analyse the results and, giving reasons, explain whether you think the spreadsheet has generated the numbers randomly or whether it is biased.

SOLUTIONS

(i) Ordering *page 1*

Exercise A

1. **a)** $<$ **b)** $=$ **c)** $>$ **d)** $<$ **e)** $=$
   f) $>$

2. **a)**
   0 1 2 3

   b)
   2 3 4 5

   c)
   −6 −5 −4 −3

   d)
   −3 −2 −1 0

   e)
   1 2 3 4 5 6

   f)
   −4 −3 −2 −1 0 1

   g)
   −2 −1 0 1 2 3

   h)
   −2 −1 0 1 2 3

3. **a)** $x > 0$ **b)** $x \leqslant 3$ **c)** $0 \leqslant x \leqslant 4$
   d) $-4 < x \leqslant -1$

4. **a)** $x \leqslant 20\,000$ **b)** $135 \leqslant x \leqslant 180$
   c) $5x + 3 < 20$ **d)** $x \leqslant 25$
   e) $350 \leqslant x \leqslant 400$ **f)** $11 \leqslant x \leqslant 28$

Exercise B

1. 0.06 0.6 0.606 0.66 6.0 6.6 6.606

2. $\frac{4}{5}$ $\frac{1}{2}$ $\frac{6}{13}$ $\frac{7}{18}$ $\frac{1}{3}$ $\frac{2}{19}$

3. 60 cm 0.75 m 800 mm 180 cm 2 m

4. 4 kg 3500 g 1 kg $\frac{3}{4}$ kg 700 g

5. 150 cm³ 430 ml 800 cm³ 1 litre 120 cl

Student Assessment 1

1. **a)** $=$ **b)** $>$ **c)** $<$ **d)** $<$

2. **a)**
   28 29 30 31

   b)
   28 29 30 31

   c)
   $1\frac{1}{2}$ $1\frac{7}{12}$ $1\frac{2}{3}$ $1\frac{3}{4}$

   d)
   2.40 2.41 2.42 2.43 2.44 2.45

3. **a)** $x \leqslant 52$ **b)** $x \geqslant 11$ **c)** $24 \leqslant x \leqslant 38$
   d) $3x + 6 < 50$ **e)** $x \geqslant 0.03$

4. **a)**
   4 5 6 7 8

   b)
   −6 −5 −4 −3 −2

   c)
   −4 −3 −2 −1 0

   d)
   −3 −2 −1 0 1

5. $\frac{7}{6}$ $\frac{13}{18}$ $\frac{2}{3}$ $\frac{7}{12}$ $\frac{1}{6}$

Student Assessment 2

1. **a)** $=$ **b)** $<$ **c)** $>$ **d)** $>$

2. **a)**
   32 33 34 35 36

   b)
   20 21 22 23 24 25

   c)
   9.7 9.8 9.9 10

   d)
   15 16 17 18

3. **a)** $x \geqslant -1$ **b)** $x < 2$ **c)** $-2 \leqslant x < 2$
   d) $-1 \leqslant x \leqslant 1$

4. **a)**
   2 3 4 5

   b)
   1 2 3 4 5

   c)
   0 1 2 3 4

   d)
   −3 −2 −1 0 1

5. $\frac{3}{14}$ $\frac{2}{5}$ $\frac{1}{2}$ $\frac{4}{7}$ $\frac{9}{10}$

(ii) Standard Form *page 6*

Exercise A

1. d and e

2. **a)** 6×10^5 **b)** 4.8×10^7 **c)** 7.84×10^{11}
   d) 5.34×10^5 **e)** 7×10^6 **f)** 8.5×10^6

3. **a)** 6.8×10^6 **b)** 7.2×10^8 **c)** 8×10^5
   d) 7.5×10^7 **e)** 4×10^9 **f)** 5×10^7

4. **a)** 6×10^5 **b)** 2.4×10^7 **c)** 1.4×10^8
   d) 3×10^9 **e)** 1.2×10^{13} **f)** 1.8×10^7

5. 1.44×10^{11} m

6. **a)** 8.8×10^8 **b)** 2.04×10^{11}
c) 3.32×10^{11} **d)** 4.2×10^{22}
e) 5.1×10^{22} **f)** 2.5×10^{25}

7. **a)** 2×10^2 **b)** 3×10^5 **c)** 4×10^6
d) 2×10^4 **e)** 2.5×10^6 **f)** 4×10^4

8. **a)** 4.26×10^5 **b)** 8.48×10^9 **c)** 6.388×10^7
d) 3.157×10^9 **e)** 4.5×10^8 **f)** 6.01×10^7
g) 8.15×10^{10} **h)** 3.56×10^7

9. Mercury 5.8×10^7 km
Venus 1.08×10^8 km
Earth 1.5×10^8 km
Mars 2.28×10^8 km
Jupiter 7.78×10^8 km
Saturn 1.43×10^9 km
Uranus 2.87×10^9 km
Pluto 5.92×10^9 km

Exercise B

1. **a)** 6×10^{-4} **b)** 5.3×10^{-5}
c) 8.64×10^{-4} **d)** 8.8×10^{-8}
e) 7×10^{-7} **f)** 4.145×10^{-4}

2. **a)** 6.8×10^{-4} **b)** 7.5×10^{-7}
c) 4.2×10^{-10} **d)** 8×10^{-9}
e) 5.7×10^{-11} **f)** 4×10^{-11}

3. **a)** -4 **b)** -3 **c)** -8 **d)** -5
e) -7 **f)** 3

4. 6.8×10^5 6.2×10^3 8.414×10^2
6.741×10^{-4} 3.2×10^{-4} 5.8×10^{-7}
5.57×10^{-9}

Student Assessment I

1. **a)** 8×10^6 **b)** 7.2×10^{-4} **c)** 7.5×10^{10}
d) 4×10^{-4} **e)** 4.75×10^9 **f)** 6.4×10^{-7}

2. 7.41×10^{-9} 3.6×10^{-5} 5.5×10^{-3}
4.21×10^7 6.2×10^7 4.9×10^8

3. **a)** 6×10^6 8.2×10^5 4.4×10^{-3} 8×10^{-1}
5.2×10^4
b) 6×10^6 8.2×10^5 5.2×10^4 8×10^{-1}
4.4×10^{-3}

4. **a)** 2 **b)** 8 **c)** -4 **d)** 5 **e)** -5
f) -5

5. **a)** 1.2×10^8 **b)** 5.6×10^8 **c)** 2×10^5
d) 2.5×10^5

6. 6 minutes

7. 4.73×10^{15} km correct to three significant
figures (3 s.f.)

Student Assessment 2

1. **a)** 6×10^6 **b)** 4.5×10^{-3} **c)** 3.8×10^9
d) 3.61×10^{-7} **e)** 4.6×10^8 **f)** 3×10^0

2. 4.05×10^8 3.6×10^2 9×10^1 1.5×10^{-2}
7.2×10^{-3} 2.1×10^{-3}

3. **a)** 1.5×10^7 4.3×10^5 4.35×10^{-4}
4.8×10^0 8.5×10^{-3}
b) 4.35×10^{-4} 8.5×10^{-3} 4.8×10^0
4.3×10^5 1.5×10^7

4. **a)** 3 **b)** 9 **c)** -3 **d)** 6 **e)** -1
f) 8

5. **a)** 1.2×10^8 **b)** 1.48×10^{11} **c)** 6.73×10^7
d) 3.88×10^6

6. 43 minutes (2 s.f.)

7. 2.84×10^{15} km (3 s.f.)

(iii) The Four Rules *page 12*

Exercise A

1. **a)** 9 **b)** 22 **c)** 30 **d)** 2 **e)** 7
f) 3

2. **a)** $6 \times (4 + 6) \div 3 = 20$
b) $6 \times (4 + 6 \div 3) = 36$
c) $8 + 2 \times (4 - 2) = 12$
d) $(8 + 2) \times (4 - 2) = 20$
e) $(9 - 3) \times 7 + 2 = 44$
f) $(9 - 3) \times (7 + 2) = 54$

3. **a)** 1512 **b)** $33\,984$ **c)** $29\,830$
d) $41\,492$ **e)** $20\,736$ **f)** $40\,800$

4. **a)** 5 **b)** 3 **c)** 2 **d)** 7 **e)** 1 **f)** 7

5. **a)** $7, 9$ **b)** 100 **c)** $11, 16$ **d)** 6
e) 224 **f)** 28

6. **a)** 1127.4 **b)** 7603.8 **c)** 1181.5
d) 1141.0 **e)** 526.1 **f)** 3328.2

Exercise B

1. **a)** $\frac{1}{4} = \frac{2}{8} = \frac{4}{16} = \frac{16}{64} = \frac{3}{12}$
b) $\frac{2}{5} = \frac{4}{10} = \frac{8}{20} = \frac{20}{50} = \frac{16}{40}$
c) $\frac{3}{8} = \frac{6}{16} = \frac{9}{24} = \frac{15}{40} = \frac{7}{27}$
d) $\frac{4}{7} = \frac{8}{14} = \frac{12}{21} = \frac{32}{56} = \frac{36}{63}$
e) $\frac{5}{9} = \frac{16}{27} = \frac{20}{36} = \frac{50}{90} = \frac{66}{99}$

2. **a)** $\frac{1}{2}$ **b)** $\frac{1}{3}$ **c)** $\frac{2}{3}$ **d)** $\frac{4}{9}$ **e)** $\frac{3}{4}$ **f)** $\frac{9}{10}$

3. **a)** $4\frac{1}{4}$ **b)** $4\frac{3}{5}$ **c)** $2\frac{2}{3}$ **d)** $6\frac{1}{3}$ **e)** 4
f) $3\frac{7}{12}$

4. **a)** $\frac{13}{2}$ **b)** $\frac{29}{4}$ **c)** $\frac{27}{8}$ **d)** $\frac{100}{9}$ **e)** $\frac{34}{5}$ **f)** $\frac{97}{11}$

Exercise C

1. **a)** $1\frac{2}{5}$ **b)** $\frac{10}{11}$ **c)** $\frac{11}{12}$ **d)** $1\frac{2}{45}$ **e)** $1\frac{1}{65}$
f) $1\frac{11}{12}$

2. a) $\frac{1}{7}$ b) $\frac{1}{10}$ c) $\frac{5}{9}$ d) $\frac{1}{12}$ e) $\frac{9}{40}$ f) $1\frac{1}{20}$

3. a) $5\frac{3}{4}$ b) $5\frac{3}{10}$ c) $3\frac{1}{10}$ d) $6\frac{7}{24}$ e) $1\frac{1}{8}$
   f) $\frac{25}{26}$

Exercise D

1. a) $\frac{4}{3}$ b) $\frac{9}{5}$ c) $\frac{1}{7}$ d) 9 e) $\frac{4}{11}$ f) $\frac{8}{37}$

2. a) $\frac{3}{5}$ b) $\frac{7}{12}$ c) $\frac{9}{70}$ d) $\frac{21}{25}$ e) $\frac{3}{8}$ f) $1\frac{25}{56}$

3. a) $\frac{3}{5}$ b) $3\frac{107}{120}$ c) $\frac{8}{15}$ d) $12\frac{1}{4}$

Exercise E

1. a) 0.75 b) 0.8 c) 0.45 d) 0.34 e) $0.\dot{3}$ f) 0.375 g) 0.4375 h) $0.\dot{2}$ i) $0.6\dot{3}$

2. a) 2.75 b) 3.6 c) 4.35 d) 6.22 e) $5.\dot{6}$ f) 6.875 g) 5.5625 h) $4.\dot{2}$ i) $5.\dot{4}2857\dot{1}$

Exercise F

1. a) $\frac{1}{2}$ b) $\frac{7}{10}$ c) $\frac{3}{5}$ d) $\frac{3}{4}$ e) $\frac{33}{40}$ f) $\frac{1}{20}$ g) $\frac{1}{20}$ h) $\frac{201}{500}$ i) $\frac{1}{5000}$

2. a) $2\frac{2}{5}$ b) $6\frac{1}{2}$ c) $8\frac{1}{5}$ d) $3\frac{3}{4}$ e) $10\frac{11}{20}$ f) $9\frac{51}{250}$ g) $15\frac{91}{200}$ h) $30\frac{1}{1000}$ i) $1\frac{41}{2000}$

Student Assessment 1

1. a) 23 b) 18

2. 6, 15

3. 9000

4. 22 977

5. 360.2

6. $\frac{8}{18} = \frac{4}{9} = \frac{16}{36} = \frac{56}{126} = \frac{40}{90}$

7. a) $2\frac{1}{16}$ b) 9

8. a) 0.4 b) 1.75 c) $0.\dot{8}\dot{1}$ d) $1.\dot{6}$

9. a) $4\frac{1}{5}$ b) $\frac{3}{50}$ c) $1\frac{17}{20}$ d) $2\frac{1}{200}$

Student Assessment 2

1. a) 0 b) 19

2. 6, 12

3. 294

4. 18 032

5. 340.7

6. $\frac{24}{36} = \frac{8}{12} = \frac{4}{6} = \frac{20}{30} = \frac{60}{90}$

7. a) $1\frac{7}{10}$ b) 2

8. a) 0.875 b) 1.4 c) $0.\dot{8}$ d) $3.\dot{2}8571\dot{4}$

9. a) $6\frac{1}{2}$ b) $\frac{1}{25}$ c) $3\frac{13}{20}$ d) $3\frac{1}{125}$

(iv) Estimation page 18

Exercise A

1. a) 69 000 b) 74 000 c) 89 000 d) 4000 e) 100 000 f) 1 000 000

2. a) 78 500 b) 6900 c) 14 100 d) 8100 e) 1000 f) 3000

3. a) 490 b) 690 c) 8850 d) 80 e) 0 f) 1000

Exercise B

1. a) 5.6 b) 0.7 c) 11.9 d) 157.4 e) 4.0 f) 15.0 g) 3.0 h) 1.0 i) 12.0

2. a) 6.47 b) 9.59 c) 16.48 d) 0.09 e) 0.01 f) 9.30 g) 100.00 h) 0.00 i) 3.00

Exercise C

1. a) 50 000 b) 48 600 c) 7000 d) 7500 e) 500 f) 2.57 g) 1000 h) 2000 i) 15.0

2. a) 0.09 b) 0.6 c) 0.94 d) 1 e) 0.95 f) 0.003 g) 0.0031 h) 0.0097 i) 0.01

Exercise D

1. a) 419.6 b) 5.0 c) 166.3 d) 23.8 e) 57.8 f) 4427.10 g) 1.9 h) 4.1 i) 0.6

   Answers to Q.2–4 may vary slightly from those given below:

2. a) 1200 b) 3000 c) 3000 d) 150 000 e) 0.8 f) 100

3. a) 200 b) 200 c) 30 d) 550 e) 500 f) 3000

4. a) 130 b) 80 c) 9 d) 4 e) 200 f) 250

5. c) and e) are incorrect.

Answers to Q.6 and 7 may vary slightly from those given below:

6. **a)** 120 m² **b)** 40 m² **c)** 400 cm²

7. **a)** 200 cm³ **b)** 4000 cm³ **c)** 2000 cm³

Student Assessment 1

1. **a)** 2800 **b)** 7290 **c)** 49 000 **d)** 1000

2. **a)** 3.8 **b)** 6.8 **c)** 0.85 **d)** 1.58
   e) 10.0 **f)** 0.008

3. **a)** 4 **b)** 6.8 **c)** 0.8 **d)** 10 **e)** 830
   f) 0.005

Answers to Q.4–6 may vary from those given below:

4. 18 000 yds

5. 40 m²

6. **a)** 25 **b)** 4 **c)** 4

7. 92.3 cm³ (1 d.p.)

Student Assessment 2

1. **a)** 6470 **b)** 88 500 **c)** 65 000 **d)** 10

2. **a)** 6.8 **b)** 4.44 **c)** 8.0 **d)** 63.08
   e) 0.057 **f)** 3.95

3. **a)** 40 **b)** 5.4 **c)** 0.06 **d)** 49 000
   e) 700 000 **f)** 687 000

Answers to Q.4–6 may vary from those given below:

4. 34 000 yds

5. 150 m²

6. **a)** 20 **b)** 1 **c)** 3

7. 30.1 m³ (1 d.p.)

1 Number, Set Notation and Language

Core Section page 24

Exercise 1.1

1. **a)** Rational **b)** Rational **c)** Irrational
   d) Rational **e)** Rational **f)** Rational
   g) Irrational **h)** Rational **i)** Rational

2. **a)** Irrational **b)** Irrational **c)** Rational
   d) Rational **e)** Rational **f)** Rational

d) i) 216, 343 ii) n^3
e) i) 217, 344 ii) $n^3 + 1$
f) i) 226, 353 ii) $n^3 + 10$
g) i) 213, 340 ii) $n^3 - 3$
h) i) 56, 72 ii) $n^2 +$

Student Ass
1. **a)** Ratio **d)** Ra
2. **a)**

j) 1) Sequence of powers of 5 ii) 5^{10} or 9 765 625

Exercise 1.3

a) 53, 71 **b)** 67, 131 **c)** 39, 63

d) 173, 275 **e)** 170, 357 **f)** 127, 221

g) 27, 29

Exercise 1.4

a) i) 20, 23 ii) $3n + 2$

b) i) 25, 29 ii) $4n + 1$

c) i) 29, 34 ii) $5n - 1$

d) i) 18, 20 ii) $2n + 6$

e) i) 36, 43 ii) $7n - 6$

f) i) 24, 28 ii) $4n - 4$

g) i) 46, 55 ii) $9n - 8$

h) i) 65, 75 ii) $10n + 5$

i) i) 64, 75 ii) $11n - 2$

j) i) 13.5, 15.5 ii) $2n - 0.5$

k) i) 5.25, 6.25 ii) $n - 0.75$

l) i) 6, 7 ii) $n - 1$

Exercise 1.5

a) i) 50, 65 ii) $n^2 + 1$

b) i) 43, 56 ii) $n^2 + 7$

c) i) 48, 63 ii) $n^2 - 1$

...essment 1

...nal   **b)** Irrational   **c)** Rational
...tional   **e)** Rational   **f)** Irrational

   a) i) $45, 54$ ii) Terms increasing by 9
   b) i) $30, 24$ ii) Terms decreasing by 6
   c) i) $2.25, 1.125$ ii) Terms halving
   d) i) $-12, -18$ ii) Terms decreasing by 6
   e) i) $27, 8$ ii) Descending order of cube numbers
   f) i) $81, 243$ ii) Terms multiplied by 3

3.   **a)** $4n + 2$ **b)** $6n + 7$ **c)** $6n - 3$
    d) $n^2 + 3$ **e)** $10n - 10$ **f)** $n^3 - 1$

Student Assessment 2

1.   **a)** $\frac{5}{8}$ **b)** 3 **c)** $\frac{11}{25}$

2.   **a)** i) $30, 36$ ii) Terms increasing by 6
    b) i) $12, 9$ ii) Terms decreasing by 3
    c) i) $-5, -10$ ii) Terms decreasing by 5
    d) i) $64, 81$ ii) Ascending order of square numbers
    e) i) $1000, 10\,000$ ii) Ascending order of powers of 10
    f) i) $\frac{1}{16}, \frac{1}{32}$ ii) Terms halving

3.   **a)** $2n + 1$ **b)** $6n + 1$ **c)** $10n - 2$
    d) $8n - 7$ **e)** $8n - 12$ **f)** $n^2 + 1$

Extended Section   page 30

Exercise 1.6

1.   **a)** i) Continents of the world
      ii) Student's own answers

    b) i) Even numbers
      ii) Student's own answers

    c) i) Days of the week
      ii) Student's own answers

    d) i) Months with 31 days
      ii) Student's own answers

    e) i) Triangle numbers
      ii) Student's own answers

    f) i) Boy's names beginning with the letter m
      ii) Student's own answers

    g) i) Odd numbers
      ii) Student's own answers

    h) i) Vowels   ii) o, u

    i) i) Planets of the solar system
      ii) Student's own answers

    j) i) Numbers between 3 and 12
      ii) Student's own answers

    k) i) Numbers between -5 and 5
      ii) Student's own answers

2.   **a)** 7   **c)** 7   **d)** 7   **f)** Unquantifiably finite, though theoretically infinite   **h)** 5   **i)** 9

Exercise 1.7

1.   **a)** $Q = \{2, 4, 6, 8, 10, 12, 14, 16, 18, 20, 22, 24, 26, 28\}$
    b) $R = \{1, 3, 5, 7, 9, 11, 13, 15, 17, 19, 21, 23, 25, 27, 29\}$
    c) $S = \{2, 3, 5, 7, 11, 13, 17, 19, 23, 29\}$
    d) $T = \{1, 4, 9, 16, 25\}$
    e) $U = \{1, 3, 6, 10, 15, 21, 28\}$

2.   **a)** $B = \{55, 60, 65\}$
    b) $C = \{51, 54, 57, 60, 63, 66, 69\}$
    c) $D = \{64\}$

3.   **a)** $\{p, q, r\}, \{p, q\}, \{p, r\}, \{q, r\}, \{p\}, \{q\}, \{r\}, \{\ \}$
    b) $\{p, q\}, \{p, r\}, \{q, r\}, \{p\}, \{q\}, \{r\}$

4.   **a)** True   **b)** True   **c)** True   **d)** False
    e) False   **f)** True   **g)** True   **h)** False

Exercise 1.8

1.   **a)** True   **b)** True   **c)** False   **d)** False
    e) False   **f)** True

2.   **a)** $A \cap B = \{4, 6\}$ **b)** $A \cap B = \{4, 9\}$
    c) $A \cap B = \{$yellow, green$\}$

3.   **a)** $A \cup B = \{2, 3, 4, 6, 8, 9, 10, 13, 18\}$
    b) $A \cup B = \{1, 4, 5, 6, 7, 8, 9, 16\}$
    c) $A \cup B = \{$red, orange, blue, indigo, violet, yellow, green, purple, pink$\}$

4.   **a)** $\mathscr{E} = \{a, b, p, q, r, s, t\}$ **b)** $A' = \{a, b\}$

5.   **a)** $\mathscr{E} = \{1, 2, 3, 4, 5, 6, 7, 8\}$
    b) $A' = \{1, 4, 6, 8\}$ **c)** $A \cap B = \{2, 3\}$
    d) $A \cup B = \{1, 2, 3, 4, 5, 7, 8\}$
    e) $(A \cap B)' = \{1, 4, 5, 6, 7, 8\}$
    f) $A \cap B' = \{5, 7\}$

6.   **a)** i) $A = \{$even numbers from 2 to 14$\}$
      ii) $B = \{$multiples of 3 from 3 to 15$\}$
      iii) $C = \{$multiples of 4 from 4 to 20$\}$

b) i) $A \cap B = \{6, 12\}$
ii) $A \cap C = \{4, 8, 12\}$
iii) $B \cap C = \{12\}$
iv) $A \cap B \cap C = \{12\}$
v) $A \cup B = \{2, 3, 4, 6, 8, 9, 10, 12, 14, 15\}$
vi) $C \cup B = \{3, 4, 6, 8, 9, 12, 15, 16, 20\}$

7. a) i) $A = \{1, 2, 4, 5, 6, 7\}$
ii) $B = \{3, 4, 5, 8, 9\}$
iii) $C' = \{1, 2, 3, 4, 5, 8, 9\}$
iv) $A \cap B = \{4, 5\}$
v) $A \cup B = \{1, 2, 3, 4, 5, 6, 7, 8, 9\}$
vi) $(A \cap B)' = \{1, 2, 3, 6, 7, 8, 9\}$
b) $C \subset A$

8. a) i) $W = \{1, 2, 4, 5, 6, 7, 9, 10\}$
ii) $X = \{2, 3, 6, 7, 8, 9\}$
iii) $Z' = \{1, 4, 5, 6, 7, 8, 10\}$
iv) $W \cap Z = \{2, 9\}$ v) $W \cap X = \{2, 6, 7, 9\}$
vi) $Y \cap Z = \{ \ \}$ or $\emptyset$
b) Z

Exercise 1.9

1. a)

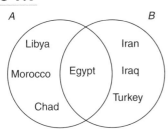

b) i) $A \cap B = \{Egypt\}$
ii) $A \cup B = \{Libya, Morocco, Chad, Egypt,$
   $Iran, Iraq, Turkey\}$

2. a)

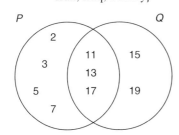

b) i) $P \cap Q = \{11, 13, 17\}$
ii) $P \cup Q = \{2, 3, 5, 7, 11, 13, 15, 17, 19\}$

3.

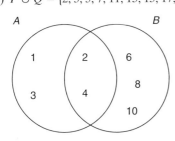

4.

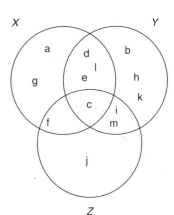

5.

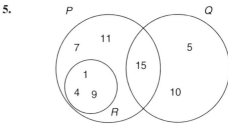

Exercise 1.10

1. a) 5   **b)** 14   **c)** 13

2. 45

3. a) 10   **b)** 50

4. a)

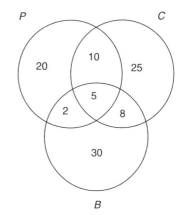

b) 100

Student Assessment 1

1. a) {even numbers from 2 to 8}
b) {even numbers}   **c)** {square numbers}
d) {oceans}

2. **a)** 7   **b)** 2   **c)** 6   **d)** 366

3. **a)**

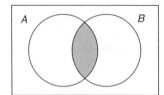

   b)

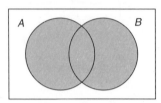

4. {a, b}, {a}, {b}, { }

5. $A' = \{m, t, h\}$

Student Assessment 2

1. **a)** {odd numbers from 1 to 7}
    b) {odd numbers}   **c)** {triangle numbers}
    d) {countries in South America}

2. **a)** 12   **b)** 3   **c)** 7
    d) Student's own answer

3. **a)**

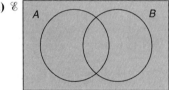

   b) ℰ

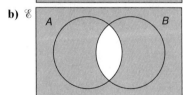

4. {o, r, k}, {w, r, k}, {w, o, k}, {w, o, r}, {w, o, r, k}

5. $P' = \{1, 3, 5, 7\}$

Student Assessment 3

1. {2, 4}, {2, 6}, {2, 8}, {4, 6}, {4, 8}, {6, 8}, {2, 4, 6}, {2, 4, 8}, {2, 6, 8}, {4, 6, 8}

2. **a)**

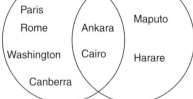

   b) {Ankara, Cairo}   **c)** {Maputo, Harare}

3. **a)**

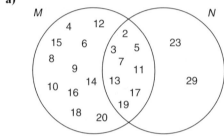

   b) {2, 3, 5, 7, 11, 13, 17, 19}
   c) {4, 6, 8, 9, 10, 12, 14, 15, 16, 18, 20, 23, 29}

4. **a)** ℰ

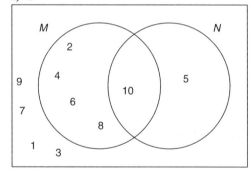

   b) $X = \{$multiples of 10$\}$

5. **a)**

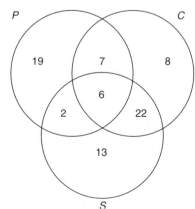

   b) 6

Student Assessment 4

1. **a)** 32
    b) {a, e, i, o, u}, {a, e, i, o}, {a, e, i, u}, {a, e, o, u}, {a, i, o, u}, {e, i, o, u}

2. **a)**

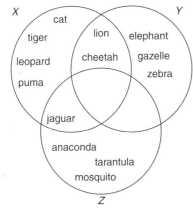

    b) {lion, cheetah}　　**c)** Ø　　**d)** Ø

3. **a)** Let the number liking only tennis be *x*.

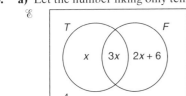

    b) 15　　**c)** 5　　**d)** 16

4. **a)** 5　　**b)** 35　　**c)** 40　　**d)** 50　　**e)** 15
    f) 12　　**g)** 10　　**h)** 78　　**i)** 78

ICT Section page 40

1. **b)** 6th term = 12
    10th term = 20
    20th term = 40

2. **a)** 18, 30, 60　　　**b)** 14, 22, 42
    c) 17, 29, 59　　　**d)** 3, 5, 10
    e) 35, 99, 399　　　**f)** 18, 50, 200
    g) −14, −26, −56　　**h)** 42, 110, 420

2 Limits of Accuracy

Core Section page 41

Exercise 2.1

1. **a)** i) Lower bound = 5.5
        Upper bound = 6.5
        ii) $5.5 \leqslant x < 6.5$
    b) i) Lower bound = 82.5
        Upper bound = 83.5
        ii) $82.5 \leqslant x < 83.5$

c) i) Lower bound = 151.5
    Upper bound = 152.5
    ii) $151.5 \leqslant x < 152.5$
d) i) Lower bound = 999.5
    Upper bound = 1000.5
    ii) $999.5 \leqslant x < 1000.5$
e) i) Lower bound = −0.5
    Upper bound = 0.5
    ii) $-0.5 < x < 0.5$
f) i) Lower bound = −4.5
    Upper bound = −3.5
    ii) $-4.5 < x \leqslant -3.5$

2. **a)** i) Lower bound = 3.75
        Upper bound = 3.85
        ii) $3.75 \leqslant x < 3.85$
    b) i) Lower bound = 15.55
        Upper bound = 15.65
        ii) $15.55 \leqslant x < 15.65$
    c) i) Lower bound = 0.95
        Upper bound = 1.05
        ii) $0.95 \leqslant x < 1.05$
    d) i) Lower bound = 9.95
        Upper bound = 10.05
        ii) $9.95 \leqslant x < 10.05$
    e) i) Lower bound = 0.25
        Upper bound = 0.35
        ii) $0.25 \leqslant x < 0.35$
    f) i) Lower bound = −0.25
        Upper bound = −0.15
        ii) $-0.25 < x \leqslant -0.15$

3. **a)** i) Lower bound = 4.15
        Upper bound = 4.25
        ii) $4.15 \leqslant x < 4.25$
    b) i) Lower bound = 0.835
        Upper bound = 0.845
        ii) $0.835 \leqslant x < 0.845$
    c) i) Lower bound = 415
        Upper bound = 425
        ii) $415 \leqslant x < 425$
    d) i) Lower bound = 4950
        Upper bound = 5050
        ii) $4950 \leqslant x < 5050$
    e) i) Lower bound = 0.0445
        Upper bound = 0.0455
        ii) $0.0445 \leqslant x < 0.0455$
    f) i) Lower bound = 24 500
        Upper bound = 25 500
        ii) $24\,500 \leqslant x < 25\,500$

4. **a)**

5.3	5.35	5.4	5.45	5.5

    b) $5.35 \leqslant M < 5.45$

5. **a)**

11.7	11.75	11.8	11.85	11.9

    b) $11.75 \leqslant T < 11.85$

6. **a)** Lower bound = 615 m³
Upper bound = 625 m³
b) $615 \leqslant x < 625$

7. **a)** Lower bound = 625 m
Upper bound = 635 m
b) $395 \leqslant W < 405$

Student Assessment 1

1. **a)**
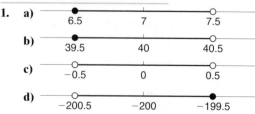

2. **a)** $205 \leqslant x < 215$ **b)** $63.5 \leqslant x < 64.5$
c) $2.95 \leqslant x < 3.05$ **d)** $0.875 \leqslant x < 0.885$

3. Length: $349.5 \leqslant L < 350.5$
Width: $199.5 \leqslant W < 200.5$

4.
58.85   58.9   58.95

5. $0.0035 \leqslant x < 0.0045$

6. **a)** $4.825 \leqslant x < 4.835$ **b)** $5.045 \leqslant y < 5.055$
c) $9.95 \leqslant z < 10.05$
d) $-100.005 < p \leqslant -99.995$

Student Assessment 2

1. **a)**
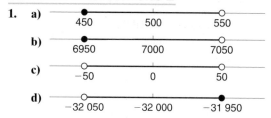

2. **a)** $253.5 \leqslant x < 254.5$ **b)** $40.45 \leqslant x < 40.55$
c) $0.4095 \leqslant x < 0.4105$
d) $99.95 \leqslant x < 100.5$

3. $20.25 \leqslant L < 20.75$ $9.75 \leqslant W < 10.25$

4.
245.5   245.6   245.7

5. Lower bound = 365.245
Upper bound = 365.255

6. **a)** $10.895 \leqslant x < 10.905$
b) $2.995 \leqslant x < 3.005$ **c)** $0.45 \leqslant x < 0.55$
d) $-175.005 < x \leqslant -174.995$

Extended Section page 45

Exercise 2.2

1. **a)** Lower bound = 263.25
Upper bound = 297.25
b) Lower bound = 3295.25
Upper bound = 3455.25
c) Lower bound = 4925.25
Upper bound = 5075.25
d) Lower bound = 3.76
Upper bound = 4.26 (2 d.p.)
e) Lower bound = 2.83
Upper bound = 3.19 (2 d.p.)
f) Lower bound = 8.03
Upper bound = 8.66 (2 d.p.)
g) Lower bound = 44.95
Upper bound = 52.82 (2 d.p.)
h) Lower bound = 39.77
Upper bound = 42.23 (2 d.p.)
i) Lower bound = 16.14
Upper bound = 18.88 (2 d.p.)
j) Lower bound = 3.55
Upper bound = 7.12 (2 d.p.)
k) Lower bound = 1.47
Upper bound = 1.63 (2 d.p.)
l) Lower bound = 18.51
Upper bound = 28.59 (2 d.p.)

2. **a)** Lower bound = 6.7
Upper bound = 6.9
b) Lower bound = 29.69
Upper bound = 30.80 (2 d.p.)
c) Lower bound = 147.76
Upper bound = 150.25 (2 d.p.)
d) Lower bound = 13.3
Upper bound = 13.5
e) Lower bound = 1.75
Upper bound = 1.81 (2 d.p.)
f) Lower bound = 0.39
Upper bound = 0.46 (2 d.p.)
g) Lower bound = 34.10
Upper bound = 40.03 (2 d.p.)
h) Lower bound = 0.98
Upper bound = 1.02 (2 d.p.)
i) Lower bound = 0
Upper bound = 0.04

3. **a)** Lower bound = 20 002.5
Upper bound = 20 962.5
b) Lower bound = 1.06
Upper bound = 1.15 (2 d.p.)
c) Lower bound = 1 116 250
Upper bound = 1 188 250
d) Lower bound = 88.43
Upper bound = 91.60 (2 d.p.)
e) Lower bound = 131.75
Upper bound = 139.34 (2 d.p.)

f) Lower bound = 18.95
Upper bound = 20.39 (2 d.p.)
g) Lower bound = 2.24
Upper bound = 2.53 (2 d.p.)
h) Lower bound = 3.61
Upper bound = 3.97 (2 d.p.)
i) Lower bound = 60.34
Upper bound = 68.52 (2 d.p.)

Exercise 2.3

1. Lower bound = 3.5 kg
Upper bound = 4.5 kg

2. Lower bound = 21.8 cm
Upper bound = 22.2 cm

3. Lower bound = 13.7 m
Upper bound = 13.74 m

4. Lower bound = 74.13 cm^2
Upper bound = 75.88 cm^2 (2 d.p.)

5. Lower bound = 68 425 m^2
Upper bound = 75 625 m^2

6. Lower bound = 13.1
Upper bound = 13.5 (1 d.p.)

7. Lower bound = 9.5 cm
Upper bound = 10.6 cm (1 d.p.)

8. **a)** Lower bound = 53.1 cm
Upper bound = 53.7 cm (1 d.p.)
b) Lower bound = 224.3 cm^2
Upper bound = 229.7 cm^2 (1 d.p.)

9. **a)** Lower bound = 11.2 cm
Upper bound = 11.4 cm (1 d.p.)
b) Lower bound = 70.5 cm
Upper bound = 71.3 cm (1 d.p.)

10. Lower bound = 11.5 g/cm^3
Upper bound = 13.6 g/cm^3 (1 d.p.)

11. Least = 93.3 h   Greatest = 138.5 h (1 d.p.)

Student Assessment 1

1. **a)** Lower bound = 965.25
Upper bound = 1035.25
b) Lower bound = 6218.75
Upper bound = 6381.75
c) Lower bound = 6.2
Upper bound = 6.3 (1 d.p.)
d) Lower bound = 47.7
Upper bound = 54.7 (1 d.p.)
e) Lower bound = 0.8
Upper bound = 1.2 (1 d.p.)
f) Lower bound = 0.5
Upper bound = 2.1 (1 d.p.)

2. Lower bound = 118.7 cm^2
Upper bound = 121.1 cm^2 (1 d.p.)

3. Lower bound = 10.5 cm
Upper bound = 13.5 cm

4. $0.8 \leqslant x < 1$

5. Lower bound = 0.46 kg
Upper bound = 0.48 kg (2 d.p.)

Student Assessment 2

1. **a)** Lower bound = 110.25
Upper bound = 140.25
b) Lower bound = 945.25
Upper bound = 1055.25
c) Lower bound = 6.6
Upper bound = 6.7 (1 d.p.)
d) Lower bound = 53.8
Upper bound = 199.5 (1 d.p.)
e) Lower bound = 0.4
Upper bound = 0.7 (1 d.p.)
f) Lower bound = 0.2
Upper bound = 0.6 (1 d.p.)

2. Lower bound = 4.1 cm
Upper bound = 4.5 cm (1 d.p.)

3. Lower bound = 15.45 cm
Upper bound = 15.75 cm (2 d.p.)

4. $8.9 \leqslant x < 9.1$

5. Lower bound = 0.69 kg
Upper bound = 0.71 kg (2 d.p.)

Student Assessment 3

1. 255 kg

2. 31 575 cm^3 (5 s.f.)

3. **a)** $124.5 \leqslant V < 125.5$ cm^3
b) $4.99 \leqslant L < 5.01$ cm (2 d.p.)

4. **a)** $25.10 \leqslant C < 25.16$ (2 d.p.)
b) $50.14 \leqslant A < 50.39$ cm^2 (2 d.p.)

5. 22 cups

6. **a)** Lower bound = 4.5 cm
Upper bound = 5.5 cm
b) Lower bound = 5.43 cm
Upper bound = 5.44 cm

Student Assessment 4

1. Lower bound = 0.85 kg
Upper bound = 1.15 kg

2. **a)** Lower bound = 1251 cm^3
Upper bound = 1647 cm^3 (4 s.f.)
b) Lower bound = 723.5 cm^2
Upper bound = 863.5 cm^2

3. **a)** $63.5 \leqslant A < 64.5$ cm²
   b) $15.9 \leqslant P < 16.1$ cm (1 d.p.)

4. Lower bound = 0.28 g/cm³
   Upper bound = 0.31 g/cm³ (2 d.p.)

5. **a)** Lower bound = 7.5 g
   Upper bound = 8.5 g
   b) Lower bound = 8.07 g
   Upper bound = 8.08 g (2 d.p.)

6. **a)** $37.67 \leqslant C < 37.73$ cm (2 d.p.)
   b) $112.91 \leqslant A < 113.29$ cm² (2 d.p.)

ICT Section page 50

Circle	Radius (cm)	Accuracy (decimal places)	Radius (cm)		Area (cm²)	
			Lower bound	Upper bound	Lower bound	Upper bound
2a	10	0	9.5	10.5	283.5	346.4
2b	6.6	1	6.55	6.65	134.78	138.93
2c	20.4	1	20.35	20.45	1301.00	1313.82
2d	12.50	2	12.495	12.505	490.481	491.267
2e	10.00	2	9.995	10.005	313.845	314.474

3 Ratio, Proportion and Measures of Rate

Core Section page 51

Exercise 3.1

1. 48

2. 16 h 40 min

3. 11 units

4. **a)** 7500 bricks   **b)** 53 h

5. **a)** 6250 litres   **b)** 128 km

6. 1111 km (4 s.f.)

7. **a)** 450   **b)** 75   **c)** 120

Exercise 3.2

1. **a)** 450 kg   **b)** 1250 kg

2. **a)** Butter 600 g Flour 2 kg Sugar 200 g
      Currants 400 g
   b) 120 cakes

3. **a)** 16.8 litres
   b) Red 1.2 litres   White 14.3 litres

4. **a)** 125   **b)** Red 216 Yellow 135   **c)** 20

5. **a)** 42 litres
   b) Orange juice 495 litres
      Mango juice 110 litres

Exercise 3.3

1. 60 : 90

2. 16 : 24 : 32

3. 3.25 : 1.75

4. 18 : 27

5. 10 : 50

6. 7 : 1

7. Orange 556 ml (3 s.f.)   Water 444 ml (3 s.f.)

8. **a)** 55 : 45   **b)** 440 boys   360 girls

9. $\frac{3}{5}$

10. 32 cm

11. 4 km and 3 km

12. 40°, 80°, 120°, 120°

13. 45°, 75°, 60°

14. 24 yr old $400 000   28 yr old $466 667
    32 yr old $533 333

15. Alex £2000   Maria £3500   Ahmet £2500

Exercise 3.4

1. 4

2.

Speed (km/h)	60	40	30	120	90	50	10
Time (h)	2	3	4	1	$1\frac{1}{3}$	$2\frac{2}{5}$	12

3. 30

4. **a)** i) 12 h   ii) 4 h   iii) 48 h
   b) i) 16   ii) 3   iii) 48

5. **a)** 30 rows   **b)** 42 chairs

6. 6 h 40 min

7. 4

8. 18 h

Student Assessment 1

1. 16 cm and 14 cm

2. 1200 g

3. 200 g

4. **a)** 2 km   **b)** 48 cm

5. **a)** 26 litres of petrol and 4 litres of oil
   b) 3250 ml of oil

6. **a)** 1 : 40   **b)** 13.75 cm

7. Girl £1040   Boy £960

8. $24°, 60°, 96°$

9. **a)** 15 s   **b)** 8 copiers

10. 6 h

Student Assessment 2

1. **a)** $\frac{7}{10}$ **b)** 45 cm

2. **a)** 375 g   **b)** 625 g

3. **a)** 450 m   **b)** 80 cm

4. **a)** 1 : 25   **b)** 1.75 m

5. 300 : 750 : 1950

6. $60°, 90°, 90°, 120°$

7. $150°$

8. **a)** 13.5 h   **b)** 12 pumps

9.

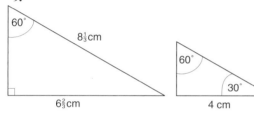

10. **a)** 4 min 48 s   **b)** 1.6 litres/min

Extended Section page 59

Exercise 3.5

1. **a)** 3   **b)** 21   **c)** 27   **d)** 3   **e)** 10

2. **a)** 0.5   **b)** 8   **c)** 24.5   **d)** 8   **e)** 16

3. **a)** 24   **b)** $\frac{3}{8}$ **c)** $\frac{1}{9}$ **d)** 1

4. **a)** 0.25   **b)** 25   **c)** 4   **d)** $\frac{1}{16}$

Exercise 3.6

1. **a)** i) $y \propto x^3$ ii) $y = kx^3$

   b) i) $y \propto \dfrac{1}{x^3}$ ii) $y = \dfrac{k}{x^3}$

   c) i) $t \propto P$ ii) $t = kp$

   d) i) $s \propto \dfrac{1}{t}$ ii) $s = \dfrac{k}{t}$

   e) i) $A \propto r^2$ ii) $A = kr^2$

   f) i) $T \propto \dfrac{1}{\sqrt{g}}$ ii) $T = \dfrac{k}{\sqrt{g}}$

2. 10.5

3. **a)** $\frac{1}{2}$ **b)** 2

4. 32

5. **a)** $\frac{1}{8}$ **b)** 0.4

6. 75

Exercise 3.7

1. **a)** $h = kt^2$ **b)** $k = 5$ **c)** 45 m   **d)** 6 s

2. **a)** $v = k\sqrt{e}$ **b)** $x = 3$ **c)** 49 J

3. **a)** $l = k(m)^{\frac{1}{3}}$ **b)** $k = 3$ **c)** $l = 6$ cm

4. **a)** $P = kI^2$ **b)** $I = 5$ amps

Exercise 3.8

1. **a)** 160   **b)** 250   **c)** 175   **d)** 110
   e) 225   **f)** 128

2. **a)** $93\frac{1}{3}$ **b)** $116\frac{2}{3}$ **c)** 80   **d)** $157\frac{1}{2}$
   e) 154   **f)** 85

3. **a)** 40   **b)** 50   **c)** 35   **d)** 36   **e)** 15
   f) 52

4. **a)** $22\frac{1}{2}$ **b)** $6\frac{2}{3}$ **c)** $17\frac{1}{2}$ **d)** $5\frac{5}{8}$ **e)** $18\frac{3}{4}$
   f) $13\frac{1}{2}$

5. 50

6. 32

7. 210

8. 90

Exercise 3.9

1. 22 cm by 16.5 cm

2. 28 cm by 21 cm

3. 5 : 2

4. Min. = 32 : 1   Max. = 35 : 1

5. **a)** i) 28 cm² ii) 112 cm²   **b)** 4 : 1

6. **a)** i) 9 cm² ii) 81 cm²   **b)** 9 : 1

7. **a)** i) 30 cm³ ii) 240 cm³   **b)** 8 : 1

8. **a)** i) 64 cm³ ii) 1728 cm³   **b)** 27 : 1

9. **a)** 16 cm²   **b)** 4 cm²   **c)** 1 : 4

10. Student's own answer

Student Assessment 1

1. **a)** 3   **b)** 12   **c)** 3   **d)** 6

2. **a)** 4   **b)** $\frac{1}{4}$ **c)** 4   **d)** 8

3. **a)** 0.5   **b)** 32   **c)** $\frac{1}{2}$ **d)** 6

4. **a)** 2.5   **b)** $\frac{1}{2}$ **c)** $\frac{1}{4}$ **d)** $\frac{1}{16}$

5. **a)** $\frac{1}{4}$ **b)** $\frac{1}{4}$ **c)** ± 1 **d)** ± 5

Student Assessment 2

1. **a)** 1.5 **b)** 15 **c)** 3 **d)** 12
2. **a)** 10 **b)** 2.5 **c)** 1 **d)** 20
3. **a)** $\frac{1}{3}$ **b)** 72 **c)** $\frac{1}{3}$ **d)** 12
4. **a)** 5 **b)** $\frac{5}{4}$ **c)** $\frac{1}{2}$ **d)** 1
5. **a)** $\frac{1}{3}$ **b)** $\frac{4}{3}$ **c)** ± 2 **d)** ± 1

Student Assessment 3

1. **a)**

x	1	2	3	4	5
y	5	10	15	20	25

b)

x	1	2	3	4	5
y	30	15	10	7.5	6

c)

x	1	2	3	4	5
y	20	5	2.$\dot{2}$	1.25	0.8

2. **a)** 28 m (2 s.f.) **b)** 137 km/h (3 s.f.)
3. **a)** $V = kr^3$ **b)** 4.19 (2 d.p.)
   c) 113.13 cm^3 (2 d.p.) **d)** 4.57 cm (2 d.p.)
4. **a)** $V = kah$ **b)** 12 cm^2
5. 2
6. **a)** 27 : 8 **b)** 9 : 4

Student Assessment 4

1. **a)**

x	1	2	4	8	16	32
y	32	16	8	4	2	1

b) 1.6

2. **a)**

x	1	2	4	5	10
y	5	10	20	25	50

b)

x	1	2	4	5	10
y	20	10	5	4	2

c)

x	4	16	25	36	64
y	4	8	10	12	16

3. **a)** 0.8 **b)** 0.8
4. **a)** 1.6 (1 d.p.) **b)** 2
5. 21.6 cm by 9 cm
6. **a)** 12.5 cm **b)** 1 : 64

ICT Section page 67

2. =B3*B3
3. Total surface area: =C3*6 Volume: =B3^3
5. B7 =B3*F10

   C7 =B7*B7

   D7 =C7*6

   E7 =B7^3
6. F11=C7/C3 or F11=D7/D3
7. F12=E7/E3
9. If scale factor of enlargement is n, area factor of enlargement is n^2 and volume factor of enlargement is n^3.

4 Percentages

Core Section page 69

Exercise 4.1

1. White = 47% Blue = 23% Red = 30%
2. 70%
3. **a)** 60% **b)** 40%
4. **a)** $\frac{73}{100}$ **b)** $\frac{28}{100}$ **c)** $\frac{10}{100}$ **d)** $\frac{25}{100}$
5. **a)** 27% **b)** 30% **c)** 14% **d)** 25%
6. **a)** 0.39 **b)** 0.47 **c)** 0.83 **d)** 0.07
   e) 0.02 **f)** 0.2
7. **a)** 31% **b)** 67% **c)** 9% **d)** 5%
   e) 20% **f)** 75%

Exercise 4.2

1. **a)** 25% **b)** 66.$\dot{6}$% **c)** 62.5%
   d) 180% **e)** 490% **f)** 387.5%
2. **a)** 0.75 **b)** 0.8 **c)** 0.2 **d)** 0.07
   e) 1.875 **f)** 0.1$\dot{6}$
3. **a)** 20 **b)** 100 **c)** 50 **d)** 36
   e) 4.5 **f)** 7.5
4. **a)** 8.5 **b)** 8.5 **c)** 52 **d)** 52
   e) 17.5 **f)** 17.5
5. **a)** Black 6 **b)** Blonde 3 **c)** Brown 21
6. Beef 66 Chicken 24 Pork 12 Lamb 18
7. English 143 Pakistani 44 Greek 11 Other 22
8. Newspapers 69 Pens 36 Books 18 Other 27

Exercise 4.3

1. **a)** 48%   **b)** 36.8%   **c)** 35%   **d)** 50%
    e) 45%   **f)** 40%   **g)** $33\frac{1}{3}$%
    h) 57% (2 s.f.)

2. Win 50%   Lose $33\frac{1}{3}$%   Draw $16\frac{2}{3}$%

3. A = 34.5% (1 d.p.)   B = 25.6% (1 d.p.)
    C = 23.0% (1 d.p.)   D = 16.9% (1 d.p.)

4. Red 35.5%   Blue 31.0%   White 17.7%
    Silver 6.6%   Green 6.0%   Black 3.2%

Exercise 4.4

1. **a)** 187.5   **b)** 322   **c)** 7140   **d)** 245
    e) 90   **f)** 121.5

2. **a)** 90   **b)** 38   **c)** 9   **d)** 900   **e)** 50
    f) 43.5

3. **a)** 20%   **b)** 80%   **c)** 110%   **d)** 5%
    e) 85%   **f)** 225%

4. **a)** 50%   **b)** 30%   **c)** 5%   **d)** 100%
    e) 36%   **f)** 5%

5. 7475 tonnes

6. 6825 BRL

7. **a)** £75   **b)** £93.75

8. **a)** 43   **b)** 17.2%

9. 1100

Exercise 4.5

1. £11 033 750

2. 520 875 euro

3. £10 368

4. 1331 students

5. 3 276 800 tonnes

6. 2 years

7. 5 years

8. 3 years

Student Assessment 1

1.

Fraction	Decimal	Percentage
$\frac{3}{4}$	0.75	75%
$\frac{4}{5}$	0.8	80%
$\frac{1}{3}$	0.$\dot{3}$	$33\frac{1}{3}$%
$\frac{5}{8}$	0.625	62.5%
$\frac{3}{2}$	1.5	150%

2. 640 m

3. £345.60

4. $10 125

5. **a)** 20%   **b)** 41.7% (1 d.p.)   **c)** 22.5%
    d) 85.7% (1 d.p.)   **e)** 7%   **f)** 30%

6. 16% profit

7. **a)** £36   **b)** 25%

8. 2001

9. 499 200 euro

10. 4 years

Student Assessment 2

1.

Fraction	Decimal	Percentage
$\frac{1}{4}$	0.25	25%
$\frac{3}{5}$	0.6	60%
$\frac{5}{8}$	0.625	62.5%
$\frac{2}{3}$	0.$\dot{6}$	$66\frac{2}{3}$%
$2\frac{1}{4}$	2.25	225%

2. 750 m

3. €1400

4. £97 200

5. **a)** 29.2% (1 d.p.)   **b)** 21.7% (1 d.p.)
    c) 125%   **d)** 8.3% (1 d.p.)   **e)** 20%
    f) 10%

6. 8.3%

7. **a)** ¥6500   **b)** 61.8% (1 d.p.)

8. 8 days

9. 955 080 euro

10. 5 years

Extended Section page 77

Exercise 4.6

1. **a)** 600   **b)** 350   **c)** 900   **d)** 250
    e) 125   **f)** 1.5

2. **a)** 56   **b)** 65   **c)** 90   **d)** 20
    e) 0.25   **f)** −38

3. 280 pages

4. 12 500 families

5. 22

6. 12 200 000 m³

Student Assessment I

1. £3500   £12 000   £1   £56

2. £500   £250   £20 000   £137 500

3. 15

4. £15 000

5. 40 000 tonnes

6. 46 500 units

Student Assessment 2

1. $200 $25 $524 $10

2. $462 $4000 $4500 $5500

3. 15 marks

4. 35 000

5. 25 000 units

6. 470 tonnes

ICT Section page 79

3. e.g. in cell C4 enter =B4/B$3*100

5 Graphs In Practical Situations

Core Section page 80

Exercise 5.1

1.

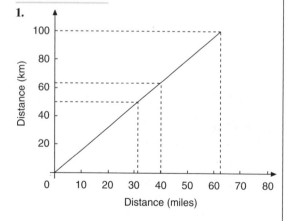

a) 50 km = 31 miles

b) 40 miles = 64 km, therefore 80 miles = 128 km

c) 100 km/h = 62 mph

d) 40 mph = 64 km/h

2.

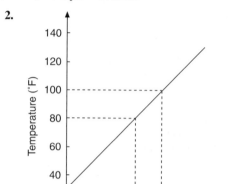

a) 25 °C = 80 °F   **b)** 100 °F = 35 °C

c) 0 °C = 30 °F   **d)** 100 °F = 35 °C, therefore 200 °F = 70 °C

3.

i) a) 25 °C = 77 °F   **b)** 100 °F = 38 °C

c) 0 °C = 32 °F   **d)** 100 °F = 38 °C, therefore 200 °F = 76 °C

ii) The rough conversion is most useful at lower temperatures (i.e. between 0 and 20 °C).

4. a)

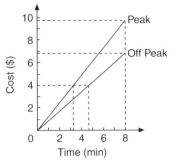

b) 8 min = $6.80
c) 8 min = $9.60
d) Extra time = 1 min 20 s

5.

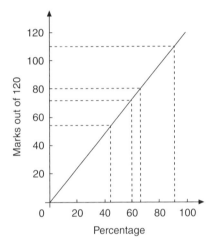

a) 80 = 67%
b) 110 = 92%
c) 54 = 45%
d) 72 = 60%

Exercise 5.2

1. a) 6 m/s   **b)** 4 m/s   **c)** 39 km/h
d) 20 km/h   **e)** 160 km/h   **f)** 50 km/h

2. a) 400 m   **b)** 182 m   **c)** 210 km
d) 255 km   **e)** 10 km   **f)** 79.2 km

3. a) 5 s   **b)** 50 s   **c)** 4 min
d) 71 min 26 s (nearest second)   **e)** 5 s
f) 4 min

Exercise 5.3

1.

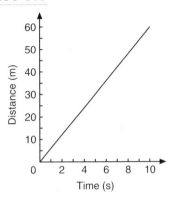

2.

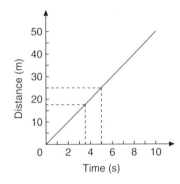

a) 25 m = 5s
b) 3.5 s = 17 m

3. a) Speed A = 40 m/s   Speed B = $13\frac{1}{3}$ m/s
b) Distance apart = $133\frac{1}{3}$ m

4. a) $\frac{2}{3}$ m/s   **b)** 6 m/s, $\frac{2}{3}$ m/s   **c)** 1 m/s
d) $\frac{1}{2}$ m   **e)** $7\frac{1}{3}$ m

Exercise 5.4

1. a) 45 km/h   **b)** 20 km/h
c) Paul has arrived at the restaurant and is waiting for Helena.

2.

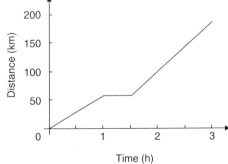

3.

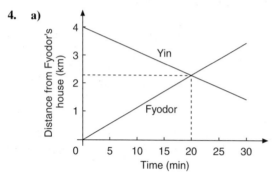

4. **a)**

b) After 20 min   **c)** Distance $= 2\frac{1}{3}$ km

5. **a)**

b) Time ≈ 6.57 p.m.
c) Distance from Q ≈ 79 km
d) The 6.10 p.m. train from station Q arrives first.

6. **a)** a: $133\frac{1}{3}$ km/h   b: 0 km/h   c: 200 km/h
b) d: 100 km/h   e: 200 km/h

Student Assessment 1

1. **a)**

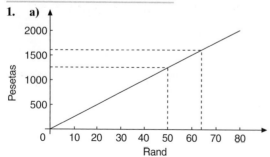

b) 50 rand = 1250 pesetas
c) 1600 pesetas = 64 rand

2. **a)**

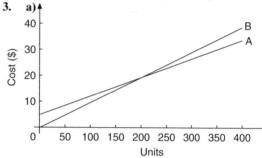

    i) 5 km: 50 rand   ii) 8.5 km: 71 rand
b) 80 rand: 10 km

3. **a)**

b) If the customer uses under 200 units/quarter then he should use account type B.

4. **a)**

b) 180 km

5. **a)**

b) Distance ≈ 122 km

c) Time ≈ 1 h 13 min after start

Student Assessment 2

1.

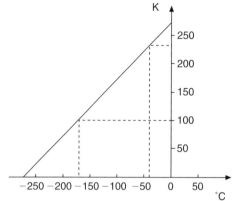

a) −40 °C = 233 K   **b)** 100 K = −173 °C

2. **a)**

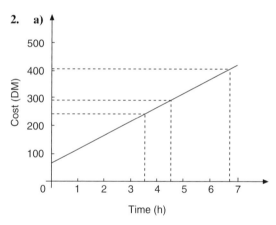

b) 295 DM   **c)** ≈ 408 DM   **d)** $3\frac{1}{2}$ h

3. **a)**

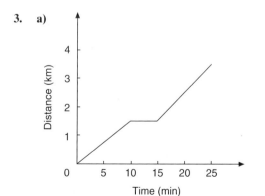

b) 25 min

4. **a)** B, C

b) B because it illustrates going back in time,
C because it illustrates infinite speed

Extended Section page 87

Exercise 5.5

1. Acceleration is 0.375 m/s²

2. Acceleration is 0.2 m/s²

3. Acceleration is 7 m/s²

4. Acceleration is 1.75 m/s²

5. Deceleration is 0.25 m/s²

6. Deceleration is 1 m/s²

7.

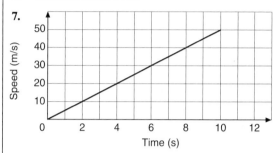

8.

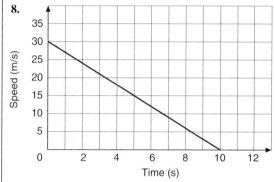

Exercise 5.6

1. **a)** 1.5 m/s²   **b)** 0 m/s²   **c)** 0.5 m/s²

2. **a)** Cheetah   **b)** 7.5 m/s²   **c)** 5 m/s²
d) 15 m/s²

3. **a)** 0.5 m/s²   **b)** 0.25 m/s²
c) 0.10 m/s² (1 d.p.)
d) Travelling at a constant speed of 30 m/s
e) Stationary

Exercise 5.7

1. a)

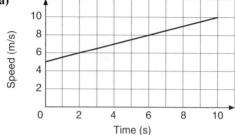

b) 0.5 m/s² **c)** 75 m

2. a)

Time (s)	0	0.5	1	1.5	2	2.5	3
Speed (m/s)	6	5	4	3	2	1	0

b)

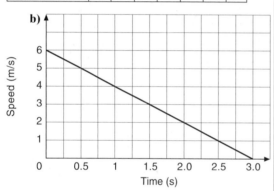

c) 9 m

3. a) 1.5 m/s² **b)** 2400 m **c)** 40 s

4. a) 390 m **b)** 240 m

5. 21.45 km

6. 720 m

7. a) 0.37 m/s² **b)** 2.16 m/s² **c)** 208 m
d) 204 m **e)** 4 m

Student Assessment 1

1. a) 2 m/s² **b)** 225 m **c)** 10.6 s (1 d.p.)

2. a) 4 m/s² **b)** 3 m/s² **c)** 102 m
d) 9.83 s

3. a)

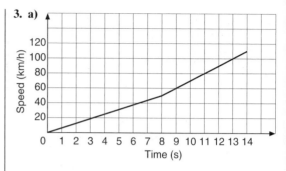

b) 189 m (3 s.f.)

4. a) 1 m/s² **b)** 750 m **c)** 1237.5 m

Student Assessment 2

1. a) 2.5 m/s² **b)** 180 m **c)** 3.5 s (1 d.p.)

2. a) 0.28 m/s² (1 d.p.) **b)** 93.8 m (1 d.p.)
c) 97.2 m (1 d.p.)

3. a)

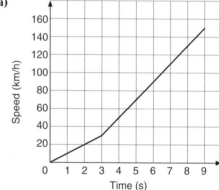

b) 162.5 m

4. a) 3.3 m/s² (1 d.p.) **b)** 240 m **c)** 212.5 m

6 Algebraic Representation and Manipulation

Core Section page 96

Exercise 6.1

1. a) $4x - 12$ **b)** $10p - 20$ **c)** $-42x + 24y$
d) $6a - 9b - 12c$ **e)** $-14m + 21n$
f) $-16x + 6y$

2. **a)** $3x^2 - 9xy$ **b)** $a^2 + ab + ac$
 c) $8m^2 - 4mn$ **d)** $-15a^2 + 20ab$
 e) $4x^2 - 4xy$ **f)** $24p^2 - 8pq$

3. **a)** $-2x^2 + 3y^2$ **b)** $a - b$ **c)** $7p - 2q$
 d) $3x - 4y + 2z$ **e)** $3x - \frac{3}{2}y$
 f) $2x^2 - 3xy$

4. **a)** $12r^3 - 15rs + 6rt$ **b)** $a^3 + a^2b + a^2c$
 c) $6a^3 - 9a^2b$ **d)** $p^2q + pq^2 - p^2q^2$
 e) $m^3 - m^2n + m^3n$ **f)** $a^6 + a^5b$

Exercise 6.2

1. **a)** $-a - 8$ **b)** $4x - 20$ **c)** $3p - 16$
 d) $21m - 6n$ **e)** 3 **f)** $-p - 3p^2$

2. **a)** $8m^2 + 28m + 2$ **b)** $x - 4$ **c)** $2p + 22$
 d) $m - 12$ **e)** $a^2 + 6a + 2$
 f) $7ab - 16ac + 3c$

3. **a)** $4x + 4$ **b)** $5x - \frac{3}{2}y$ **c)** $\frac{9}{4}x - \frac{5}{2}y$
 d) $\frac{9}{2}x - \frac{1}{2}y$ **e)** $7x - 4y$ **f)** 0

Exercise 6.3

1. **a)** $2(2x - 3)$ **b)** $6(3 - 2p)$ **c)** $3(2y - 1)$
 d) $2(2a - 3b)$ **e)** $3(p - q)$
 f) $4(2m + 3n + 4r)$

2. **a)** $a(3b + 4c - 5d)$ **b)** $2p(4q + 3r - 2s)$
 c) $a(a - b)$ **d)** $2x(2x - 3y)$
 e) $ab(c + d + f)$ **f)** $3m(m + 3)$

3. **a)** $3pq(r - 3s)$ **b)** $5m(m - 2n)$
 c) $4xy(2x - y)$ **d)** $b^2(2a^2 - 3c^2)$
 e) $12(p - 3)$ **f)** $6(7x - 9)$

4. **a)** $6(3 + 2y)$ **b)** $7(2a - 3b)$
 c) $11x(1 + y)$ **d)** $4(s - 4t + 5r)$
 e) $5q(p - 2r + 3s)$ **f)** $4y(x + 2y)$

5. **a)** $m(m + n)$ **b)** $3p(p - 2q)$
 c) $qr(p + s)$ **d)** $ab(1 + a + b)$
 e) $p^3(3 - 4p)$ **f)** $b^2c(7b + c)$

6. **a)** $m(m^2 - mn + n^2)$ **b)** $2r^2(2r - 3 + 4s)$
 c) $7xy(8x - 4y)$ **d)** $18mn(4m + 2n - mn)$

Exercise 6.4

1. **a)** 0 **b)** 30 **c)** 14 **d)** 20 **e)** -13
 f) -4

2. **a)** -3 **b)** -30 **c)** 20 **d)** -16
 e) -40 **f)** 42

3. **a)** -160 **b)** -23 **c)** 42 **d)** -17
 e) -189 **f)** 113

4. **a)** 48 **b)** -8 **c)** 15 **d)** 16 **e)** -5
 f) 9

5. **a)** 12 **b)** -5 **c)** -5 **d)** 7 **e)** 7
 f) 36

Exercise 6.5

1. **a)** $n = r - m$ **b)** $m = p - n$
 c) $n = 3p - 2m$ **d)** $q = 3x - 2p$

 e) $a = \dfrac{cd}{b}$ **f)** $d = \dfrac{ab}{c}$

2. **a)** $x = \dfrac{4m}{3y}$ **b)** $r = \dfrac{7pq}{5}$ **c)** $x = \dfrac{c}{3}$

 d) $x = \dfrac{y - 7}{3}$ **e)** $y = \dfrac{3r + 9}{5}$

 f) $x = \dfrac{5y - 9}{3}$

3. **a)** $b = \dfrac{2a - 5}{6}$ **b)** $a = \dfrac{6b + 5}{2}$

 c) $z = \dfrac{3x - 7y}{4}$ **d)** $x = \dfrac{4x + 7y}{3}$

 e) $y = \dfrac{3x - 4z}{7}$ **f)** $p = \dfrac{8 + q}{2r}$

4. **a)** $p = 4r$ **b)** $p = \dfrac{4}{3r}$ **c)** $p = \dfrac{n}{10}$

 d) $n = 10p$ **e)** $p = \dfrac{2t}{q + r}$ **f)** $q = \dfrac{2t}{p} - r$

5. **a)** $r = \dfrac{3m - n}{t(p + q)}$ **b)** $t = \dfrac{3m - n}{r(p + q)}$

 c) $m = \dfrac{rt(p + q) + n}{3}$

 d) $n = 3m - rt(p + q)$ **e)** $p = \dfrac{3m - n}{rt} - q$

 f) $q = \dfrac{3m - n}{rt} - p$

6. **a)** $d = \dfrac{ab}{ce}$ **b)** $a = \dfrac{dec}{b}$ **c)** $c = \dfrac{ab}{de}$

 d) $a = cd - b$ **e)** $b = d - \dfrac{a}{c}$

 f) $c = \dfrac{a}{d - b}$

Student Assessment I

1. **a)** $10a - 30b + 15c$ **b)** $15x^2 - 27x$
 c) $-15xy^2 - 5y^3$ **d)** $15x^3y + 9x^2y^2 - 3x^5$
 e) $12 - p$ **f)** $14m^2 - 14m$ **g)** $4x + 3$
 h) $\frac{13}{2}m^2 - 3m$

2. **a)** $4(3a - b)$ **b)** $x(x - 4y)$
c) $4p^2(2p - q)$ **d)** $8xy(3 - 2x + y)$

3. **a)** -21 **b)** 26 **c)** 43 **d)** 7 **e)** 12
f) 12

4. **a)** $q = x - 3p$ **b)** $n = \dfrac{3m - 8r}{5}$

c) $y = \dfrac{2mt}{3}$ **d)** $w = \dfrac{2y}{x} - y$ **e)** $p = \dfrac{xyt}{2rs}$

f) $x = w(m + n) - y$

Student Assessment 2

1. **a)** $6x - 9y + 15z$ **b)** $8pm - 28p$
c) $-8m^2n + 4mn^2$ **d)** $20p^3q - 8p^2q^2 - 8p^3$
e) $-2x - 2$ **f)** $22x^2 - 14x$ **g)** 2
h) $\frac{5}{2}x^2 - x$

2. **a)** $8(2p - q)$ **b)** $p(p - 6q)$
c) $5pq(p - 2q)$ **d)** $3pq(3 - 2p + 4q)$

3. **a)** 0 **b)** -7 **c)** 29 **d)** 7 **e)** 7
f) 35

4. **a)** $n = p - 4m$ **b)** $y = \dfrac{4x - 5z}{3}$

c) $y = \dfrac{10px}{3}$ **d)** $y = \dfrac{3w}{m} - x$

e) $r = \dfrac{pqt}{4mn}$ **f)** $q = r(m - n) - p$

Extended Section page 101

Exercise 6.6

1. **a)** $x^2 + 5x + 6$ **b)** $x^2 + 7x + 12$
c) $x^2 + 7x + 10$ **d)** $x^2 + 7x + 6$
e) $x^2 + x - 6$ **f)** $x^2 + 5x - 24$

2. **a)** $x^2 + 2x - 24$ **b)** $x^2 - 3x - 28$
c) $x^2 - 2x - 35$ **d)** $x^2 - 2x - 15$
e) $x^2 - 2x - 3$ **f)** $x^2 + 2x - 63$

3. **a)** $x^2 - 5x + 6$ **b)** $x^2 - 7x + 10$
c) $x^2 - 12x + 32$ **d)** $x^2 + 6x + 9$
e) $x^2 - 6x + 9$ **f)** $x^2 - 12x + 35$

4. **a)** $x^2 - 9$ **b)** $x^2 - 49$ **c)** $x^2 - 64$
d) $x^2 - y^2$ **e)** $a^2 - b^2$ **f)** $p^2 - q^2$

5. **a)** $2y^2 + 7y + 6$ **b)** $3y^2 + 25y + 28$
c) $2y^2 + 17y + 8$ **d)** $4y^2 + 6y + 2$
e) $6y^2 + 23y + 20$ **f)** $18y^2 + 15y + 3$

6. **a)** $2p^2 + 13p - 24$ **b)** $4p^2 + 23p - 35$
c) $6p^2 + p - 12$ **d)** $12p^2 + 13p - 35$
e) $18p^2 - 2$ **f)** $28p^2 + 44p - 24$

7. **a)** $4x^2 - 4x + 1$ **b)** $9x^2 + 6x + 1$
c) $16x^2 - 16x + 4$ **d)** $25x^2 - 40x + 16$
e) $4x^2 + 24x + 36$ **f)** $4x^2 - 9$

8. **a)** $-4x^2 + 9$ **b)** $16x^2 - 9$ **c)** $-16x^2 + 9$
d) $-25y^2 + 49$ **e)** $8y^2 - 18$
f) $25y^2 - 70y + 49$

Exercise 6.7

1. **a)** $(x + y)(a + b)$ **b)** $(x - y)(a + b)$
c) $(3 + x)(m + n)$ **d)** $(m + n)(4 + x)$
e) $(m - n)(3 + x)$ **f)** $(x + z)(6 + y)$

2. **a)** $(p + q)(r - s)$ **b)** $(p + 3)(q - 4)$
c) $(q - 4)(p + 3)$ **d)** $(r + 2t)(s + t)$
e) $(s + t)(r - 2t)$ **f)** $(b + c)(a - 4c)$

3. **a)** $(y + x)(x + 4)$ **b)** $(x - 2)(x - y)$
c) $(a - 7)(b + 3)$ **d)** $(b - 1)(a - 1)$
e) $(p - 4)(q - 4)$ **f)** $(m - 5)(n - 5)$

4. **a)** $(m - 3)(n - 2)$ **b)** $(m - 3r)(n - 2r)$
c) $(p - 4q)(r - 4)$ **d)** $(a - c)(b - 1)$
e) $(x - 2y)(x - 2z)$ **f)** $(2a + b)(a + b)$

Exercise 6.8

1. **a)** $(a - b)(a + b)$ **b)** $(m - n)(m + n)$
c) $(x - 5)(x + 5)$ **d)** $(m - 7)(m + 7)$
e) $(9 - x)(9 + x)$ **f)** $(10 - y)(10 + y)$

2. **a)** $(12 - y)(12 + y)$ **b)** $(q - 13)(q + 13)$
c) $(m - 1)(m + 1)$ **d)** $(1 - t)(1 + t)$
e) $(2x - y)(2x + y)$ **f)** $(5p - 8q)(5p + 8q)$

3. **a)** $(3x - 2y)(3x + 2y)$
b) $(4p - 6q)(4p + 6q)$ **c)** $(8x - y)(8x + y)$
d) $(x - 10y)(x + 10y)$
e) $(pq - 2p)(pq + 2p)$
f) $(ab - cd)(ab + cd)$

4. **a)** $(mn - 3y)(mn + 3y)$
b) $(\frac{1}{2}x - \frac{1}{3}y)(\frac{1}{2}x + \frac{1}{3}y)$ **c)** $(p^2 - q^2)(p^2 + q^2)$
d) $(2m^2 - 6y^2)(2m^2 + 6y^2)$
e) $(4x^2 - 9y^2)(4x^2 + 9y^2)$
f) $(2x - 9y^2)(2x + 9y^2)$

Exercise 6.9

1. **a)** 60 **b)** 240 **c)** 2400 **d)** 280
e) 7600 **f)** 9200

2. **a)** 2000 **b)** 9800 **c)** 200 **d)** 3200
e) $998\,000$ **f)** 161

3. **a)** 68 **b)** 86 **c)** 1780 **d)** 70
e) 55 **f)** 5

4. **a)** 72.4 **b)** 0.8 **c)** 65 **d)** 15
e) $1\,222\,000$ **f)** 231

Exercise 6.10

1. **a)** $(x+4)(x+3)$ **b)** $(x+6)(x+2)$
   c) $(x+12)(x+1)$ **d)** $(x-3)(x-4)$
   e) $(x-6)(x-2)$ **f)** $(x-12)(x-1)$

2. **a)** $(x+5)(x+1)$ **b)** $(x+4)(x+2)$
   c) $(x+3)^2$ **d)** $(x+5)^2$ **e)** $(x+11)^2$
   f) $(x-6)(x-7)$

3. **a)** $(x+12)(x+2)$ **b)** $(x+8)(x+3)$
   c) $(x-6)(x-4)$ **d)** $(x+12)(x+3)$
   e) $(x+18)(x+2)$ **f)** $(x-6)^2$

4. **a)** $(x+5)(x-3)$ **b)** $(x-5)(x+3)$
   c) $(x+4)(x-3)$ **d)** $(x-4)(x+3)$
   e) $(x+6)(x-2)$ **f)** $(x-12)(x-3)$

5. **a)** $(x-4)(x+2)$ **b)** $(x-5)(x+4)$
   c) $(x+6)(x-5)$ **d)** $(x+6)(x-7)$
   e) $(x+7)(x-9)$ **f)** $(x+9)(x-6)$

6. **a)** $(2x+2)(x+1)$ **b)** $(2x+3)(x+2)$
   c) $(2x-3)(x+2)$ **d)** $(2x-3)(x-2)$
   e) $(3x+2)(x+2)$ **f)** $(3x-1)(x+4)$
   g) $(2x+3)^2$ **h)** $(3x-1)^2$
   i) $(3x+1)(2x-1)$

Exercise 6.11

1. **a)** $x=\dfrac{p}{2m}$ **b)** $x=\sqrt{\dfrac{T}{3}}$ **c)** $x=\sqrt{\dfrac{y^2}{m}}$

   d) $x=\sqrt{p^2-q^2-y^2}$

   e) $x=\sqrt{y^2-n^2-m^2}$

   f) $x=\sqrt{\dfrac{p^2-q^2+y^2}{4}}$

2. **a)** $x=\dfrac{P}{Qr}$ **b)** $x=\sqrt{\dfrac{P}{Qr}}$ **c)** $x=\sqrt{\dfrac{Pr}{Q}}$

   d) $x=\sqrt{\dfrac{n}{m}}$ **e)** $x=\sqrt{\dfrac{wst}{r}}$

   f) $x=\sqrt{\dfrac{wr}{p+q}}$

3. **a)** $x=(rp)^2$ **b)** $x=\left(\dfrac{mn}{p}\right)^2$ **c)** $x=\dfrac{k}{g^2}$

   d) $x=\dfrac{rg}{4\pi^2}$ **e)** $x=\dfrac{4m^2r}{p^2}$

   f) $x=\dfrac{4m^2r}{p^2}$

Exercise 6.12

1. **a)** $t=\dfrac{v-u}{a}$ **b)** $u=\sqrt{v^2-2as}$

c) $s=\dfrac{v^2-u^2}{2a}$ **d)** $u=\dfrac{s-\frac{1}{2}at^2}{t}$

e) $a=\dfrac{2(s-ut)}{t^2}$ **f)** $t=\sqrt{\dfrac{2(s-ut)}{a}}$

2. **a)** $r=\dfrac{A}{\pi\sqrt{s^2+t^2}}$ **b)** $h=\sqrt{\left(\dfrac{A}{\pi r}\right)^2-r^2}$

c) $u=\dfrac{vf}{v-f}$ **d)** $v=\dfrac{fu}{u-f}$

e) $l=\left(\dfrac{t}{2\pi}\right)^2 g$ **f)** $g=l\left(\dfrac{2\pi}{t}\right)^2$

Exercise 6.13

1. **a)** $\dfrac{xp}{yq}$ **b)** $\dfrac{q}{y}$ **c)** $\dfrac{p}{r}$ **d)** $\dfrac{d}{c}$ **e)** $\dfrac{bd}{c^2}$

   f) p

2. **a)** m^2 **b)** r^5 **c)** x^6 **d)** xy^2

   e) abc^3 **f)** $\dfrac{r^3}{pq}$

3. **a)** $\dfrac{2x}{y}$ **b)** $4q^2$ **c)** $5n$ **d)** $3x^3y$

   e) $3p$ **f)** $\dfrac{2}{3mn}$

4. **a)** $\dfrac{2a}{3b}$ **b)** $\dfrac{2y}{x}$ **c)** 2 **d)** $3xy$ **e)** 3

   f) $6xy$

5. **a)** $\dfrac{8y}{3}$ **b)** $\dfrac{5p}{2}$ **c)** $\dfrac{p^3}{5}$ **d)** $\dfrac{e}{c^2}$

   e) $\dfrac{4qm}{9p}$ **f)** $\dfrac{x^3}{yz}$

Exercise 6.14

1. **a)** $\dfrac{4}{7}$ **b)** $\dfrac{a+b}{7}$ **c)** $\dfrac{11}{13}$ **d)** $\dfrac{c+d}{13}$

   e) $\dfrac{x+y+z}{3}$ **f)** $\dfrac{p^2+q^2}{5}$

2. **a)** $\dfrac{3}{11}$ **b)** $\dfrac{c-d}{11}$ **c)** $\dfrac{4}{a}$ **d)** $\dfrac{2a-5b}{3}$

   e) $\dfrac{2x-3y}{7}$ **f)** $-\dfrac{1}{2x}$

3. **a)** $\dfrac{1}{2}$ **b)** $\dfrac{3}{2a}$ **c)** $\dfrac{5}{3c}$ **d)** $\dfrac{7}{2x}$

e) $\dfrac{3}{2p}$ **f)** $-\dfrac{1}{2w}$

4. a) $\dfrac{3p - q}{12}$ **b)** $\dfrac{x - 2y}{4}$ **c)** $\dfrac{3m - n}{9}$

d) $\dfrac{x - 2y}{12}$ **e)** $\dfrac{5r + m}{10}$ **f)** $\dfrac{5s - t}{15}$

5. a) $\dfrac{7x}{12}$ **b)** $\dfrac{9x - 2y}{15}$ **c)** $\dfrac{m}{2}$ **d)** $\dfrac{m}{p}$

e) $\dfrac{x}{2y}$ **f)** $\dfrac{4r}{7s}$

Exercise 6.15

1. a) $\dfrac{5}{6}$ **b)** $\dfrac{8}{15}$ **c)** $\dfrac{11}{28}$ **d)** $\dfrac{11}{15}$

e) $\dfrac{29}{36}$ **f)** $\dfrac{24}{35}$

2. a) $\dfrac{3a + 2b}{6}$ **b)** $\dfrac{5a + 3b}{15}$ **c)** $\dfrac{7p + 4q}{28}$

d) $\dfrac{6a + 3b}{15}$ **e)** $\dfrac{9x + 20y}{36}$ **f)** $\dfrac{10x + 14y}{35}$

3. a) $\dfrac{a}{6}$ **b)** $\dfrac{2a}{15}$ **c)** $\dfrac{11p}{28}$ **d)** $\dfrac{11a}{15}$

e) $\dfrac{29x}{36}$ **f)** $\dfrac{24x}{35}$

4. a) $\dfrac{m}{10}$ **b)** $\dfrac{r}{10}$ **c)** $-\dfrac{x}{4}$ **d)** $\dfrac{29x}{28}$

e) $\dfrac{23x}{6}$ **f)** $\dfrac{p}{6}$

5. a) $\dfrac{p}{2}$ **b)** $\dfrac{2c}{3}$ **c)** $\dfrac{4x}{5}$ **d)** $\dfrac{m}{3}$ **e)** $\dfrac{q}{5}$

f) $\dfrac{w}{4}$

6. a) $\dfrac{3m}{2}$ **b)** $\dfrac{7m}{3}$ **c)** $\dfrac{-m}{2}$ **d)** $\dfrac{5m}{2}$

e) $\dfrac{p}{3}$ **f)** $\dfrac{36q}{7}$

7. a) $\dfrac{pr - p}{r}$ **b)** $\dfrac{x + xy}{y}$ **c)** $\dfrac{mn + m}{n}$

d) $\dfrac{a + ab}{b}$ **e)** $\dfrac{2xy - x}{y}$ **f)** $\dfrac{2pq - 3p}{q}$

Exercise 6.16

1. a) $\dfrac{3x + 4}{(x + 2)(x + 1)}$ **b)** $\dfrac{m - 7}{(m + 2)(m - 1)}$

c) $\dfrac{3p - 7}{(p - 3)(p - 2)}$ **d)** $\dfrac{w + 11}{(w - 1)(w + 3)}$

e) $\dfrac{3y}{(y + 4)(y + 1)}$ **f)** $\dfrac{12 - m}{(m - 2)(m + 3)}$

2. a) $\dfrac{x}{x + 2}$ **b)** $\dfrac{y}{y + 3}$ **c)** $\dfrac{m + 2}{m - 3}$

d) $\dfrac{p}{p - 5}$ **e)** $\dfrac{m}{m + 4}$ **f)** $\dfrac{m + 1}{m + 2}$

3. a) $\dfrac{x}{x + 3}$ **b)** $\dfrac{x}{x + 4}$ **c)** $\dfrac{y}{y - 1}$

d) $\dfrac{x}{x + 3}$ **e)** $\dfrac{x}{x + 2}$ **f)** $\dfrac{x}{x + 1}$

4. a) $\dfrac{x}{x + 1}$ **b)** $\dfrac{x}{x + 3}$ **c)** $\dfrac{x}{x - 3}$

d) $\dfrac{x}{x + 2}$ **e)** $\dfrac{x}{x - 3}$ **f)** $\dfrac{x}{x + 7}$

Student Assessment 1

1. a) $(m - 5n)(x - 5)$ **b)** $(2x - 9y)(2x + 9y)$
  c) 7600 **d)** $(x^2 - y^2)(x^2 + y^2)$

2. a) $x^2 + 8x + 15$ **b)** $x^2 - 14x + 49$
  c) $x^2 + 10x + 25$ **d)** $x^2 - 5x + 14$
  e) $6x^2 + 13x - 8$ **f)** $25y^2 - 70y + 49$

3. a) $(x - 16)(x - 2)$ **b)** $(x + 4)(x - 6)$
  c) $(x - 6)(x - 3)$ **d)** $(x - 1)^2$
  e) $(2x - 1)(x + 3)$ **f)** $(3x - 2)^2$

4. a) $a = \dfrac{v^2 - u^2}{2s}$ **b)** $h = \sqrt{p^2 - r^2}$

  c) $s = \sqrt{\dfrac{rn}{m}}$ **d)** $g = \dfrac{4\pi^2 l}{t^2}$

5. a) b^2 **b)** x^4 **c)** $\dfrac{4n^2}{m}$ **d)** $\dfrac{p^3 q^3}{r^2}$

6. a) $\dfrac{11p}{5}$ **b)** $\dfrac{12m + 5n}{16}$ **c)** $\dfrac{3m}{4y}$

  d) $\dfrac{5r}{12x}$

7. a) $\dfrac{10x}{21}$ **b)** $\dfrac{8x - 9y}{12}$ **c)** $\dfrac{6r}{7}$

8. **a)** $\dfrac{5m + 13}{(m + 2)(m + 3)}$ **b)** $\dfrac{y + 3}{y - 3}$

c) $\dfrac{x}{x + 7}$

Student Assessment 2

1. **a)** $(q + r)(p - 3r)$ **b)** $(1 - t^2)(1 + t^2)$
c) $750\,000$ **d)** 50

2. **a)** $x^2 - 2x - 8$ **b)** $x^2 - 16x + 64$
c) $x^2 + 2xy + y^2$ **d)** $x^2 - 121$
e) $6x^2 - 13x + 6$ **f)** $9x^2 - 30x + 25$

3. **a)** $(x - 11)(x + 7)$ **b)** $(x - 3)(x - 3)$
c) $(x - 12)(x + 12)$ **d)** $3(x - 2)(x + 3)$
e) $(2x - 3)(x + 4)$ **f)** $(2x - 5)^2$

4. **a)** $f = \sqrt{\dfrac{p}{m}}$ **b)** $t = \sqrt{\dfrac{m}{5}}$

c) $p = \left(\dfrac{A}{\pi r}\right)^2 - q$ **d)** $x = \dfrac{ty}{y - t}$

5. **a)** x^4 **b)** nq **c)** y^3 **d)** $\dfrac{4}{q}$

6. **a)** $\dfrac{2m}{11}$ **b)** $\dfrac{-3p}{16}$ **c)** $\dfrac{3x}{4y}$

d) $\dfrac{6m + 13n}{30p}$

7. **a)** $\dfrac{9p}{20}$ **b)** $\dfrac{m}{10}$ **c)** $\dfrac{-p}{12}$

8. **a)** $\dfrac{7x - 23}{(x - 5)(x - 2)}$ **b)** $\dfrac{a - b}{a + b}$ **c)** $\dfrac{1}{x + 3}$

Student Assessment 3

1. **a)** 0.20 m^3 (2 d.p.) **b)** $r = \sqrt{\dfrac{V}{\pi h}}$

c) $r = 5.40$ cm

2. **a)** 2400 cm^2 **b)** $h = \dfrac{A}{2\pi r} - r$ **c)** 10.9 cm

3. **a)** 1994.30 fr **b)** Mario **c)** £50
d) \$452.64

4. **a)** 5.4 cm (1 d.p.) **b)** 3.7 m (1 d.p.)
c) $x = \sqrt{d^2 - y^2 - z^2}$ **d)** 0.71 m (2 d.p.)

5. **a)** 4.4 s (1 d.p.) **b)** $l = \dfrac{T^2 g}{4\pi^2}$

c) 2.28 m (2 d.p.)

Student Assessment 4

1. **a)** 0.53 m^2 (2 d.p.) **b)** $r = \sqrt{\dfrac{V}{\pi r}}$

c) 5.05 cm

2. **a)** $66\,°\text{C}$ (2 s.f.) **b)** $-11\,°\text{C}$ (2 s.f.)

c) $F = \dfrac{9C}{5} + 32$ **d)** $320\,°\text{F}$

3. **a)** 15 hours **b)** $H = 1200(T - k)$
c) 5000 m

4. **a)** 524 cm^3 (3 s.f.) **b)** $r = \sqrt[3]{\dfrac{3V}{4\pi}}$

c) $r = 8.4$ m (1 d.p.)

5. **a)** £251.50 **b)** $n = \dfrac{x - 1.5}{0.05}$ **c)** $n = 470$

7 Equations and Inequalities

Core Section page 116

Exercise 7.1

1. **a)** $x = -4$ **b)** $y = 5$ **c)** $y = -5$
d) $p = -4$ **e)** $y = 8$ **f)** $x = -5.5$

2. **a)** $x = 4\frac{1}{3}$ **b)** $x = 5$ **c)** $x = 6$
d) $y = -8$ **e)** $y = 4$ **f)** $m = 10$

3. **a)** $m = 1$ **b)** $p = 3$ **c)** $k = -1$
d) $x = -21$ **e)** $x = 2$ **f)** $y = 3$

4. **a)** $x = 6$ **b)** $y = 14$ **c)** $x = 4$
d) $m = 12$ **e)** $x = 35$ **f)** $p = 20$

5. **a)** $x = 15$ **b)** $x = -5$ **c)** $x = 7.5$
d) $x = 8$ **e)** $x = 2.5$ **f)** $x = 10$

6. **a)** $x = 5$ **b)** $x = 14$ **c)** $x = 22$
d) $x = 5$ **e)** $x = 8$ **f)** $x = 2$

7. **a)** $y = 10$ **b)** $x = 17$ **c)** $x = 13$
d) $y = -5$ **e)** $x = 4$ **f)** $x = 6.5$

Exercise 7.2

1. **a)** i) $3x + 60 = 180$ ii) $x = 40°$
iii) $40°, 60°, 80°$
b) i) $3x = 180$ ii) $x = 60°$ iii) $20°, 80°, 80°$
c) i) $18x = 180$ ii) $x = 10°$
iii) $20°, 50°, 110°$
d) i) $6x = 180$ ii) $x = 30°$ iii) $30°, 60°, 90°$

e) i) $7x - 30 = 180$ ii) $x = 30°$
iii) $10°, 40°, 130°$
f) i) $9x - 45 = 180$ ii) $x = 25°$
iii) $25°, 55°, 100°$

2. **a)** i) $12x = 360$ ii) $x = 30°$
iii) $90°, 120°, 150°$
b) i) $11x + 30 = 360$ ii) $x = 30°$
iii) $90°, 135°, 135°$
c) i) $12x + 60 = 360$ ii) $x = 25°$
iii) $35°, 80°, 90°, 155°$
d) i) $10x + 30 = 360$ ii) $x = 33°$
iii) $33°, 94°, 114°, 119°$

3. **a)** i) $11x - 80 = 360$ ii) $x = 40°$
iii) $40°, 80°, 80°, 160°$
b) i) $10x + 60 = 360$ ii) $x = 30°$
iii) $45°, 90°, 95°, 130°$
c) i) $9x + 45 = 360$ ii) $x = 35°$
iii) $35°, 55°, 120°, 150°$
d) i) $16x + 8 = 360$ ii) $x = 22°$
iii) $44°, 96°, 100°, 120°$
e) i) $6x + 18 = 180$ ii) $x = 27°$
iii) $50°, 50°, 130°, 130°$
f) i) $14x - 6 = 540$ ii) $x = 39°$
iii) $77°, 86°, 105°, 119°, 153°$

4. **a)** $20°$ **b)** $25°$ **c)** 14 **d)** $25°$
e) 31 **f)** $40°$

5. **a)** $50°$ **b)** $13°$ **c)** $40°$ **d)** $40°$

6. **a)** 5 **b)** 2 **c)** 7 **d)** 1.1 **e)** 25
f) 15

Exercise 7.3

1. **a)** $x = 4$ $y = 2$ **b)** $x = 6$ $y = 5$
c) $x = 6$ $y = -1$ **d)** $x = 5$ $y = 2$
e) $x = 5$ $y = 2$ **f)** $x = 4$ $y = 9$

2. **a)** $x = 3$ $y = 2$ **b)** $x = 7$ $y = 4$
c) $x = 1$ $y = 1$ **d)** $x = 1$ $y = 5$
e) $x = 1$ $y = 10$ **f)** $x = 8$ $y = 2$

3. **a)** $x = 5$ $y = 4$ **b)** $x = 4$ $y = 3$
c) $x = 10$ $y = 5$ **d)** $x = 6$ $y = 4$
e) $x = 4$ $y = 4$ **f)** $x = 10$ $y = -2$

4. **a)** $x = 5$ $y = 4$ **b)** $x = 4$ $y = 2$
c) $x = 5$ $y = 3$ **d)** $x = 5$ $y = -2$
e) $x = 1$ $y = 5$ **f)** $x = -3$ $y = -3$

5. **a)** $x = -5$ $y = -2$ **b)** $x = -3$ $y = -4$
c) $x = 4$ $y = 3\frac{2}{3}$ **d)** $x = 2$ $y = 7$
e) $x = 1$ $y = 1$ **f)** $x = 2$ $y = 9$

6. **a)** $x = 2$ $y = 3$ **b)** $x = 5$ $y = 10$
c) $x = 4$ $y = 6$ **d)** $x = 4$ $y = 4$
e) $x = 5$ $y = 1$ **f)** $x = -3$ $y = -3$

7. **a)** $x = 1$ $y = -1$ **b)** $x = 11\frac{2}{3}$ $y = 8$
c) $x = 4$ $y = 0$ **d)** $x = 3$ $y = 4$
e) $x = 2$ $y = 8$ **f)** $x = 1$ $y = 1$

Exercise 7.4

1. **a)** $x = 2$ $y = 3$ **b)** $x = 1$ $y = 4$
c) $x = 5$ $y = 2$ **d)** $x = 3$ $y = 3$
e) $x = 4$ $y = 2$ **f)** $x = 6$ $y = 1$

2. **a)** $x = 1$ $y = 4$ **b)** $x = 5$ $y = 2$
c) $x = 3$ $y = 3$ **d)** $x = 6$ $y = 1$
e) $x = 2$ $y = 3$ **f)** $x = 2$ $y = 3$

3. **a)** $x = 0$ $y = 3$ **b)** $x = 5$ $y = 2$
c) $x = 1$ $y = 7$ **d)** $x = 6$ $y = 4$
e) $x = 2$ $y = 5$ **f)** $x = 3$ $y = 0$

4. **a)** $x = 1$ $y = 0.5$ **b)** $x = 2.5$ $y = 4$
c) $x = \frac{1}{5}$ $y = 4$ **d)** $x = \frac{3}{4}$ $y = \frac{1}{2}$
e) $x = 11$ $y = -\frac{5}{3}$ **f)** $x = \frac{1}{2}$ $y = 1$

Exercise 7.5

1. 10 and 7

2. 16 and 9

3. $x = 1$ $y = 4$

4. $x = 2$ $y = 5$

5. 60 and 20 years old

6. 60 and 6 years old

Student Assessment 1

1. **a)** 9 **b)** 11 **c)** -4 **d)** 6

2. **a)** 1.5 **b)** 7 **c)** 4 **d)** 3

3. **a)** -10 **b)** 12 **c)** 10 **d)** $11\frac{1}{4}$

4. **a)** 16 **b)** $-8\frac{2}{3}$ **c)** 2 **d)** 3.5

5. **a)** $x = 5$ $y = 2$ **b)** $x = 3\frac{1}{3}$ $y = 4\frac{1}{3}$
c) $x = 5$ $y = 4$ **d)** $x = 5$ $y = 1$

Student Assessment 2

1. **a)** -6 **b)** 6 **c)** 4 **d)** 2.4

2. **a)** 0.5 **b)** 4 **c)** 9.5 **d)** 5

3. **a)** 6 **b)** 15 **c)** 22 **d)** 6

4. **a)** 8.5 **b)** $4\frac{1}{3}$ **c)** 8.5 **d)** 12

5. **a)** $x = 7$ $y = 4$ **b)** $p = 1$ $q = 2$
c) $x = 7$ $y = 1$ **d)** $m = 3$ $n = 5$

Extended Section page 126

Exercise 7.6

1. **a)** 7 **b)** 3 **c)** 6 **d)** 7 **e)** 10
   f) 15
2. **a)** 7 **b)** 6 **c)** 6 **d)** 10 **e)** 3
3. **a)** 8 **b)** 5 **c)** 8 **d)** 6 **e)** 6
4. **a)** 6, 18, 26 **b)** 160, 214, 246, 246
   c) 50°, 80°, 100°, 140°, 170°
   d) 80°, 80°, 80°, 160°, 160°, 160°
   e) 150°, 150°, 150°, 150°, 120°, 120°, 120°, 120°

Exercise 7.7

1. **a)** −4 and −3 **b)** −2 and −6
   c) −1 and −12 **d)** 2 and 5 **e)** 2 and 3
   f) 2 and 4
2. **a)** −5 and 2 **b)** −2 and 5 **c)** −7 and 2
   d) −2 and 7 **e)** −5 and 3 **f)** −3 and 5
3. **a)** −3 and −2 **b)** −3 **c)** −8 and −3
   d) 4 and 6 **e)** −4 and 3 **f)** −2 and 6
4. **a)** −2 and 4 **b)** −4 and 5 **c)** −6 and 5
   d) −6 and 7 **e)** −7 and 9 **f)** −9 and 6

Exercise 7.8

1. **a)** −3 and 3 **b)** −4 and 4 **c)** −5 and 5
   d) −11 and 11 **e)** −12 and 12
   f) −15 and 15
2. **a)** −2.5 and 2.5 **b)** −2 and 2
   c) −1.6 and 1.6 **d)** −$\frac{1}{2}$ and $\frac{1}{2}$
   e) −$\frac{1}{3}$ and $\frac{1}{3}$ **f)** −$\frac{1}{20}$ and $\frac{1}{20}$
3. **a)** −4 and −1 **b)** −5 and −2
   c) −4 and −2 **d)** 2 and 4 **e)** 2 and 5
   f) −4 and 2
4. **a)** −2 and 5 **b)** −5 and 2 **c)** −3 and 6
   d) −6 and 3 **e)** −4 and 6 **f)** −6 and 8
5. **a)** −4 and 3 **b)** −6 and −2
   c) −9 and 4 **d)** −1 **e)** −2
   f) −9 and −8
6. **a)** 0 and 8 **b)** 0 and 7 **c)** −3 and 0
   d) −4 and 0 **e)** 0 and 9 **f)** 0 and 4
7. **a)** −1.5 and −1 **b)** −1 and 2.5
   c) −1 and $\frac{1}{3}$ **d)** −5 and −$\frac{1}{2}$ **e)** 1.5 and 5
   f) −1$\frac{1}{3}$ and $\frac{1}{2}$
8. **a)** −12 and 0 **b)** −9 and −3
   c) −8 and 4 **d)** −7 and 2 **e)** −6 and 6
   f) −10 and 10

Exercise 7.9

1. −4 and 3
2. −6 and 7
3. 5
4. 4
5. 6
6. 10
7. 6
8. **a)** $9x + 14 = 50$ **b)** $x = 4$
   c) 11 m × 6 m

Exercise 7.10

1. **a)** −3.14 and 4.14 **b)** −5.87 and 1.87
   c) −6.14 and 1.14 **d)** −4.73 and −1.27
   e) −6.89 and 1.89 **f)** 3.38 and 5.62
2. **a)** −5.30 and −1.70 **b)** −5.92 and 5.92
   c) −3.79 and 0.79 **d)** −1.14 and 6.14
   e) −4.77 and 3.77 **f)** −2.83 and 2.83
3. **a)** −0.73 and 2.73 **b)** −1.87 and 5.87
   c) −1.79 and 2.79 **d)** −3.83 and 1.83
   e) 0.38 and 2.62 **f)** 0.39 and 7.61
4. **a)** −0.85 and 2.35 **b)** −1.40 and 0.90
   c) 0.14 and 1.46 **d)** −2.28 and −0.22
   e) −0.39 and 1.72 **f)** −1.54 and 1.39

Exercise 7.11

1. **a)** $x < 4$

   b) $x > 1$

   c) $x \leqslant 3$

   d) $x \geqslant 7$

   e) $x < 2$

   f) $x > 2$

2. **a)** $x < 7$

   b) $x \geqslant -2$

   c) $x > -3$

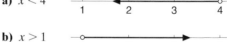

d) $x \geqslant -12$

e) $x > -24$

f) $x \geqslant -3$

3. **a)** $x < 2$

b) $x \leqslant 12$

c) $x \geqslant 2$

d) $x \geqslant -2$

e) $x > -0.5$

f) $x \geqslant 2$

Exercise 7.12

1. **a)** $2 < x \leqslant 4$

b) $1 \leqslant x < 5$

c) $3.5 \leqslant x < 5$

d) $2 \leqslant x < 4.2$

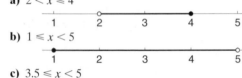

2. **a)** $1 < x \leqslant 5$

b) $-1 \leqslant x < 1$

c) $2 < x < 3$

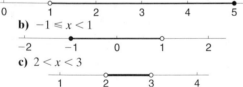

d) No solution

Student Assessment 1

1. **a)** $4x + 40 = 180$ **b)** $x = 35°$
c) $35°, 70°, 75°$

2. 9

3. $30°, 30°, 30°, 30°, 30°, 30°, 30°, 45°, 45°, 45°, 45°$

4. -5 and -1

5. 7.16 and 0.84 (2 d.p.)

6. $6 \leqslant x < 7$

7. None

Student Assessment 2

1. **a)** $9x - 90 = 360$ **b)** $x = 50°$
c) $50°, 60°, 100°, 150°$

2. 2

3. $135°, 85°, 95°, 95°, 130°$

4. -4 and 5

5. 2.19 and -0.69 (2 d.p.)

6. $3 < x \leqslant 5$

7. All values

Student Assessment 3

1. **a)**

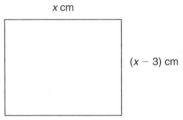

x cm

$(x - 3)$ cm

Perimeter = 54 cm

b) $4x - 6 = 54$
c) Length $= 15$ cm   Width $= 12$ cm

2. **a)** $x, x - 8, x - 23$ **b)** $55, 47, 32$

3. **b)** -0.32 and 6.32 (2 d.p.)

4. **a)** $x - y = 18$ $x + y + 70 + 40 = 360$
b) $x = 134$ $y = 116$

5. **a)**

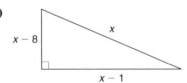

$x - 8$

x

$x - 1$

c) 5 cm, 12 cm, 13 cm

Student Assessment 4

1. **a)** $x - y = 30$ $x + y + 40 = 180$
b) $x = 85$ $y = 55$

2. **b)** $5x + 100$ **c)** $92°$
e) $92° + 82° + 72° + 62° + 52° = 360°$

3. **b)** $120 - 4x^2$ **c)** $4x^2 - 64 = 0$ $x = 4$

4. **b)** 0.44 and 4.56 (2 d.p.)

5. a)

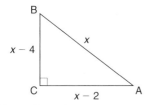

c) 10 cm, 8 cm, 6 cm

ICT Section page 138

1. a) $(x-3)(x-2)$
b) the curve crosses the x axis at 2 and 3

2. a) $(x+4)(x-1)$
b) $(x-5)(x-2)$
c) $(x-3)(x+1)$
d) $(x-6)(x-4)$
e) $(x-2)(x-2)$ or $(x-2)^2$
f) $(1-x)(1+x)$

8 Straight Line Graphs

Exercise 8.1 page 139

1. a) 1 **b)** 1 **c)** 1 **d)** 2 **e)** 3
f) 2 **g)** 4 **h)** $\frac{1}{2}$ **i)** 0 **j)** Infinite
k) $\frac{1}{4}$ **l)** $\frac{3}{2}$

2. a) -1 **b)** -1 **c)** -2 **d)** $-\frac{1}{2}$
e) -1 **f)** -2 **g)** $-\frac{3}{2}$ **h)** $\frac{2}{3}$ **i)** $-\frac{1}{4}$
j) -1 **k)** 0 **l)** -4

Exercise 8.2

1. a) i) 5.7 units **ii)** (3, 4)
b) i) 4.2 units **ii)** (4.5, 2.5)
c) i) 5.7 units **ii)** (3, 6)
d) i) 8.9 units **ii)** (2, 4)
e) i) 6.3 units **ii)** (3, 4)
f) i) 6.7 units **ii)** (−1.5, 4)
g) i) 8.2 units **ii)** (−2, 1)
h) i) 8.9 units **ii)** (0, 0)
i) i) 7 units **ii)** (0.5, 5)
j) i) 6 units **ii)** (2, 3)
k) i) 8.2 units **ii)** (0, 4)
l) i) 10.8 units **ii)** (0, 1.5)

2. a) i) 4.2 units **ii)** (2.5, 2.5)
b) i) 5.7 units **ii)** (5, 4)
c) i) 8.9 units **ii)** (4, 2)
d) i) 8.9 units **ii)** (5, 0)
e) i) 4.2 units **ii)** (−1.5, 4.5)
f) i) 4.5 units **ii)** (−4, −3)
g) i) 7.2 units **ii)** (0, 3)
h) i) 7.2 units **ii)** (5, −1)

i) i) 12.4 units **ii)** (0, 2.5)
j) i) 8.5 units **ii)** (1, −1)
k) i) 11 units **ii)** (−0.5, −3)
l) i) 8.2 units **ii)** (4, 2)

Exercise 8.3

1. a) $y=7$ **b)** $y=2$ **c)** $x=7$
d) $x=3$ **e)** $y=x$ **f)** $y=\frac{1}{2}x$
g) $y=-x$ **h)** $y=-2x$

Exercise 8.4

1. a) $y=x+1$ **b)** $y=x+3$ **c)** $y=x-2$
d) $y=2x+2$ **e)** $y=\frac{1}{2}x+5$
f) $y=\frac{1}{2}x-1$

2. a) $y=-x+4$ **b)** $y=-x-2$
c) $y=-2x-2$ **d)** $y=-\frac{1}{2}x+3$
e) $y=-\frac{3}{2}x+2$ **f)** $y=-4x+1$

3. a) 1) a) 1 **b)** 1 **c)** 1 **d)** 2 **e)** $\frac{1}{2}$
f) $\frac{1}{2}$
2) a) -1 **b)** -1 **c)** -2 **d)** $-\frac{1}{2}$
e) $-\frac{3}{2}$ **f)** -4
b) The gradient is equal to the coefficient of x.
c) The constant being added/subtracted indicates where the line intersects the y-axis.

4. Only the intercept c is different.

5. The lines are parallel.

Exercise 8.5

1. a) $m=2$ $c=1$ **b)** $m=3$ $c=5$
c) $m=1$ $c=-2$ **d)** $m=\frac{1}{2}$ $c=4$
e) $m=-3$ $c=6$ **f)** $m=-\frac{2}{3}$ $c=1$
g) $m=-1$ $c=0$ **h)** $m=-1$ $c=-2$
i) $m=-2$ $c=2$

2. a) $m=3$ $c=1$ **b)** $m=-\frac{1}{2}$ $c=2$
c) $m=-2$ $c=-3$ **d)** $m=-2$ $c=-4$
e) $m=\frac{1}{4}$ $c=6$ **f)** $m=3$ $c=2$
g) $m=1$ $c=-2$ **h)** $m=-8$ $c=6$
i) $m=3$ $c=1$

3. a) $m=2$ $c=-3$ **b)** $m=\frac{1}{2}$ $c=4$
c) $m=2$ $c=-4$ **d)** $m=-8$ $c=12$
e) $m=2$ $c=0$ **f)** $m=-3$ $c=3$
g) $m=2$ $c=1$ **h)** $m=-\frac{1}{2}$ $c=2$
i) $m=2$ $c=-\frac{1}{2}$

4. a) $m=2$ $c=-4$ **b)** $m=1$ $c=6$
c) $m=-3$ $c=-1$ **d)** $m=-1$ $c=4$
e) $m=10$ $c=-2$ **f)** $m=-3$ $c=\frac{3}{2}$
g) $m=-9$ $c=2$ **h)** $m=6$ $c=-14$
i) $m=2$ $c=-\frac{3}{2}$

5. a) $m = 2$, $c = -2$ b) $m = 2$ $c = 3$
 c) $m = 1$ $c = 0$ d) $m = \frac{3}{2}$ $c = 6$
 e) $m = -1$ $c = \frac{2}{3}$ f) $m = -4$ $c = 2$
 g) $m = 3$ $c = -12$ h) $m = 0$ $c = 0$
 i) $m = -3$ $c = 0$

6. a) $m = 1$ $c = 0$ b) $m = -\frac{1}{2}$ $c = -2$
 c) $m = -3$ $c = 0$ d) $m = 1$ $c = 0$
 e) $m = -2$ $c = -\frac{2}{3}$ f) $m = \frac{2}{3}$ $c = -4$
 g) $m = -\frac{2}{5}$ $c = 0$ h) $m = \frac{1}{3}$ $c = -\frac{5}{6}$
 i) $m = 3$ $c = 0$ j) $m = -1$ $c = -4$

Exercise 8.6

1. a) $y = 2x - 1$ b) $y = 3x + 1$
 c) $y = 2x + 3$ d) $y = x - 4$
 e) $y = 4x + 2$ f) $y = -x + 4$
 g) $y = -2x + 2$ h) $y = -3x - 1$
 i) $y = \frac{1}{2}x$

2. a) $y = \frac{1}{7}x + \frac{26}{7}$ b) $y = \frac{6}{7}x + \frac{4}{7}$
 c) $y = \frac{3}{2}x + \frac{15}{2}$ d) $y = 9x - 13$
 e) $y = -\frac{1}{2}x + \frac{5}{2}$ f) $y = -\frac{3}{13}x + \frac{70}{13}$
 g) $y = 2$ h) $y = -3x$ i) $x = 6$

Exercise 8.7

1. a)

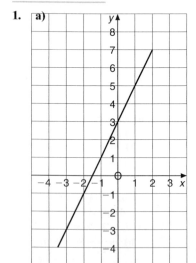

b)

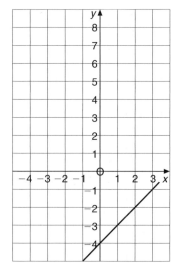

c)

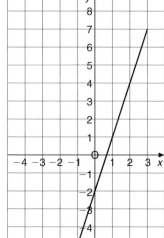

d)

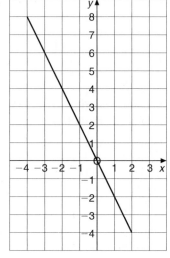

e)

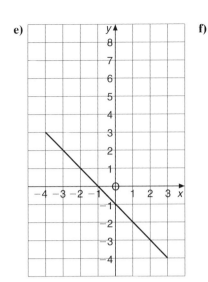

f)

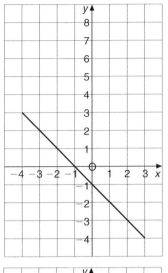

g)

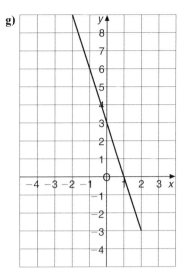

h)

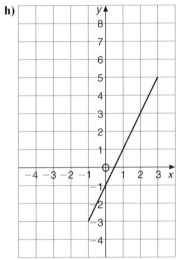

i)

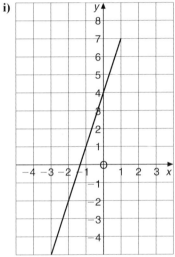

2.

a)

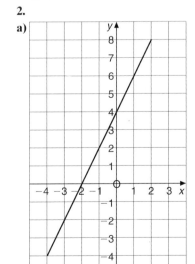

b)

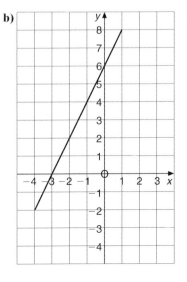

c)

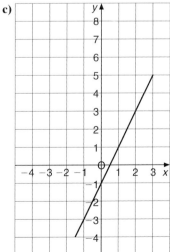

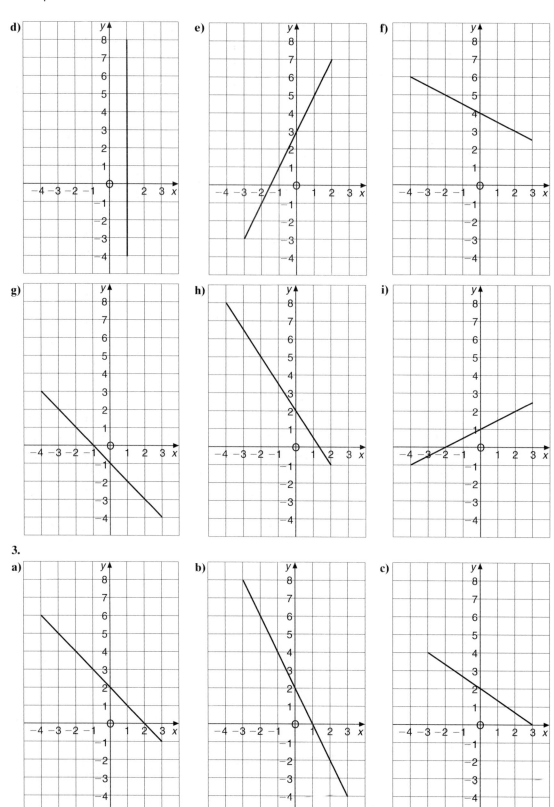

d)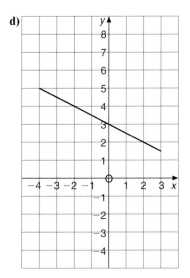

e)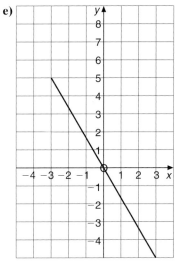

f)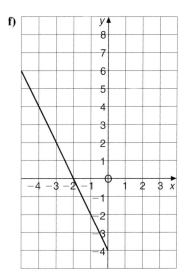

g)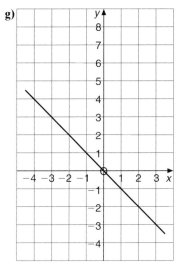

h)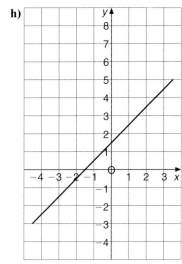

i)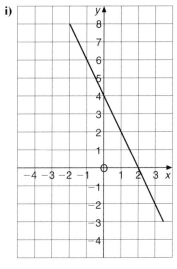

Exercise 8.8

1. **a)** $(3, 2)$ **b)** $(5, 2)$ **c)** $(2, 1)$
    d) $(2, 1)$ **e)** $(-4, 1)$ **f)** $(4, -2)$

2. **a)** $(3, -2)$ **b)** $(-1, -1)$ **c)** $(-2, 3)$
    d) $(-3, -3)$ **e)** Infinite solutions
    f) No solution

Student Assessment I

1. **a) b) c) d)**

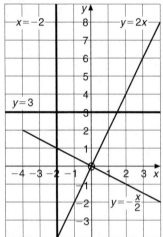

2. **a)** i) $m = 1$ $c = 1$
    ii)

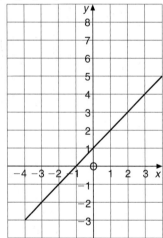

b) i) $m = -3$ $c = 3$
ii)

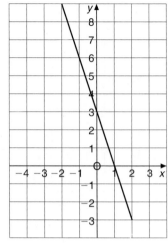

c) i) $m = 2$ $c = 4$
ii)

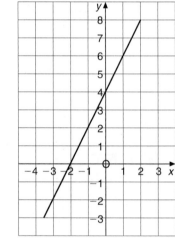

d) i) $m = \frac{5}{2}$ $c = 4$
ii)

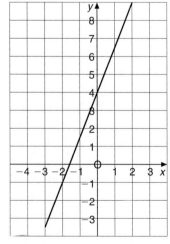

3. **a)** $y = 3x - 4$ **b)** $y = -2x + 7$

4. **a)** 13 units   **b)** 10 units

5. **a)**

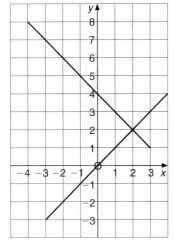

Solution is (2, 2)

b)

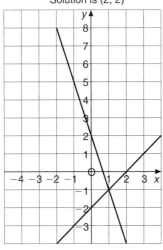

Solution is (1, −1)

c)

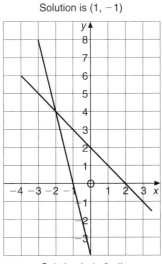

Solution is (−2, 4)

d)

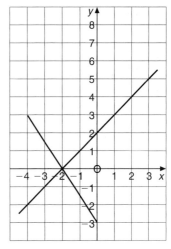

Solution is (−2, 0)

Student Assessment 2

1. **a) b) c) d)**

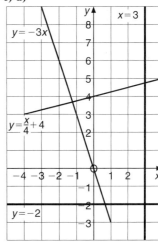

2. **a)** i) $m = 2$ $c = 3$

   ii)

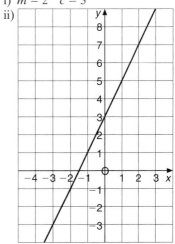

b) i) $m = -1 \quad c = 4$

ii)

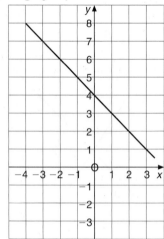

c) i) $m = 2 \quad c = -3$

ii)

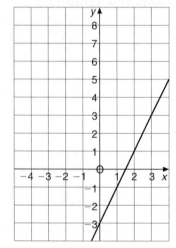

d) i) $m = \frac{3}{2} \quad c = \frac{5}{2}$

ii)

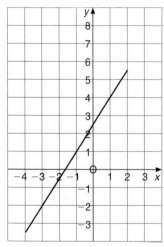

3. **a)** $y = 2x - 5$ **b)** $y = -4x + 3$

4. **a)** 5 units **b)** 26 units

5. **a)**

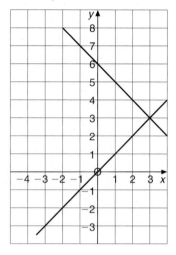

b)

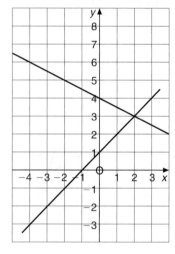

c)

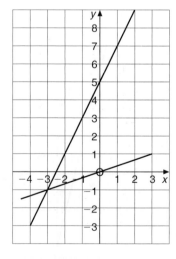

d)

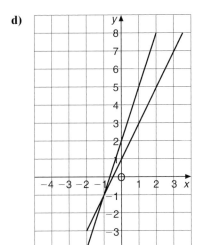

9 Functions

Exercise 9.1 page 153

1. **a)** 6 **b)** 10 **c)** 3
   d) 5 **e)** 2 **f)** −2
   g) −10 **h)** 1

2. **a)** 10 **b)** 22 **c)** 8
   d) −4 **e)** −5 **f)** −18
   g) −23 **h)** −6

3. **a)** 2 **b)** −28 **c)** −20.5
   d) −14 **e)** 1.5 **f)** 12
   g) 34.5 **h)** 13.5

4. **a)** −19 **b)** −26.5 **c)** −7
   d) −8.2 **e)** 20 **f)** 8
   g) −1 **h)** 3.5

Exercise 9.2

1. **a)** 2 **b)** 6.5 **c)** 2.375
   d) 0.5 **e)** 0.125 **f)** −4
   g) −2.5 **h)** −0.7

2. **a)** 4 **b)** 9 **c)** −1
   d) −6 **e)** −3.5 **f)** −16
   g) $-\frac{4}{3}$ **h)** $-\frac{7}{6}$

3. **a)** 0.5 **b)** 2 **c)** −4
   d) −0.25 **e)** 5 **f)** 2.75
   g) 35 **h)** 4.25

4. **a)** 4 **b)** 1.5 **c)** 2.75
   d) 0.25 **e)** −3.5 **f)** 0.5
   g) 0.375 **h)** 0.875

Exercise 9.3

1. **a)** 19 **b)** 52 **c)** 4 **d)** 3 **e)** 4
   f) 3.25 **g)** 12 **h)** 5

2. **a)** 70 **b)** 187 **c)** −2 **d)** −5 **e)** 7
   f) 4 **g)** −4.25 **h)** $-4\frac{2}{3}$

3. **a)** −14 **b)** 3.5 **c)** 4 **d)** −0.5
   e) −28 **f)** 2 **g)** −6 **h)** −68

4. **a)** −5 **b)** 32.5 **c)** 0 **d)** −6.875
   e) −7.5 **f)** 15 **g)** −7 **h)** −8.125

5. **a)** 0 **b)** 24 **c)** 0 **d)** 10.5
   e) −13.5 **f)** $-\frac{25}{6}$ **g)** −97.5
   h) 21.9 (1 d.p.)

6. **a)** 9 **b)** 0 **c)** −5 **d)** Infinite
   e) 0 **f)** 11.2 **g)** $-\frac{11}{6}$ **h)** 9

Exercise 9.4

1. **a)** $2x + 3$ **b)** $4x − 5$ **c)** $2x^2 + 1$
   d) $x + 1$ **e)** $\dfrac{x}{2} + 3$ **f)** $x + 1$

2. **a)** $12x^2 − 4$ **b)** $\dfrac{3x^2}{16} − 4$ **c)** $6x^2 − 4$
   d) $27x^2$ **e)** $3x^2 − 6x − 1$
   f) $12x^2 + 24x + 8$

3. **a)** $4x^2 + 3x + 2$ **b)** $16x^2 + 6x$
   c) $4x^2 + 19x$ **d)** $4x^2 − 5x$
   e) $\dfrac{2x^2 + 3x − 4}{2}$ **f)** $36x^2 + 57x + 20$

Exercise 9.5

1. **a)** $x − 3$ **b)** $x − 6$ **c)** $x + 5$
   d) x **e)** $\dfrac{x}{2}$ **f)** $3x$

2. **a)** $\dfrac{x}{4}$ **b)** $\dfrac{x − 5}{2}$ **c)** $\dfrac{x + 6}{3}$
   d) $2x − 4$ **e)** $\dfrac{4x + 2}{3}$ **f)** $\dfrac{5x − 7}{8}$

3. **a)** $2(x − 3)$ **b)** $4(x + 2)$
   c) $\dfrac{x + 24}{12}$ **d)** $\dfrac{x − 18}{6}$
   e) $\dfrac{x + 4}{6}$ **f)** $\dfrac{3x + 10}{8}$

Exercise 9.6

1. **a)** 6 **b)** 4 **c)** −1
2. **a)** 2 **b)** −0.5 **c)** −6
3. **a)** 3 **b)** 1.5 **c)** 2
4. **a)** 4 **b)** −2 **c)** −11
5. **a)** 4.5 **b)** 6 **c)** 0
6. **a)** 8 **b)** −2 **c)** 0.5

Exercise 9.7

1. **a)** $x + 2$ **b)** $x + 3$ **c)** $2x$ **d)** x
2. **a)** $2x + 8$ **b)** $6x + 1$ **c)** $8x + 2$
   d) $-x + 2$
3. **a)** $\dfrac{2x - 1}{2}$ **b)** $\dfrac{3x - 5}{2}$ **c)** $\dfrac{x + 7}{3}$
   d) $\dfrac{8x - 13}{20}$
4. **a)** 1 **b)** 8 **c)** 39 **d)** −2
5. **a)** 50 **b)** −2.9 **c)** 10 **d)** −1.6

Student Assessment 1

1. **a)** 9 **b)** −1 **c)** −16
2. **a)** 5 **b)** −1 **c)** −5.5
3. **a)** −6 **b)** 0 **c)** 15
4. **a)** $-x + 4$ **b)** $\dfrac{2x + 18}{3}$
5. **a)** 5 **b)** 2
6. $-4x + 14$

Student Assessment 2

1. **a)** 13 **b)** −2 **c)** 1
2. **a)** −2 **b)** 1 **c)** −1
3. **a)** −2 **b)** 0 **c)** 18
4. **a)** $\dfrac{x - 9}{-3}$ **b)** $4x + 6$
5. **a)** 1 **b)** 2
6. $32x - 6$

ICT Section page 159

1. **a)** $f^{-1}(x) = \dfrac{x}{3}$ **b)** $f^{-1}(x) = \dfrac{x + 2}{2}$
   c) $f^{-1}(x) = 2(x - 1)$ **d)** $f^{-1}(x) = \dfrac{3x - 1}{2}$
   e) $f^{-1}(x) = \dfrac{2x + 3}{4}$ **f)** $f^{-1}(x) = \dfrac{8x - 5}{3}$
3. Reflections in the line $y = x$

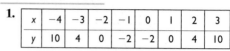

10 Graphs of Functions

Core Section page 160

Exercise 10.1

1.

x	−4	−3	−2	−1	0	1	2	3
y	10	4	0	−2	−2	0	4	10

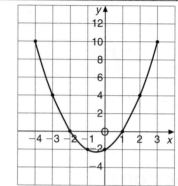

2.

x	−3	−2	−1	0	1	2	3	4	5
y	−12	−5	0	3	4	3	0	−5	−12

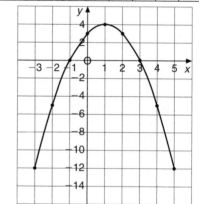

3.

x	−1	0	1	2	3	4	5
y	9	4	1	0	1	4	9

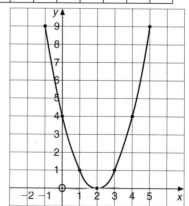

4.

x	−4	−3	−2	−1	0	1	2
y	−9	−4	−1	0	−1	−4	−9

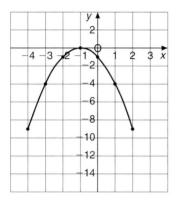

7.

x	−3	−2	−1	0	1	2	3
y	−15	−4	3	6	5	0	−9

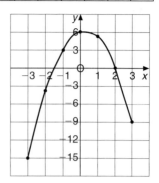

5.

x	−4	−3	−2	−1	0	1	2	3	4	5	6
y	9	0	−7	−12	−15	−16	−15	−12	−7	0	9

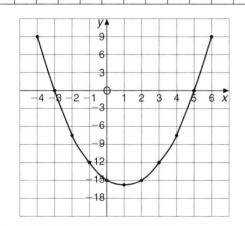

8.

x	−2	−1	0	1	2	3
y	12	0	−6	−6	0	12

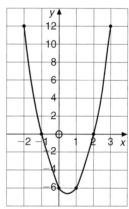

6.

x	−2	−1	0	1	2	3
y	9	1	−3	−3	1	9

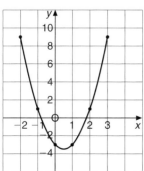

9.

x	−1	0	1	2	3
y	7	−4	−7	−2	11

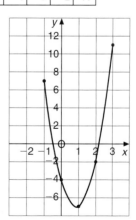

10.

x	−2	−1	0	1	2	3
y	−25	−9	−1	−1	−9	−25

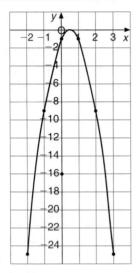

Exercise 10.2

1. −2 and 3

2. −1 and 1

3. 3

4. −4 and 3

5. 2

6. 0.5 and 3

7. 1

8. $-\frac{1}{3}$ and 2

Exercise 10.3

1. −1.6 and 2.6

2. No solution

3. 2 and 4

4. −3.5 and 2.5

5. 0.3 and 3.7

6. 0 and 3.5

7. −0.2 and 2.2

8. $-\frac{1}{3}$ and 2

Exercise 10.4

1.

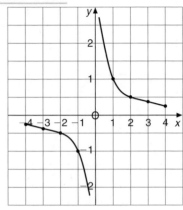

2.

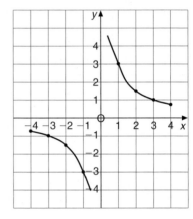

3.

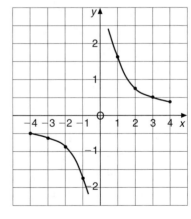

Student Assessment 1

1. **a)** 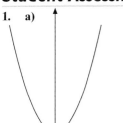 **b)**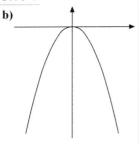

2. **a)**

x	−7	−6	−5	−4	−3	−2	−1	0	1	2
y	8	3	0	−1	0	3	8	15	24	35

b)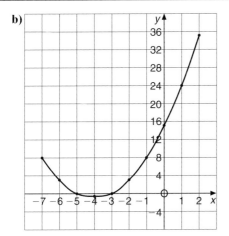

3. **a)**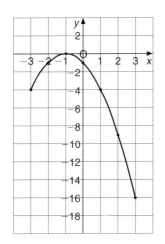

b)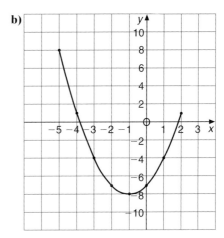

4. **a)**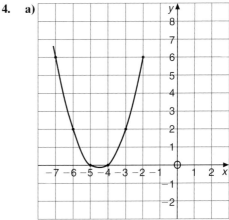

b) $x = -7$ and -2

5. **a)**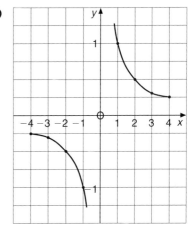

b) $x = 0.4$ and 2.6

Student Assessment 2

1.

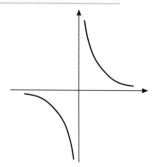

2. a)

x	−7	−6	−5	−4	−3	−2	−1	0	1	2
y	−12	−6	−2	0	0	−2	−6	−12	−20	−30

b)

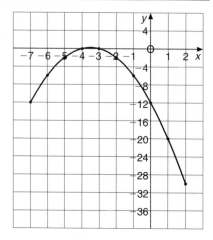

3. a)

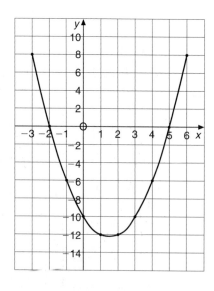

b)

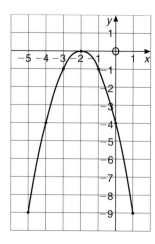

4. a)

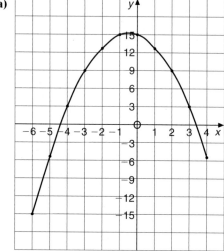

b) i) $x = -3.7$ and 2.7 ii) $x = -1.8$ and 2.8

5. a)

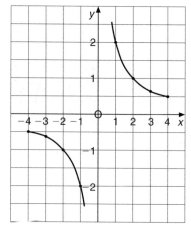

b) $x = -2$ and 1

Extended Section page 165

Exercise 10.5

1. i)

x	−3	−2	−1	0	1	2	3
f(x)	−7	−4	−1	2	5	8	11

ii)

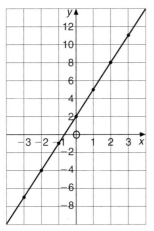

2. i)

x	−3	−2	−1	0	1	2	3
f(x)	2.5	3	3.5	4	4.5	5	5.5

ii)

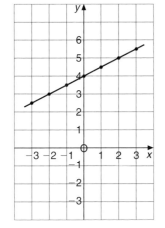

3. i)

x	−4	−3	−2	−1	0	1	2
f(x)	5	3	1	−1	−3	−5	−7

ii)

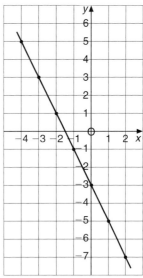

4. i)

x	−3	−2	−1	0	1	2	3
f(x)	17	7	1	−1	1	7	17

ii)

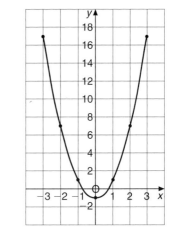

5. i)

x	−5	−4	−3	−2	−1	0	1	2	3
f(x)	5.5	2	−0.5	−2	−2.5	−2	−0.5	2	5.5

ii)

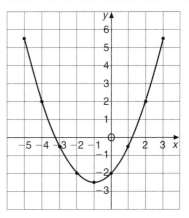

6. i)

x	−2	−1	0	1	2
f(x)	17	6	1	2	9

ii)

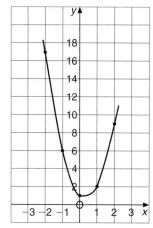

7. i)

x	−2	−1	0	1	2
f(x)	−16	−2	0	2	16

ii)

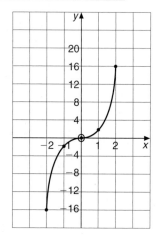

8. i)

x	−2	−1	0	1	2
f(x)	14	1	0	−1	−14

ii)

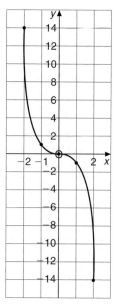

9. i)

x	−3	−2	−1	0	1	2	3
f(x)	−4.5	3	4.5	3	1.5	3	10.5

ii)

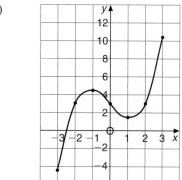

10. i)

x	−3	−2	−1	0	1	2	3
f(x)	−1	−1.5	−3	−	3	1.5	1

ii)

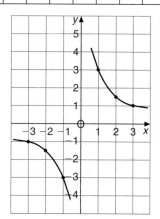

11. i)

x	−3	−2	−1	0	1	2	3
f(x)	0.22	0.5	2	−	2	0.5	0.22

ii)

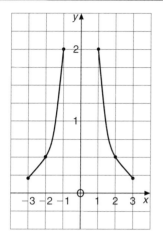

12. i)

x	−3	−2	−1	0	1	2	3
f(x)	−8.88	−5.75	−2	−	4	6.25	9.11

ii)

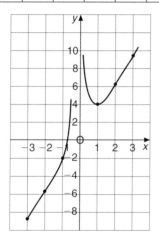

Exercise 10.6

1. i)

x	−3	−2	−1	0	1	2	3
f(x)	0.04	0.11	0.33	1	3	9	27

ii)

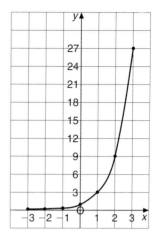

2. i)

x	−3	−2	−1	0	1	2	3
f(x)	1	1	1	1	1	1	1

ii)

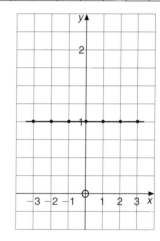

3. i)

x	−3	−2	−1	0	1	2	3
f(x)	3.125	3.25	3.5	4	5	7	11

ii)

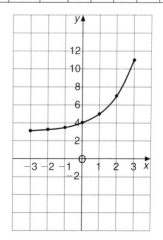

4. i)

x	−3	−2	−1	0	1	2	3
f(x)	−2.875	−1.75	−0.5	1	3	6	11

ii)

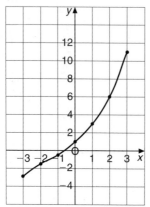

5. i)

x	−3	−2	−1	0	1	2	3
f(x)	3.125	2.25	1.5	1	1	2	5

ii)

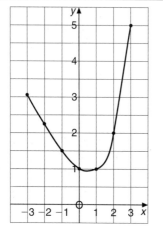

6. i)

x	−3	−2	−1	0	1	2	3
f(x)	−8.96	−3.89	−0.67	1	2	5	18

ii)

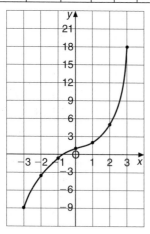

Exercise 10.7

1. i)   ii) 2

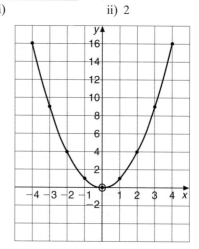

2. i)   ii) −2

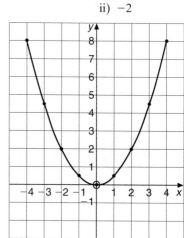

3. i)   ii) 3

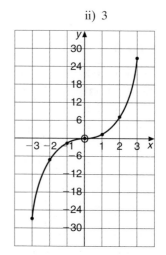

4. i)

ii) 24

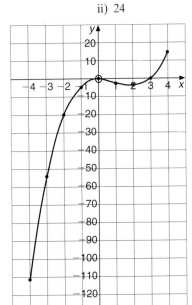

Exercise 10.8

1. **a)**   **b)** $x = \pm 2.8$

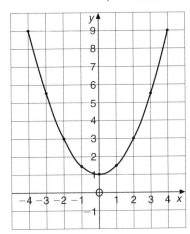

5. i)

ii) -4

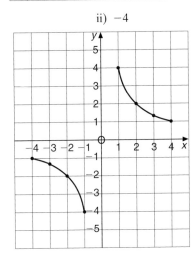

2. **a)**   **b)** $x = 1.7$

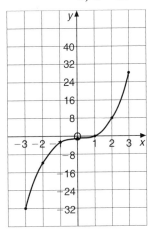

6. i)

ii) 0.7

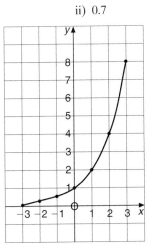

3. **a)**   **b)** $x = 1.5$

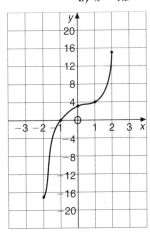

4. **a)**   **b)** $x = -0.4, 0.5, 2.4$

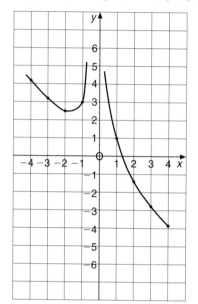

5. **a)**   **b)** $x = -0.7, 3$

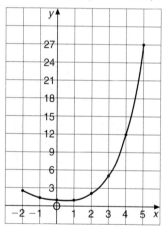

Student Assessment I

1. **a)** i) Linear   ii) Quadratic
 b) Student's own answer

2. **a)** i)

x	-5	-4	-3	-2	-1	0	1	2
$f(x)$	10	4	0	-2	-2	0	4	10

ii)

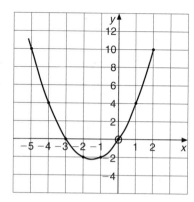

b) i)

x	-3	-2	-1	0	1	2	3
$f(x)$	-9.3	-6.5	-4	$-$	4	6.5	9.3

ii)

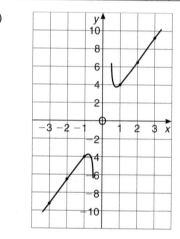

3. **a)**   **b)** i) 5.5 ii) -2.5

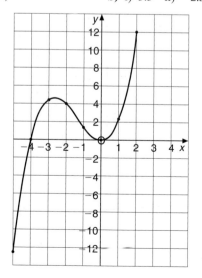

4. **a)**

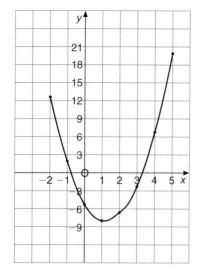

b) $x = -0.8$ and 3.3 **c)** $x = -1.6$ and 3.1

Student Assessment 2

1. **a)** i) Reciprocal   ii) Exponential
   b) Student's own answers

2. **a)** i)

x	-3	-2	-1	0	1	2	3
$f(x)$	-2.9	-1.8	-0.5	1	3	6	11

   ii)

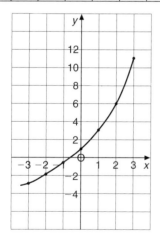

b) i)

x	-3	-2	-1	0	1	2	3
$f(x)$	-9.0	-3.9	-0.7	1	2	5	18

   ii)

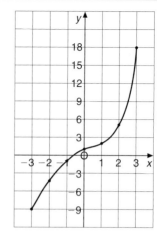

3. **a)**

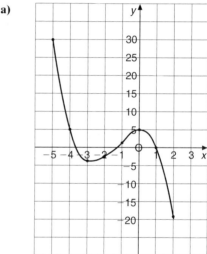

b) i) 0   ii) 4

4. a)

x	−3	−2	−1	−0.5	−0.25	0	0.25	0.5	1	2	3
y	−4.9	−4.8	−4	−1	11	−	11	−1	−4	−4.8	−4.9

b)

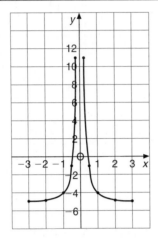

c) $x = \pm 0.4$ **d)** $x = \pm 0.4$ and ± 2.6

ICT Section page 172

2. Student's results may differ slightly

x	−4	−3	−2	−1	0	1	2	3	4
Gradient	−8	−6	−4	−2	0	2	4	6	8

3. For the function $y = x^2$, the gradient is $2x$

11 Indices

Core Section page 174

Exercise 11.1

1. **a)** 3^3 **b)** 2^5 **c)** 4^2 **d)** 6^4 **e)** 8^6
   f) 5^1

2. **a)** $2^3 \times 3^2$ **b)** $4^5 \times 5^2$ **c)** $3^2 \times 4^3 \times 5^2$
   d) 2×7^4 **e)** 6^2 **f)** $3^3 \times 4^2 \times 6^5$

3. **a)** 4×4 **b)** $5 \times 5 \times 5 \times 5 \times 5 \times 5 \times 5$
   c) $3 \times 3 \times 3 \times 3 \times 3$
   d) $4 \times 4 \times 4 \times 6 \times 6 \times 6$
   e) $7 \times 7 \times 2 \times 2 \times 2 \times 2 \times 2 \times 2 \times 2$
   f) $3 \times 3 \times 4 \times 4 \times 4 \times 2 \times 2 \times 2 \times 2$

4. **a)** 32   **b)** 81   **c)** 64   **d)** 216
   e) 1 000 000   **f)** 256   **g)** 72
   h) 125 000

Exercise 11.2

1. **a)** 3^6 **b)** 8^7 **c)** 5^9 **d)** 4^{10} **e)** 2^4
   f) $3^5 \times 6^6$ **g)** $4^8 \times 5^9 \times 6^2$
   h) $2^4 \times 5^{10} \times 6^8$

2. **a)** 4^4 **b)** 5^3 **c)** 2   **d)** 6^3 **e)** 6^3
   f) 8   **g)** 4^3 **h)** 3^7

3. **a)** 5^4 **b)** 4^{12} **c)** 10^{10} **d)** 3^{15} **e)** 6^8
   f) 8^6

4. **a)** 2^3 **b)** 3   **c)** 5^3 **d)** 4^5 **e)** 2^8
   f) $6^4 \times 8^5$ **g)** $4^3 \times 5^2$ **h)** 4×6^7

5. **a)** c^8 **b)** m^2 **c)** b^9 **d)** m^3n^6

   e) $2a^4b$ **f)** $3x^3y^2$ **g)** $\dfrac{uv^3}{2}$ **h)** $\dfrac{x^2y^3z^2}{3}$

6. **a)** $12a^5$ **b)** $8a^5b^3$ **c)** $8p^6$ **d)** $16m^4n^6$
   e) $200p^{13}$ **f)** $32m^5n^{11}$ **g)** $24xy^3$
   h) $a^{(d+e)}b^{(d+e)}$

Exercise 11.3

1. **a)** 8   **b)** 25   **c)** 1   **d)** 1   **e)** 1
   f) 0.25

2. **a)** $\frac{1}{4}$ **b)** $\frac{1}{9}$ **c)** $\frac{3}{50}$ **d)** $\frac{1}{200}$ **e)** 1
   f) $\frac{1}{1000}$

3. **a)** 1   **b)** 2   **c)** 4   **d)** $\frac{1}{2}$ **e)** $\frac{1}{6}$
   f) 10

4. **a)** 12   **b)** 32   **c)** 225   **d)** 80   **e)** 7
   f) 64

Exercise 11.4

1. **a)** 2   **b)** 4   **c)** 3   **d)** 3   **e)** 4
   f) 0

2. **a)** 4   **b)** 1   **c)** $\frac{3}{2}$ **d)** −1   **e)** 2
   f) 0

3. **a)** −3   **b)** −4   **c)** −5   **d)** $-\frac{2}{3}$
   e) −2   **f)** $-\frac{1}{2}$

4. **a)** −3   **b)** −7   **c)** −3   **d)** 2   **e)** 8
   f) 5

Exercise 11.5

1. **a)** 7 cm   **b)** 0 cm   **c)** 5 hours
   e) Approx. $5\frac{1}{2}$ hours

2. **a)** Approx. 2.5   **b)** $-\frac{1}{2}$

3. **a)** Approx. 4.3   **b)** Approx. 3.3
   c) Approx. 2.3

4. **a)** Approx. 0.3   **b)** Approx. 2.4

Student Assessment 1

1. **a)** $2^3 \times 5^2$ **b)** $2^2 \times 3^5$

2. **a)** $4 \times 4 \times 4$ **b)** $6 \times 6 \times 6 \times 6$

3. **a)** 800   **b)** 27

4. **a)** 3^7 **b)** $6^5 \times 3^9$ **c)** 2^7 **d)** 6
   e) $3^2 \times 4^2$ **f)** 1

5. **a)** 4   **b)** 9   **c)** 5   **d)** 1

6. **a)** 7   **b)** -2 **c)** -1 **d)** $\frac{1}{3}$

Student Assessment 2

1. **a)** $2^2 \times 3^5$ **b)** 2^{14}

2. **a)** $6 \times 6 \times 6 \times 6 \times 6$

   b) $\dfrac{1}{2 \times 2 \times 2 \times 2 \times 2}$

3. **a)** 27 000   **b)** 125

4. **a)** 2^7 **b)** $7^7 \times 3^{12}$ **c)** 2^6 **d)** 3^3
   e) 4^{-1} **f)** 2^2

5. **a)** 5   **b)** 16   **c)** 49   **d)** 48

6. **a)** 2.5   **b)** -0.5 **c)** 0   **d)** $\frac{2}{9}$

Extended Section page 181

Exercise 11.6

1. **a)** 4   **b)** 5   **c)** 10   **d)** 3   **e)** 9   **f)** 10

2. **a)** 2   **b)** 3   **c)** 2   **d)** 2   **e)** 6   **f)** 4

3. **a)** 8   **b)** 32   **c)** 27   **d)** 64   **e)** 1   **f)** 9

4. **a)** 25   **b)** 8   **c)** 32   **d)** 100   **e)** 32   **f)** 27

5. **a)** $\frac{1}{2}$ **b)** $\frac{1}{3}$ **c)** $\frac{1}{2}$ **d)** $\frac{1}{3}$ **e)** $\frac{1}{2}$ **f)** $\frac{1}{6}$

6. **a)** $\frac{1}{3}$ **b)** $\frac{1}{2}$ **c)** $\frac{1}{4}$ **d)** $\frac{1}{3}$ **e)** $\frac{1}{6}$ **f)** $\frac{1}{3}$

Exercise 11.7

1. **a)** 1   **b)** 7   **c)** 2   **d)** 1   **e)** 81   **f)** 6

2. **a)** 25   **b)** 2   **c)** 2   **d)** 27   **e)** 4   **f)** 2

3. **a)** 4   **b)** 2   **c)** 64   **d)** 9   **e)** 3   **f)** 27

Student Assessment 1

1. **a)** 9   **b)** 3   **c)** 3   **d)** 125   **e)** 49
   f) 0.5   **g)** 5   **h)** 16

2. **a)** 1   **b)** 9   **c)** 4   **d)** 25   **e)** 2
   f) 8   **g)** 1   **h)** 45

3. **a)**

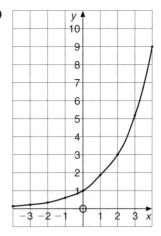

   b) Approx. 2.9

Student Assessment 2

1. **a)** 2   **b)** 81   **c)** $\frac{1}{3}$ **d)** 64   **e)** 3
   f) 2   **g)** 6   **h)** 32

2. **a)** 15   **b)** 4   **c)** 3   **d)** 3125   **e)** 4
   f) 1   **g)** 27   **h)** 10

3. **a)**

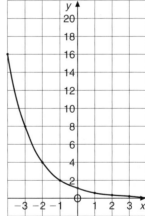

   b) Approx. -2.6

ICT Section page 184

1. $x \approx 2.7$
2. $x \approx 2.6$
3. $x \approx 2.1$
4. $x \approx 0.6$

12 Linear Programming

Exercise 12.1 page 185

1. **a)** $x < 2$ **b)** $y \geqslant 2$ **c)** $x \leqslant 2$
     d) $y \leqslant 6$ **e)** $t > 0$ **f)** $p \leqslant -3$

2. **a)** $2 < y \leqslant 4$ **b)** $1 \leqslant p < 5$
     c) $5 \leqslant m < 7$ **d)** $3 < x < 4$

Exercise 12.2

1. **a)**

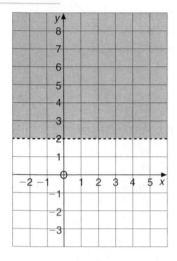

b)

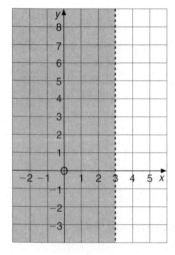

c)

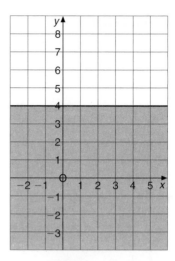

d)

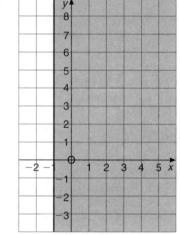

e)

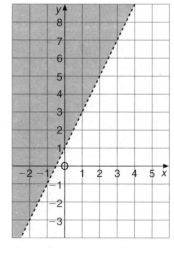

f)

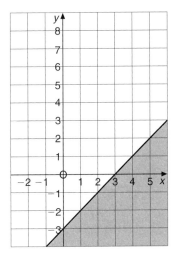

2. **a)**

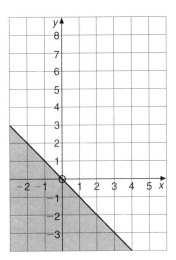

b)

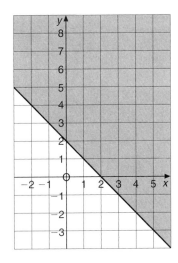

c)

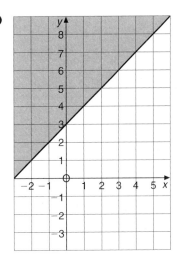

d)

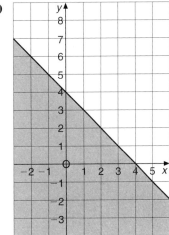

e)

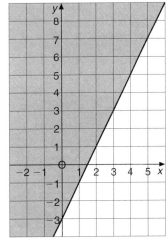

f)

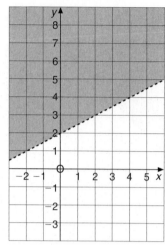

Exercise 12.3

1.

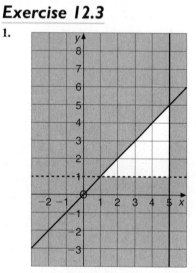

2.

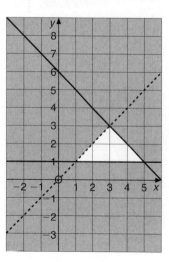

3.

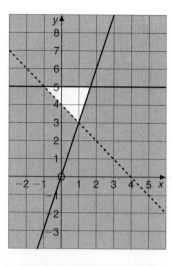

4.

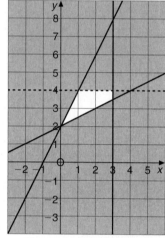

Exercise 12.4

1. **a)** $x \geqslant 5$ $2 < y < 8$ $x + y \leqslant 12$
   b)

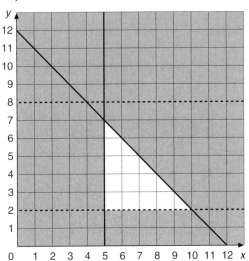

   c) Student's own answer

2. **a)** $p \geqslant 5$ $q \geqslant 2$ $p + q \leqslant 10$
   b)

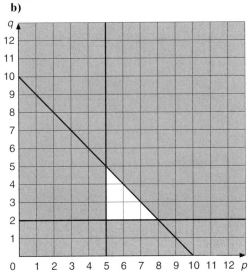

   c) Student's own answer

3. **a)** $m \geqslant 2$ $n \geqslant 2m$ $m + n \leqslant 11$
   b)

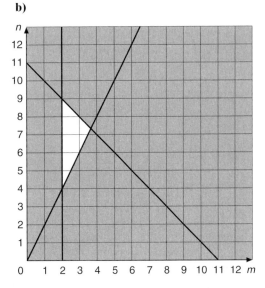

   c) Student's own answer

4. **a)** $3 \leqslant L < 9$ $S < 6$ $L + S \leqslant 10$
   b)

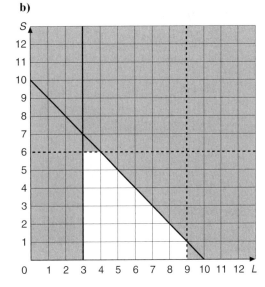

   c) Student's own answer

Student Assessment 1

1. **a)** $x \leqslant 5$ **b)** $y \leqslant 3$

2. **a)** $2 < y \leqslant 4$ **b)** $4 < p \leqslant 8$

3. **a)** $r < 10$ $3 < s < 9$ $s + r < 12$
    b)

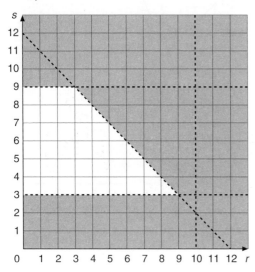

    c) Student's own answer

4. **a)** $12 < A + E < 20$ $A < 10$ $E \leqslant A + 3$
    b)

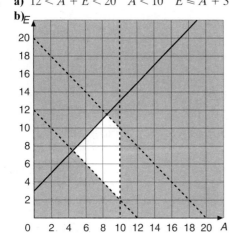

    c) Student's own answers

Student Assessment 2

1. **a)** $x \leqslant 7$ **b)** $y \leqslant 9$

2. **a)** $1 \leqslant p < 4$ **b)** $4 < x \leqslant 7$

3. **a)** $x < 4$ $y > 2$ $x + y \leqslant 8$
    b)

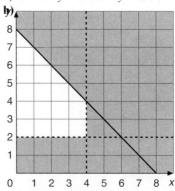

    c) Student's own answers

4. **a)** $c + e \leqslant 40$ $c \geqslant 16$ $e \geqslant 5$ $c > 2e$
    b)

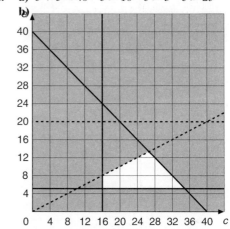

    c) Student's own answers

ICT Section page 190

1.

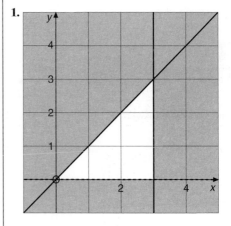

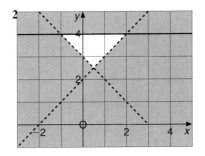

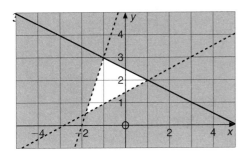

Loci page 191

Exercise A

1.

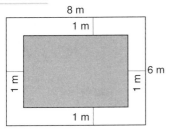

2.

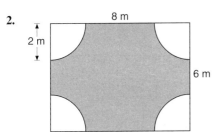

3.

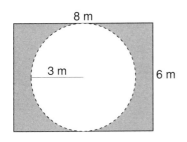

4.

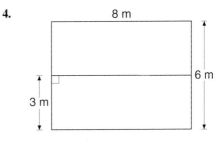

5. **a)**　　**b)**　　**c)**　　**d)**

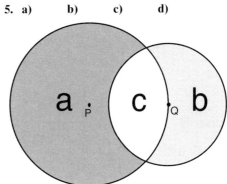

6.

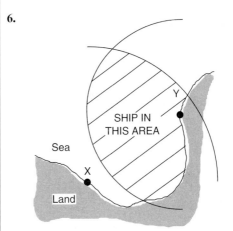

7. **a)** Student's own diagrams. L, M and N will all lie on the circumference of a circle, the centre of the circle being the point equidistant from L, M and N.

   b) There would be no point equidistant from all three (except in the infinite!).

8. C is on the circumference of a circle with AB as its diameter.

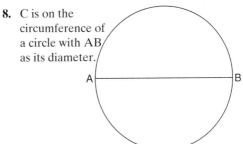

9. Student's own construction

10.

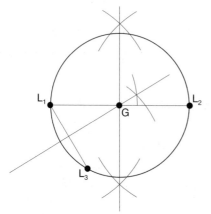

Exercise B

1. **a)**

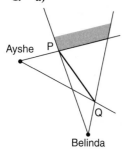

b)

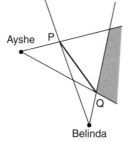

c)

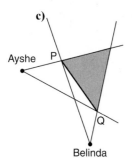

d)

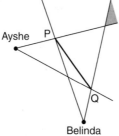

2.

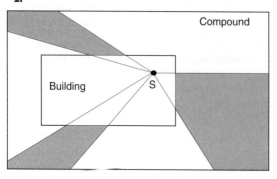

3.

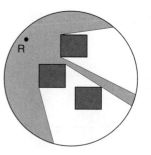

Exercise C

1.

2.

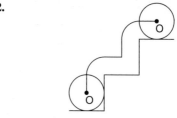

3.

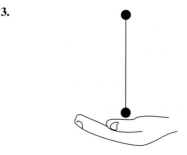

4.

5.

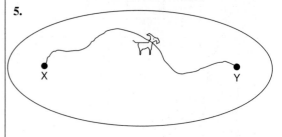

Student Assessment 1

1.

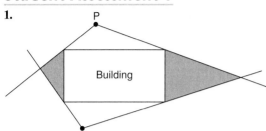

2. a) b)

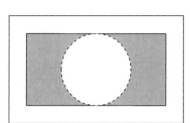

3.

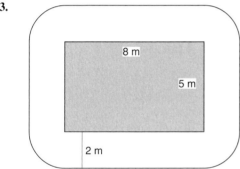

4. Ship's path is perpendicular bisector of AB.

5.

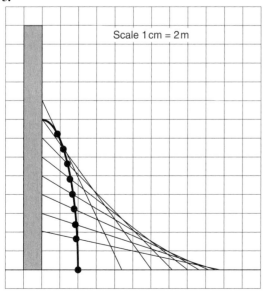

6.

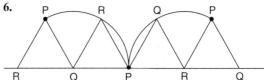

Student Assessment 2

1.

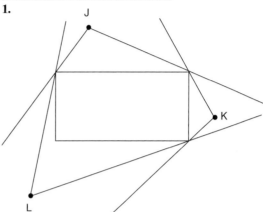

None of the friends can see each other, as shown above.

2.

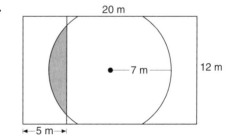

3.

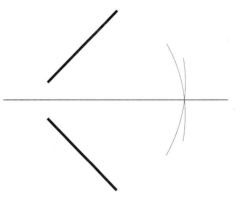

4.

5.

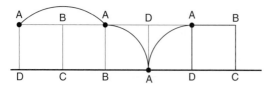

13 Geometrical Relationships

Exercise 13.1 page 198

1. **a)** Interior angles are the same, i.e. 60°, 30° and 90°.
   b) $\frac{5}{8}$ **c)** $x = 6.25$ cm $y = 3.75$ cm

2. A, C and F are similar. B and D are similar.

3. **a)** 6 cm   **b)** 9 cm

4. $p = 4.8$ cm $q = 4.5$ cm $r = 7.5$ cm

5. $e = 5$ cm $f = 2\frac{2}{3}$ cm

6. **a)** 10 cm²   **b)** 1.6   **c)** 25.6 cm²

7. **a)** 10 cm   **b)** 2.5   **c)** 150 cm²

8. **a)** $9\sqrt{3}$ and $\sqrt{3}$ cm²   **b)** 9 : 1

9. **a)** $33\frac{1}{3}$ cm²   **b)** $6\frac{2}{3}$ cm

Exercise 13.2

1. 50 cm²

2. 10 cm²

3. **a)** i) 455.6 cm² (1 d.p.)   ii) 90 cm²
   iii) 40 cm²
   b) Triangle I

4. 43.56 cm²

5. 56.25 cm²

6. 18.1 cm² (1 d.p.)

Exercise 13.3

1. **a)** i) 8 m   ii) 4.8 m   **b)** $2\sqrt{2}$ m
   c) $16\sqrt{2}$ cm²

2. **a)** 220 cm²   **b)** 1375 cm²   **c)** 200 cm³
   d) 3125 cm³

3. **a)** 54 cm²   **b)** 3   **c)** 729 cm³

4. **a)** $1 : n^2$ **b)** $1 : n^3$

5. 112 cm³

6. 0.64 litres

Exercise 13.4

1. 20 cm

2. **a)** 1 : 8   **b)** 1 : 7

3. 16 cm³

4. **a)** Not similar. Student's own reasons
   b) 1 : 3

5. **a)** 16 cm²   **b)** 64 cm²   **c)** 144 cm²
   d) A = 48 cm³   B = 192 cm³   C = 432 cm³

6. **a)** 30 km²   **b)** 6 cm²

7. **a)** 10 cm × 20 cm × 30 cm   **b)** 100 g

Student Assessment 1

1. A and C

2. $1 : \left(\dfrac{H}{h}\right)^2$

3. **a)** Yes. Student's own explanation   **b)** 5 cm
   c) 8 cm   **d)** 6 cm

4. 15 m

5. **a)**

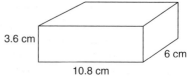

   b) 233.28 cm³   **c)** 250.56 cm²

6. **a)** 12.8 cm³   **b)** 880 cm²   **c)** 35.2 cm²

7. 1250 cm²

8. 27 000 cm³

Student Assessment 2

1. **a)** 5 cm   **b)** $x = 4.5$ cm $y = 7.5$ cm

2. **a)** $1 : \left(\dfrac{H}{h}\right)^2$ **b)** $1 : \left(\dfrac{H}{h}\right)^3$

3. $x = \sqrt{41}$ cm $y = \dfrac{3\sqrt{41}}{5}$ cm $z = 6.4$ cm

4. 156 cm² (3 s.f.)

5. **a)** 1000 cm³   **b)** 600 cm²

6. **a)** 6 cm   **b)** 18.75 cm³

7. **a)** 4 m by 3 m   **b)** 12 m²

8. 1216.8 cm² (1 d.p.)

ICT Section page 206

1. **d)** The ratios are the same.

3. **a)** The ratios remain the same.
   b) Ratios are still equal to each other (though probably of a different value from 1d).

14 Angle Properties

Core Section page 207

Exercise 14.1

1. **a)** 720°   **b)** 1260°   **c)** 900°

2. **a)** 135°   **b)** 90°   **c)** 144°   **d)** 150°

3. **a)** 72°   **b)** 30°   **c)** 51.4° (1 d.p.)

4. **a)** 18   **b)** 10   **c)** 36   **d)** 8   **e)** 20
   f) 120

5. **a)** 5   **b)** 12   **c)** 20   **d)** 15   **e)** 40
   f) 360

6. 12

7.

Number of sides	Name	Sum of exterior angles	Size of an exterior angle	Sum of interior angles	Size of an interior angle
3	Equilateral triangle	360°	120°	180°	60°
4	Square	360°	90°	360°	90°
5	Pentagon	360°	72°	540°	108°
6	Hexagon	360°	60°	720°	120°
7	Heptagon	360°	51.4°	900°	128.6°
8	Octagon	360°	45°	1080°	135°
9	Nonagon	360°	40°	1260°	140°
10	Decagon	360°	36°	1440°	144°
12	Dodecagon	360°	30°	1800°	150°

Exercise 14.2

1. 60°

2. 135°

3. 20°

4. 32°

5. 110°

6. 22.5°

Exercise 14.3

1. 35°

2. 60°

3. 40°

4. 45°

5. 24°

6. 26°

7. 13 cm

8. 8 cm

9. 17.7 cm (1 d.p.)

Student Assessment 1

1. 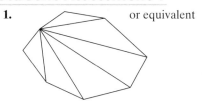 or equivalent

2. 162°

3. 1260°

4. 360°

5. 72°

6. **a)** 90°   **b)** 6.5 cm

7. 58°

8. 30°

9. 25°

10. 152°

Student Assessment 2

1. 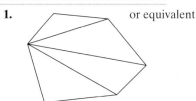 or equivalent

2. 165°

3. 1800°

4. 30°

5. **a)** 90°   **b)** 13 cm

6. 28°

7. 125°

8. 45°

9. 42°

10. 38°

Extended Section page 214

Exercise 14.4

1. **a)** $x = 54°$ **b)** 54°, 108°, 162°, 54°, 162°

2. 125°, 145°

3. 64°

4. $a = 135°, b = 125°, c = 130°, d = 110°, e = 85°$

Exercise 14.5

1. 55°

2. 80°

3. 90°

4. 100°

5. 80°

6. 20°

7. $x = 54°, y = 18°$

8. $x = 50°, y = 25°$

Exercise 14.6

1. $a = 72°$

2. $b = 33°, c = 66°$

3. $d = 48°, e = 32°$

4. $f = 30°, g = 120°, h = 120°, i = 30°, j = 30°$

5. $k = 55°, l = 55°, m = 55°, n = 55°$

6. $p = 65°, q = 40°$

Exercise 14.7

1. $a = 80°, b = 65°$

2. $c = 110°, d = 98°, e = 110°$

3. $f = 115°, g = 75°$

4. $i = 98°, j = 90°, k = 90°$

5. $l = 95°, m = 55°, n = 55°, p = 85°, q = 95°$

6. $r = 70°, s = 120°, t = 60°, u = 110°$

Student Assessment 1

1. **a)** ii) ∠OBA ∠OBC
   b) i) ∠DAB + ∠DCB ∠ADC + ∠CBA
   c) iii) ∠DAC = ∠DBC ∠ADB = ∠ACB
   d) ii) ∠CAB iii) ∠ACB = ∠ABC

2. **a)** 42°   **b)** 21°

3. ∠DAB = 117° ∠ABC = 92°

4. ∠BDC = 25° ∠DAB = 115°

5. ∠OQR = 15° ∠ORQ = ∠OPQ = 90°
   ∠RPQ = ∠PRQ = 75°
   ∠OPR = ∠ORP = 15° ∠ROP = 150°

6. 35°, 54°, 91° respectively

7. 95°, 85°, 85°, 92° respectively

Student Assessment 2

1. **a)** i) ∠DAB + ∠DCB ∠CDA + ∠CBA
   b) i) ∠AOC + ∠CBA ∠OCB + ∠OAB
   ii) ∠OAB + ∠OCB
   iii) ∠ABO = ∠CBO ∠AOB = ∠COB
   c) iii) ∠TPS = ∠TQS = ∠TRS
   ∠PTR = ∠PSR ∠PTQ = ∠PSQ
   ∠QTR = ∠QSR
   d) ii) ∠OAX + ∠OCY
   iii) ∠DCB = ∠BAD ∠CDA = ∠CBA

2. 26°

3. ∠ABC = 50° ∠OAB = 25° ∠CAO = 40°

4. ∠BCD = 40° = ∠BAD
   ∠ABC = 65° = ∠ADC

5. 28° and 56° respectively

6. 67.5°

15 Trigonometry

Core Section page 220

Exercise 15.1

1. **a)** 1.8 cm   **b)** 4.0 cm   **c)** 19.2 cm
   d) 4.9 cm   **e)** 37.3 cm   **f)** 13.9 cm

2. **a)** 14.3 cm   **b)** 9.0 cm   **c)** 9.3 cm
   d) 4.1 cm   **e)** 13.9 cm   **f)** 6.2 cm

3. **a)** 49.4°   **b)** 51.1°   **c)** 51.3°   **d)** 63.4°
   e) 50.4°   **f)** 71.6°

Exercise 15.2

1. **a)** 2.4 cm   **b)** 18.5 cm   **c)** 6.2 cm
   d) 2.4 cm   **e)** 43.8 cm   **f)** 31.8 cm

2. **a)** 38.7°   **b)** 48.6°   **c)** 38.1°   **d)** 49.8°
   e) 32.6°   **f)** 14.5°

Exercise 15.3

1. **a)** 36.0 cm   **b)** 15.1 cm   **c)** 48.2°   **d)** 81.1°
   e) 6.7 cm   **f)** 16.8 cm   **g)** 70.5°   **h)** 2.1 cm

Exercise 15.4

1. **a)** 5 cm   **b)** 11.4 mm (1 d.p.)   **c)** 12 cm
   d) 13.2 cm (1 d.p.)

2. **a)** 11.0 cm (1 d.p.)   **b)** 14.8 cm (1 d.p.)
   c) 7.9 cm (1 d.p.)   **d)** 7.3 cm (1 d.p.)
   e) 3 cm   **f)** 13.9 cm (1 d.p.)

3. 71.6 km   **4.** 66.9 km

5. **a)** 225° − 135° = 90°   **b)** 73.8 km

6. 57 009 m

7. **a)** 8.5 km   **b)** 15.5 km (1 d.p.)

8. **a)** 13.3 m (1 d.p.)   **b)** 15.0 m (1 d.p.)

Exercise 15.5

1. **a)** 43.6° (1 d.p.)   **b)** 19.5 cm (1 d.p.)
   c) 16.7 cm (1 d.p.)   **d)** 42.5° (1 d.p.)

2. **a)** i) 20.8 km (1 d.p.)   ii) 215.2° (1 d.p.)
   b) i) 228.4 km (1 d.p.)   ii) 101.7 km (1 d.p.)
      iii) 103.2 km (1 d.p.)   iv) 147.4 km (1 d.p.)
      v) 414.8 km (1 d.p.)   vi) 216.9° (1 d.p.)
   c) i) 6.7 m (1 d.p.)   ii) 19.6 m (1 d.p.)
      iii) 15.3 m (1 d.p.)
   d) i) 48.2° (1 d.p.)   ii) 41.8° (1 d.p.)
      iii) 8 cm   iv) 8.9 cm (1 d.p.)
      v) 76.0 cm² (1 d.p.)

Student Assessment 1

1. **a)** 4.0 cm   **b)** 43.9 cm   **c)** 20.8 cm
   d) 3.9 cm

2. **a)** 37°   **b)** 56°   **c)** 31°   **d)** 34°

3. **a)** 5.0 cm   **b)** 6.6 cm   **c)** 9.3 cm
   d) 28.5 cm

Student Assessment 2

1. **a)** 160.8 km   **b)** 177.5 km

2. **a)** $\tan \theta = \dfrac{5}{x}$ **b)** $\tan \theta = \dfrac{7.5}{(x + 16)}$

   c) $\dfrac{5}{x} = \dfrac{7.5}{(x + 16)}$ **d)** 32 m

   e) 8.9° (1 d.p.)

3. **a)** 285 m (3 s.f.)   **b)** 117° (3 s.f.)
   c) 297° (3 s.f.)

4. **a)** 1.96 km (2 d.p.)   **b)** 3.42 km (2 d.p.)
   c) 3.57 km (2 d.p.)

Extended Section page 230

Exercise 15.6

1. **a)** i) 12.2 km (1 d.p.)   ii) 9.5° (1 d.p.)
   b) i) 10.1 km (1 d.p.)   ii) 1.2 km (1 d.p.)
   c) i) 22.6° (1 d.p.)   ii) 130 m
   d) i) 0.34 km (2 d.p.)   ii) 0.94 km (2 d.p.)

2. **a)** i) 64.0 m (1 d.p.)   ii) 30.2 m (1 d.p.)
   b) 6.9 km (1 d.p.)
   c) i) 5.9 km (1 d.p.)   ii) 1.6 km (1 d.p.)
   d) i) 7.5 km (1 d.p.)   ii) 3.2 km (1 d.p.)

3. **a)** i) 2.9 km (1 d.p.)   ii) 6.9 km (1 d.p.)
      iii) 11.4° (1 d.p.)   iv) 20.4 km (1 d.p.)
   b) i) 2.7 km (1 d.p.)   ii) 1.0 km (1 d.p.)
      iii) 3.5° (1 d.p.)   iv) 16.83 km (2 d.p.)
   c) i) 225.2 m (1 d.p.)   ii) 48.4° (1 d.p.)

Exercise 15.7

1. **a)** sin 120°   **b)** sin 100°   **c)** sin 65°
   d) sin 40°   **e)** sin 52°   **f)** sin 13°

2. **a)** sin 145°   **b)** sin 130°   **c)** sin 150°
   d) sin 132°   **e)** sin 76°   **f)** sin 53°

3. **a)** 19°, 161°   **b)** 82°, 98°   **c)** 5°, 175°
   d) 72°, 108°   **e)** 13°, 167°   **f)** 28°, 152°

4. **a)** 70°, 110°   **b)** 9°, 171°   **c)** 53°, 127°
   d) 34°, 146°   **e)** 16°, 164°   **f)** 19°, 161°

Exercise 15.8

1. **a)** −cos 160°   **b)** −cos 95°   **c)** −cos 148°
   d) −cos 85°   **e)** −cos 33°   **f)** −cos 74°

2. **a)** −cos 82°   **b)** −cos 36°   **c)** −cos 20°
   d) −cos 37°   **e)** −cos 9°   **f)** −cos 57°

3. **a)** cos 80°   **b)** −cos 90°   **c)** cos 70°
   d) cos 135°   **e)** cos 58°   **f)** cos 155°

4. **a)** −cos 55°   **b)** −cos 73°   **c)** cos 60°
   d) cos 82°   **e)** cos 88°   **f)** cos 70°

Exercise 15.9

1. **a)** 8.9 cm   **b)** 8.9 cm   **c)** 6.0 mm   **d)** 8.6 cm

2. **a)** 33.2°   **b)** 52.7°   **c)** 77.0°   **d)** 44.0°

3. **a)** 25°, 155° (nearest degree)
b)

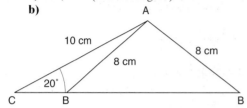

4. **a)** 75°, 105° (nearest degree)
b)

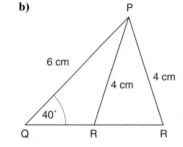

Exercise 15.10

1. **a)** 4.7 m   **b)** 12.1 cm   **c)** 9.1 cm
   d) 3.1 cm   **e)** 10.7 cm

2. **a)** 125.1°   **b)** 108.2°   **c)** 33.6°
   d) 37.0°   **e)** 122.9°

Exercise 15.11

1. **a)** 42.9 m (1 d.p.)   **b)** 116.9° (1 d.p.)
   c) 24.6° (1 d.p.)   **d)** 33.4° (1 d.p.)
   e) 35.0 m (1 d.p.)

2. 370 m

3. **a)** 669.2 m (1 d.p.)   **b)** 546.4 m (1 d.p.)
   c) 473.2 m (1 d.p.)

4. 73.9 m (1 d.p.)

Exercise 15.12

1. **a)** 70.0 cm² **b)** 70.9 mm² **c)** 121.6 cm²
   d) 17.0 cm²

2. **a)** 24.6° **b)** 13.0 cm **c)** 23.1 cm
   d) 63.2°

3. 16 800 m²

4. **a)** 3.9 m² (1 d.p.)   **b)** 222 m² (3 s.f.)

Exercise 15.13

1. **a)** 5.7 cm (1 d.p.)   **b)** 6.9 cm (1 d.p.)
   c) 54.7° (1 d.p.)

2. **a)** 5.8 cm (1 d.p.)   **b)** 6.2 cm (1 d.p.)
   c) 18.9° (1 d.p.)

3. **a)** 6.4 cm (1 d.p.)   **b)** 13.6 cm (1 d.p.)
   c) 61.9° (1 d.p.)

4. **a)** 75.3° (1 d.p.)   **b)** 56.3° (1 d.p.)

5. **a)** i) 7.2 cm (1 d.p.)   ii) 21.1° (1 d.p.)
   b) i) 33.7° (1 d.p.)   ii) 68.9° (1 d.p.)

6. **a)** i) 8.5 cm (1 d.p.)   ii) 28.3° (1 d.p.)
   b) i) 20.6° (1 d.p.)   ii) 61.7° (1 d.p.)

7. **a)** 6.5 cm   **b)** 11.3 cm (1 d.p.)
   c) 70.7 cm (1 d.p.)

8. **a)** 11.7 cm   **b)** 7.6 cm (1 d.p.)

9. **a)** TU = TQ = 10 cm QU = √72 cm
   b) 90°, 37°, 53°   **c)** 24 cm²

Exercise 15.14

1. **a)** RW   **b)** TQ   **c)** SQ   **d)** WU
   e) QV   **f)** SV

2. **a)** JM   **b)** KN   **c)** HM   **d)** HO
   e) JO   **f)** MO

3. **a)** ∠TPS   **b)** ∠UPQ   **c)** ∠VSW
   d) ∠RTV   **e)** ∠SUR   **f)** ∠VPW

4. **a)** 5.8 cm (1 d.p.)   **b)** 31.0° (1 d.p.)

5. **a)** 10.2 cm (1 d.p.)   **b)** 29.2° (1 d.p.)
   c) 51.3° (1 d.p.)

6. **a)** 6.7 cm (1 d.p.)   **b)** 61.4° (1 d.p.)

7. **a)** 7.8 cm (1 d.p.)   **b)** 11.3 cm (1 d.p.)
   c) 12.4° (1 d.p.)

8. **a)** 14.1 cm (1 d.p.)   **b)** 8.5 cm (1 d.p.)
   c) 7.5 cm (1 d.p.)   **d)** 69.3° (1 d.p.)

9. **a)** 17.0 cm (1 d.p.)   **b)** 5.7 cm (1 d.p.)
   c) 7 cm   **d)** 51.1° (1 d.p.)

Student Assessment 1

1. **a)** 4003 m   **b)** 2.3° (1 d.p.)

2. Student's graph

3. **a)** sin 130°   **b)** sin 30°   **c)** −cos 135°
   d) −cos 60°

4. 134° (3 s.f.)

5. **a)** 11.7 cm (1 d.p.)   **b)** 12.3 cm (1 d.p.)
   c) 29° (2 s.f.)

Student Assessment 2

1. **a)** 10.8 cm (1 d.p.)   **b)** 11.9 cm (1 d.p.)
   c) 30° (2 s.f.)   **d)** 49° (2 s.f.)

2. Student's graph

3. **a)** −cos 52°   **b)** cos 100°

4. **a)** 9.8 cm (1 d.p.)   **b)** 30°
    c) 19.6 cm (1 d.p.)

5. **a)** 678.4 m (1 d.p.)   **b)** 11.6° (1 d.p.)
    c) 718.0 m (1 d.p.)

Student Assessment 3

1. **a)** 18.0 m (1 d.p.)   **b)** 27° (2 s.f.)
    c) 28.8 m (1 d.p.)   **d)** 278 m² (3 s.f.)

2. **a)** 12.7 cm (1 d.p.)   **b)** 67° (2 s.f.)
    c) 93 cm²   **d)** 14.7 cm (1 d.p.)

3. **a)** 38°, 322°   **b)** 106°, 254°

4. **a)** 25° (2 s.f.)   **b)** 33° (2 s.f.)
    c) 5.9 km (1 d.p.)   **d)** 4:41 p.m.

Student Assessment 4

1. **a)** 22° (2 s.f.)   **b)** 9° (1 s.f.)   **c)** 2.2 km
    d) 10° (2 s.f.)   **e)** 15° (2 s.f.)
    f) 1.76 km (2 d.p.)

2. **a)**

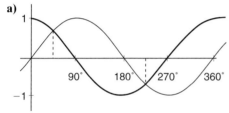

    b) $\theta = 45°, 225°$

3. **a)** 5.8 cm (1 d.p.)   **b)** 6.7 cm (1 d.p.)
    c) 7.8 cm (1 d.p.)   **d)** 47° (2 s.f.)
    e) 19.0 cm² (1 d.p.)   **f)** 37° (2 s.f.)

ICT Section page 249

1. **b)** i) sin 90 = 1
            sin 180 = 0
            sin 270 = –1
    c) $y = 0.7$ crosses the sin curve at two points
    d) $\sin^{-1} 0.5 = 30°$ and 150°

2. **b)** two
    c) 45°

3. **a)** 0° and 180°
    b) 38.2° and 141.8°

16 Mensuration

Core Section page 250

Exercise 16.1

1. **a)** 25.13 cm   **b)** 21.99 cm   **c)** 28.90 cm
    d) 1.57 m

2. **a)** 50.27 cm²   **b)** 38.48 cm²   **c)** 66.48 cm²
    d) 0.20 m²

3. **a)** 2.4 cm (1 d.p.)   **b)** 0.5 cm (exact)
    c) 0.6 m (1 d.p.)   **d)** 1.3 mm (1 d.p.)

4. **a)** 4.5 cm (1 d.p.)   **b)** 6.0 cm (1 d.p.)
    c) 3.2 m (1 d.p.)   **d)** 4.3 mm (1 d.p.)

Exercise 16.2

1. i) 1.57 m (2 d.p.)   ii) 637 times (3 s.f.)

2. 188.5 m (1 d.p.)

3. 264 mm² (3 s.f.)

4. **a)** 144 cm²   **b)** 28.3 cm² (1 d.p.)
    c) 30.9 cm² (1 d.p.)

5. **a)** 57.1 m (1 d.p.)   **b)** 178.3 m² (1 d.p.)

Exercise 16.3

1. 58.5 cm²

2. 84 cm²

3. 153.3 cm² (1 d.p.)

4. 157.5 cm²

Exercise 16.4

1. 4

2. 3

3. 23.0 m² (1 d.p.)

4. **a)** 16 m², 24 m²   **b)** 100 m²   **c)** 15

Exercise 16.5

1. **a)** 460 cm²   **b)** 208 cm²   **c)** 147.78 cm²
    d) 33.52 cm²

2. **a)** 2 cm   **b)** 4 cm   **c)** 6 cm   **d)** 5 cm

3. **a)** 100.5 cm² (1 d.p.)   **b)** 276.5 cm² (1 d.p.)
    c) 279.3 cm² (1 d.p.)   **d)** 25.6 cm² (1 d.p.)

4. **a)** 1.2 cm   **b)** 0.5 cm   **c)** 1.7 cm
    d) 7.0 cm

Exercise 16.6

1. **a)** 24 cm² **b)** 2 cm

2. **a)** 216 cm² **b)** 15.2 cm (1 d.p.)

3. **a)** 94.2 cm² (1 d.p.) **b)** 14 cm

4. 4.4 cm

Exercise 16.7

1. **a)** 24 cm³ **b)** 18 cm³ **c)** 27.6 cm³
   d) 8.82 cm³

2. **a)** 452.4 cm³ (1 d.p.) **b)** 277.1 cm³ (1 d.p.)
   c) 196.3 cm³ (1 d.p.) **d)** 0.5 cm³ (1 d.p.)

3. **a)** 108 cm³ **b)** 140 cm³ **c)** 42 cm³
   d) 6.2 cm³

4. **a)** 70 cm³ **b)** 96 cm³ **c)** 380 cm³
   d) 137.5 cm³

Exercise 16.8

1. **a)** 16 cm **b)** 4096 cm³ **c)** 3216 cm³
   d) 21.5% (1 d.p.)

2. **a)** 42 cm² **b)** 840 cm³

3. 5.7 cm

4. 2.90 m³ (2 d.p.)

Student Assessment 1

1. **a)** Circumference = 34.6 cm   Area = 95.0 cm²
   b) Circumference = 50.3 mm
      Area = 201.1 mm²

2. 9.9 cm²

3. **a)** 39.3 cm² **b)** 34 cm² **c)** 101.3 cm²

4. **a)** 10.2 cm² **b)** 282.7 cm² **c)** 633.0 cm²

5. **a)** 339.3 mm³ **b)** 9.8 cm³

Student Assessment 2

1. **a)** Circumference = 27.0 cm   Area = 58.1 cm²
   b) Circumference = 47.1 mm
      Area = 176.7 mm²

2. 325.8 cm²

3. **a)** 56.5 cm² **b)** 108 cm² **c)** 254.5 cm²

4. 418 cm³

5. **a)** 1011.6 cm² **b)** 523.1 cm²

Extended Section page 260

Exercise 16.9

1. **a)** 6.3 cm **b)** 2.1 cm **c)** 11.5 cm
   d) 23.6 cm

2. **a)** 33° (2 s.f.) **b)** 229° (3 s.f.)
   c) 57° (2 s.f.) **d)** 115° (3 s.f.)

3. **a)** 12.2 cm (1 d.p.) **b)** 4.6 cm (1 d.p.)
   c) 18.6 cm (1 d.p.) **d)** 4.0 cm (1 d.p.)

Exercise 16.10

1. **a)** 48.8 cm (1 d.p.) **b)** 104.5 cm (1 d.p.)

2. **a)** 3.7 cm (1 d.p.) **b)** 49.7 cm (1 d.p.)
   c) 68.8° (1 d.p.)

3. **a)** 12 cm **b)** 54 cm **c)** 47.7° (1 d.p.)

Exercise 16.11

1. **a)** 33.5 cm² (1 d.p.) **b)** 205.3 cm² (1 d.p.)
   c) 5.7 cm² (1 d.p.) **d)** 44.7 cm² (1 d.p.)

2. **a)** 18.5 cm (1 d.p.) **b)** 20.0 cm (1 d.p.)
   c) 1.7 cm (1 d.p.) **d)** 12.4 cm (1 d.p.)

3. **a)** 48° (2 s.f.) **b)** 34° (2 s.f.)
   c) 20° (2 s.f.) **d)** 127° (3 s.f.)

Exercise 16.12

1. 79.2 m² (1 d.p.)

2. **a)** 117.8 cm² (1 d.p.) **b)** 39.3 cm² (1 d.p.)
   c) 8.7 cm (1 d.p.)

3. **a)** 4.2 cm (1 d.p.) **b)** 114.5 cm² (1 d.p.)
   c) 62.8 cm³ (1 d.p.)

4. **a)** 9.2 cm (1 d.p.) **b)** 0.8 cm (1 d.p.)
   c) 34.9 cm² (1 d.p.)
   d) Length = 20 cm   Width = 10.8 cm (1 d.p.)
   e) 77 cm² (2 s.f.)

5. **a)** 20° **b)** 0.89 cm (2 d.p.)
   c) Length = 16 cm   Width = 11.8 cm (1 d.p.)
   d) 49 cm² (2 s.f.)

Exercise 16.13

1. **a)** 904.8 cm³ (1 d.p.) **b)** 3591.4 cm³ (1 d.p.)
   c) 2309.6 cm³ (1 d.p.) **d)** 1.4 cm³ (1 d.p.)

2. **a)** 3.1 cm **b)** 5.6 cm **c)** 0.4 m
   d) 6.2 mm

Exercise 16.14

1. 6.3 cm

2. 86.7 cm^3 (1 d.p.)

3. 11.9 cm (1 d.p.)

4. **a)** 4189 cm^3 (4 s.f.)   **b)** 8000 cm^3
   c) 48%

5. 10.0 cm

6. A = 4.1 cm, B = 3.6 cm, C = 3.1 cm

7. 3 : 2

Exercise 16.15

1. **a)** 452.4 cm^2 (1 d.p.)   **b)** 254.5 cm^2 (1 d.p.)
   c) 1885.7 cm^2 (1 d.p.)   **d)** 4 cm^2

2. **a)** 2.0 cm (1 d.p.)   **b)** 1.1 cm (1 d.p.)
   c) 3.1 mm (1 d.p.)   **d)** 0.5 cm

3. 1 : 4

4. **a)** 157.1 cm^2 (1 d.p.)   **b)** 15 cm
   c) 706.9 cm^2 (1 d.p.)

5. **a)** 804.2 cm^2 **b)** 5.9 cm (1 d.p.)

Exercise 16.16

1. 40 cm^3

2. 133 cm^3 (3 s.f.)

3. 64 cm^3

4. 70 cm^3

Exercise 16.17

1. 7 cm

2. 5 cm

3. **a)** 8 cm   **b)** 384 cm^3 **c)** 378 cm^3

4. **a)** 3.6 cm   **b)** 21.7 cm^3 (1 d.p.)
   c) 88.7 cm^3 (1 d.p.)

Exercise 16.18

1. 6.9 cm^2 (1 d.p.)

2. 157.4 cm^2 (1 d.p.)

3. 73.3 cm^2 (1 d.p.)

4. 1120.9 cm^2 (1 d.p.)

5. **a)** 692.8 cm^2 (1 d.p.)   **b)** 137.2 cm^2 (1 d.p.)
   c) 23.6 cm (1 d.p.)

Exercise 16.19

1. **a)** 56.5 cm^3 (1 d.p.)   **b)** 263.9 cm^3 (1 d.p.)
   c) 1.3 cm^3 (1 d.p.)   **d)** 165.9 cm^3 (1 d.p.)

2. **a)** 6.9 cm (1 d.p.)   **b)** 10.9 cm (1 d.p.)
   c) 0.8 cm (1 d.p.)   **d)** 51.3 cm (1 d.p.)

3. **a)** i) 8.0 cm   ii) 12.7 cm   iii) 843.2 cm^3
   b) i) 15.9 cm   ii) 8.4 cm   iii) 2230.3 cm^3
   c) i) 6.4 cm   ii) 4.0 cm   iii) 168.3 cm^3
   d) i) 3.8 cm   ii) 4.6 cm   iii) 70.7 cm^3

Exercise 16.20

1. 3.9 cm (1 d.p.)

2. **a)** 33.0 cm (1 d.p.)   **b)** 5.3 cm (1 d.p.)
   c) 7.3 cm (1 d.p.)   **d)** 211.0 cm^3 (1 d.p.)
   e) 148.4 cm^2 (1 d.p.)

3. **a)** 2304 cm^3 **b)** 603.2 cm^3 (1 d.p.)
   c) 1700.8 cm^3 (1 d.p.)

4. **a)** 81.6 cm^3 (1 d.p.)   **b)** 275.4 cm^3 (1 d.p.)
   c) 8 : 27

Exercise 16.21

1. 81.8 cm^3 (1 d.p.)

2. 770.7 cm^3 (1 d.p.)

3. 3166.7 cm^3 (1 d.p.)

4. **a)** 654.5 cm^3 **b)** 12.5 cm   **c)** 2945.2 cm^3

Exercise 16.22

1. **a)** 414.7 cm^2 (1 d.p.)   **b)** 1649.3 cm^2 (1 d.p.)

2. 1131.0 cm^2 (1 d.p.)

3. **a)** 314.2 cm^2 (1 d.p.)   **b)** 12.5

Student Assessment 1

1. **a)** 11.8 cm (1 d.p.)   **b)** 35.3 cm (1 d.p.)

2. **a)** 272° (3 s.f.)   **b)** 6° (1 s.f.)

3. 232.0 cm^2 (1 d.p.)

4. **a)** 530.9 cm^2 **b)** 1150.3 cm^3

5. **a)** 1210.7 cm^2 (1 d.p.)   **b)** 2592 cm^3

Student Assessment 2

1. **a)** 178.0 cm (1 d.p.)   **b)** 68.3 mm (1 d.p.)

2. **a)** 143° (3 s.f.)   **b)** 3° (1 s.f.)

3. **a)** 95.5 cm^2 (1 d.p.)

4. **a)** 603.2 cm^2 **b)** 1072.3 cm^3

5. **a)** 22.5 cm   **b)** 125.7 cm^3 **c)** 3267.3 cm^3

Student Assessment 3

1. **a)** 22.9 cm (1 d.p.) **b)** 229.2 cm² (1 d.p.)
   c) 985.1 cm² (1 d.p.) **d)** 1833.5 cm³ (1 d.p.)

2. **a)** 904.8 cm³ (1 d.p.) **b)** 12 cm
   c) 13.4 cm (1 d.p.) **d)** 958.2 cm²

3. **a)** 10 cm **b)** 82.1 cm³ (1 d.p.)
   c) 71.8 cm³ (1 d.p.) **d)** 30.8 cm³ (1 d.p.)
   e) 41.1 cm³ (1 d.p.)

Student Assessment 4

1. **a)** 3619.1 cm³ (1 d.p.)
   b) 3619.1 cm³ (1 d.p.) **c)** 904.8 cm² (1 d.p.)
   d) 1916.4 cm² (1 d.p.)

2. **a)** 43.3 cm² (1 d.p.) **b)** 173.2 cm² (1 d.p.)

3. **a)** 314.2 cm² (1 d.p.) **b)** 26.9 cm

ICT Section page 276

1. Possible formulae are given:
   In cell B2: =A2/360*2*PI()*10
   C2: =B2
   D2: =C2/(2*PI())
   E2: =SQRT((100-D2^2))
   F2: =1/3*PI()*D2^2*E2

17 Vectors

Core Section page 278

Exercise 17.1

1. **a)** $\begin{pmatrix} 2 \\ 4 \end{pmatrix}$ **b)** $\begin{pmatrix} 3 \\ 1 \end{pmatrix}$ **c)** $\begin{pmatrix} -2 \\ -4 \end{pmatrix}$ **d)** $\begin{pmatrix} 3 \\ -2 \end{pmatrix}$

   e) $\begin{pmatrix} -6 \\ 1 \end{pmatrix}$ **f)** $\begin{pmatrix} 6 \\ -1 \end{pmatrix}$ **g)** $\begin{pmatrix} -3 \\ -1 \end{pmatrix}$

   h) $\begin{pmatrix} -5 \\ -5 \end{pmatrix}$ **i)** $\begin{pmatrix} -1 \\ 3 \end{pmatrix}$

2. **a)** $\begin{pmatrix} 4 \\ 4 \end{pmatrix}$ **b)** $\begin{pmatrix} 0 \\ 4 \end{pmatrix}$ **c)** $\begin{pmatrix} -3 \\ 0 \end{pmatrix}$ **d)** $\begin{pmatrix} -2 \\ 6 \end{pmatrix}$

   e) $\begin{pmatrix} -4 \\ -2 \end{pmatrix}$ **f)** $\begin{pmatrix} 0 \\ -4 \end{pmatrix}$ **g)** $\begin{pmatrix} 3 \\ 0 \end{pmatrix}$ **h)** $\begin{pmatrix} 2 \\ -6 \end{pmatrix}$

   i) $\begin{pmatrix} -4 \\ -4 \end{pmatrix}$

3.

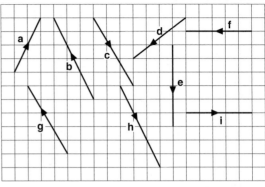

Exercise 17.2

1. **a) b) c) d) e) f)**

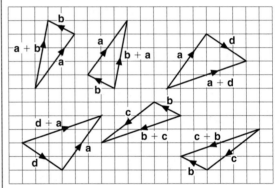

2. **a + b = b + a**, **a + d = d + a**,
   b + c = c + b

3. **a) b) c) d) e) f)**

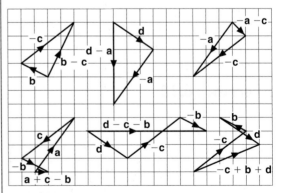

4. **a)** $\begin{pmatrix} 2 \\ 4 \end{pmatrix}$ **b)** $\begin{pmatrix} 0 \\ -6 \end{pmatrix}$ **c)** $\begin{pmatrix} 1 \\ -1 \end{pmatrix}$ **d)** $\begin{pmatrix} 1 \\ 0 \end{pmatrix}$

   e) $\begin{pmatrix} 9 \\ 0 \end{pmatrix}$ **f)** $\begin{pmatrix} 5 \\ 2 \end{pmatrix}$

Exercise 17.3

1. $d = -c$ $e = -a$ $f = 2a$ $g = \frac{1}{2}c$ $h = \frac{1}{2}b$
   $i = -\frac{1}{2}b$ $j = \frac{3}{2}b$ $k = -\frac{3}{2}a$

2. a) $\begin{pmatrix} 4 \\ 6 \end{pmatrix}$ b) $\begin{pmatrix} -12 \\ -3 \end{pmatrix}$ c) $\begin{pmatrix} 2 \\ -4 \end{pmatrix}$

   d) $\begin{pmatrix} -2 \\ 2 \end{pmatrix}$ e) $\begin{pmatrix} -2 \\ -5 \end{pmatrix}$ f) $\begin{pmatrix} -8 \\ 9 \end{pmatrix}$

   g) $\begin{pmatrix} -10 \\ -5 \end{pmatrix}$ h) $\begin{pmatrix} 3 \\ 2 \end{pmatrix}$ i) $\begin{pmatrix} 10 \\ -6 \end{pmatrix}$

3. a) $2a$ b) $-b$ c) $b + c$ d) $a - b$
   e) $2c$ f) $2c - a$

Student Assessment 1

1. a) $\begin{pmatrix} -2 \\ 3 \end{pmatrix}$ b) $\begin{pmatrix} 7 \\ 2 \end{pmatrix}$ c) $\begin{pmatrix} 6 \\ -2 \end{pmatrix}$

2. $a = \begin{pmatrix} 2 \\ 4 \end{pmatrix}$ $b = \begin{pmatrix} 4 \\ 0 \end{pmatrix}$ $c = \begin{pmatrix} 0 \\ -5 \end{pmatrix}$ $d = \begin{pmatrix} -4 \\ -2 \end{pmatrix}$

   $e = \begin{pmatrix} -2 \\ 1 \end{pmatrix}$

3.

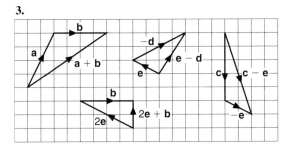

4. a) $\begin{pmatrix} -1 \\ 5 \end{pmatrix}$ b) $\begin{pmatrix} 1 \\ 5 \end{pmatrix}$ c) $\begin{pmatrix} 1 \\ 11 \end{pmatrix}$ d) $\begin{pmatrix} 0 \\ 14 \end{pmatrix}$

Student Assessment 2

1. a) $\begin{pmatrix} -2 \\ 7 \end{pmatrix}$ b) $\begin{pmatrix} -5 \\ -4 \end{pmatrix}$ c) $\begin{pmatrix} -1 \\ -3 \end{pmatrix}$

2. a) $\begin{pmatrix} 0 \\ 3 \end{pmatrix}$ b) $\begin{pmatrix} -6 \\ 0 \end{pmatrix}$ c) $\begin{pmatrix} -1 \\ -5 \end{pmatrix}$ d) $\begin{pmatrix} 3 \\ 3 \end{pmatrix}$

   e) $\begin{pmatrix} 4 \\ -1 \end{pmatrix}$

3.

4. a) $\begin{pmatrix} 7 \\ -11 \end{pmatrix}$ b) $\begin{pmatrix} -3 \\ 9 \end{pmatrix}$ c) $\begin{pmatrix} 6 \\ -6 \end{pmatrix}$

   d) $\begin{pmatrix} -18 \\ 28 \end{pmatrix}$

Extended Section page 283

Exercise 17.4

1. $|a| = 5.0$ units $|b| = 4.1$ units $|c| = 4.5$ units
   $|d| = 7.0$ units $|e| = 7.3$ units $|f| = 6.4$ units

2. a) $|AB| = 4.0$ units   b) $|BC| = 5.4$ units
   c) $|CD| = 7.2$ units   d) $|DE| = 13.0$ units
   e) $|2AB| = 8.0$ units   f) $|-CD| = 4.5$ units

3. a) 4.1 units   b) 18.4 units   c) 15.5 units
   d) 17.7 units   e) 31.8 units   f) 19.6 units

Exercise 17.5

1. $A = \begin{pmatrix} 3 \\ 3 \end{pmatrix}$ $B = \begin{pmatrix} 1 \\ 2 \end{pmatrix}$ $C = \begin{pmatrix} -3 \\ 2 \end{pmatrix}$ $D = \begin{pmatrix} -3 \\ 0 \end{pmatrix}$

   $E = \begin{pmatrix} 4 \\ -2 \end{pmatrix}$ $F = \begin{pmatrix} 2 \\ -4 \end{pmatrix}$ $G = \begin{pmatrix} -1 \\ -1 \end{pmatrix}$

   $H = \begin{pmatrix} 0 \\ -3 \end{pmatrix}$

Exercise 17.6

1. a) a b) a c) b d) $-b$ e) $2b$
   f) $-2a$ g) $a + b$ h) $b - a$
   i) $b - 2a$

2. a) $2a$ b) b c) $b - a$ d) $b - a$
   e) $-a$ f) $a - 2b$

3. a) $-2a$ b) $-a$ c) b d) $b - a$
   e) $2(b - a)$ f) $2b - a$ g) $a - 2b$
   h) $b - 2a$ i) $-a - b$

4. a) $5a$ b) $\frac{8}{3}b$ c) $\frac{1}{3}(8b - 15a)$
   d) $\frac{1}{15}(8b - 15a)$ e) $b - 2a$ f) $5a - 2b$
   g) $\frac{8}{5}b$ h) $\frac{1}{5}(10a + 8b)$ i) $\frac{1}{5}(8 - 5a)$

Exercise 17.7

1. a) i) $2a$ ii) $2a - b$ iii) $\frac{1}{4}(2a + 3b)$
   b) Proof

2. a)b) Proofs

3. a) i) $4a$ ii) $2a$ iii) $2(a + b)$ iv) $\frac{3}{2}a$
   b) Proof

4. a) i) $\frac{1}{2}(2q - p)$ ii) $\frac{2}{7}(p - q)$ b) Proof

Student Assessment 1

1. a) $|\overrightarrow{AB}| = 7.2$ units
   b) $|a| = 9.2$ units $|b| = 8.1$ units
   $|c| = 13.0$ units

2. **a)** 17.5 units **b)** 2.7 units

3. $A = \begin{pmatrix} 2 \\ 4 \end{pmatrix}$ $B = \begin{pmatrix} -1 \\ 2 \end{pmatrix}$ $C = \begin{pmatrix} -3 \\ -1 \end{pmatrix}$

$D = \begin{pmatrix} 0 \\ -3 \end{pmatrix}$ $E = \begin{pmatrix} 1 \\ -4 \end{pmatrix}$

4. **a)** Student's own vector **b)** $\overrightarrow{DF} = \frac{1}{2}\overrightarrow{BC}$
b) $\overrightarrow{CF} = -\overrightarrow{DE}$

Student Assessment 2

1. **a)** $|\overrightarrow{FG}| = 5$ units
b) $|\mathbf{a}| = 6.1$ units $|\mathbf{b}| = 12.4$ units
   $|\mathbf{c}| = 14.1$ units

2. **a)** 29.7 units **b)** 11.4 units

3. **a)** $2\mathbf{a}$ **b)** $-\mathbf{b}$ **c)** $\mathbf{b} - \mathbf{a}$

Student Assessment 3

1. **a)** i) $\frac{1}{4}\mathbf{b}$ ii) $\frac{1}{4}(4\mathbf{a} - \mathbf{b})$ iii) $\frac{2}{5}(\mathbf{b} - \mathbf{a})$
b) Proof

2. **a)** Proof **b)** Proof

3. **a)** i) $\mathbf{b} - \mathbf{a}$ ii) $\mathbf{a}$ iii) $\mathbf{a} + \mathbf{b}$
b) i) $4 : 25$ ii) $20 : 25$

Student Assessment 4

1. **a)** $\mathbf{a}$ **b)** $-\mathbf{b}$ **c)** $(1 + \sqrt{2})\mathbf{b}$

2. **a)** $5\mathbf{b}$ **b)** $5\mathbf{b} - \mathbf{a}$ **c)** $\frac{1}{2}(\mathbf{a} + 3\mathbf{b})$

3. **a)** i) $\frac{1}{2}\mathbf{a}$ ii) $\mathbf{b} - \mathbf{a}$ **b)** $2 : 3$

18 Matrics

Exercise 18.1 page 290

1. **a)** $P = 2 \times 3$ **b)** $Q = 2 \times 4$ **c)** $R = 4 \times 2$
d) $S = 4 \times 5$ **e)** $T = 5 \times 1$ **f)** $F = 1 \times 5$

2. Student's own matrices

3. $\begin{pmatrix} 6500 & 900 \\ 7200 & 1100 \\ 7300 & 1040 \end{pmatrix}$

4. $\begin{pmatrix} 3 & 4 & 2 & 1 \\ 0 & 6 & 2 & 0 \\ 1 & 3 & 0 & 2 \end{pmatrix}$

5. $(8 \ 6 \ 9 \ 3)$

6. $\begin{pmatrix} 37 & 49 \\ 74 & 58 \\ 76 & 62 \\ 89 & 56 \end{pmatrix}$

7. $\begin{pmatrix} 20 & 35 & 15 \\ 45 & 25 & 40 \\ 30 & 30 & 10 \\ 0 & 0 & 25 \\ 5 & 10 & 10 \end{pmatrix}$

8. $\begin{pmatrix} 8000 & 3000 & 5000 \\ 8000 & 6000 & 10\,000 \\ 5000 & 11\,000 & 9000 \\ 9000 & 13\,000 & 6000 \end{pmatrix}$

9. $\begin{pmatrix} 6 & 12 & 43 & 18 & 6 & 9 & 6 \\ 9 & 15 & 28 & 18 & 12 & 12 & 6 \\ 12 & 19 & 30 & 12 & 9 & 9 & 9 \end{pmatrix}$

Exercise 18.2

1. **a)** $\begin{pmatrix} 14 & 8 \\ 7 & 16 \end{pmatrix}$ **b)** $\begin{pmatrix} 9 & 3 & 24 \\ 19 & 35 & 5 \end{pmatrix}$

c) $\begin{pmatrix} 24 \\ 3 \\ 18 \end{pmatrix}$ **d)** $\begin{pmatrix} -10 & 13 & 17 \\ 2 & -13 & 18 \\ 5 & 10 & 11 \end{pmatrix}$

e) $(-3 \ 1 \ -1)$ **f)** $\begin{pmatrix} 14 & -9 \\ 15 & -5 \\ -5 & 10 \end{pmatrix}$

2. **a)** $\begin{pmatrix} 5 & 4 \\ 1 & 0 \end{pmatrix}$ **b)** $\begin{pmatrix} 2 & 2 & 2 \\ 2 & 2 & 2 \end{pmatrix}$ **c)** $\begin{pmatrix} -5 \\ 0 \\ 4 \end{pmatrix}$

d) $\begin{pmatrix} 4 & -14 & -1 \\ 2 & 0 & 4 \\ -9 & -1 & -5 \end{pmatrix}$ **e)** $\begin{pmatrix} -3 & 18 \\ -1 & -12 \end{pmatrix}$

f) $\begin{pmatrix} -3 & -4 \\ -21 & 13 \\ -3 & 2 \end{pmatrix}$

3. **a)** $\begin{pmatrix} 9 & 11 \\ 15 & 5 \\ 9 & 8 \end{pmatrix}$ **b)** $\begin{pmatrix} 9 & 11 \\ 15 & 5 \\ 9 & 8 \end{pmatrix}$

c) $\begin{pmatrix} -1 & 5 \\ 6 & 6 \\ 7 & -5 \end{pmatrix}$ **d)** $\begin{pmatrix} 10 & 3 \\ -3 & 14 \\ 3 & 4 \end{pmatrix}$

e) $\begin{pmatrix} 10 & 6 \\ 9 & -1 \\ 2 & 13 \end{pmatrix}$ **f)** $\begin{pmatrix} 11 & -2 \\ -9 & 8 \\ -4 & 9 \end{pmatrix}$

g) $\begin{pmatrix} 5 & 9 \\ 6 & 0 \\ -5 & 13 \end{pmatrix}$ **h)** $\begin{pmatrix} 3 & -1 \\ -12 & 12 \\ -5 & 0 \end{pmatrix}$

i) $\begin{pmatrix} 13 & -7 \\ -6 & 10 \\ 9 & 0 \end{pmatrix}$

4. **a)** $\begin{pmatrix} 2 & 3 & 3 \\ 3 & 1 & 4 \\ 1 & 4 & 3 \end{pmatrix} \begin{pmatrix} 5 & 1 & 1 \\ 1 & 4 & 2 \\ 1 & 3 & 3 \end{pmatrix}$ **b)** 15 races

c) $\begin{pmatrix} 7 & 4 & 4 \\ 4 & 5 & 6 \\ 2 & 7 & 6 \end{pmatrix}$

5. **a)** 444

b) $\begin{pmatrix} 265 & 312 \\ 140 & 132 \end{pmatrix} - \begin{pmatrix} 189 & 204 \\ 121 & 68 \end{pmatrix} = \begin{pmatrix} 76 & 108 \\ 19 & 64 \end{pmatrix}$

c) 267

Exercise 18.3

1. **a)** $\begin{pmatrix} 8 & 12 \\ 14 & 6 \end{pmatrix}$ **b)** $\begin{pmatrix} 21 & 6 \\ 3 & 0 \end{pmatrix}$ **c)** $\begin{pmatrix} 9 & 3 \\ 0 & 6 \end{pmatrix}$

d) $\begin{pmatrix} 10 & 5 \\ 0 & 20 \end{pmatrix}$ **e)** $\begin{pmatrix} 28 & 20 \\ 16 & 32 \end{pmatrix}$ **f)** $\begin{pmatrix} 21 & 42 \\ 14 & 7 \end{pmatrix}$

2. **a)** $\begin{pmatrix} 4 & 1 \\ 0 & 2 \end{pmatrix}$ **b)** $\begin{pmatrix} 4 & 3 \\ 2 & 1 \end{pmatrix}$ **c)** $\begin{pmatrix} 3 & 6 \\ 0.75 & 1.5 \end{pmatrix}$

d) $\begin{pmatrix} 6 & 4 \\ 0.8 & 2 \end{pmatrix}$ **e)** $\begin{pmatrix} 2.5 & 7.5 \\ 7.5 & 10 \end{pmatrix}$

f) $\begin{pmatrix} 12 & 3 \\ 3 & 6 \end{pmatrix}$

Exercise 18.4

1. **a)** $\begin{pmatrix} 18 & 78 \\ 14 & 54 \end{pmatrix}$ **b)** $\begin{pmatrix} 12 & 72 \\ 0 & 24 \end{pmatrix}$

2. **a)** $\begin{pmatrix} 52 & 12 & 18 \\ 44 & 22 & 12 \end{pmatrix}$ **b)** $\begin{pmatrix} 18 & 18 \\ 8 & 8 \\ 6 & 6 \end{pmatrix}$

3. **a)** $\begin{pmatrix} 28 & 4 & -8 \\ -31 & 19 & 40 \end{pmatrix}$ **b)** $(27 \ -17)$

4. **a)** $\begin{pmatrix} 6 & 4 & -2 \\ -3 & -2 & 1 \\ -12 & -8 & 4 \\ 18 & 12 & -6 \end{pmatrix}$ **b)** $\begin{pmatrix} 30 & 22 & 38 \\ 20 & 13 & 2 \\ 9 & 7 & 17 \\ 24 & 19 & 50 \end{pmatrix}$

Exercise 18.5

1. $\mathbf{VW} = \begin{pmatrix} 0 & 7 \\ 24 & 8 \end{pmatrix}$ $\mathbf{WV} = \begin{pmatrix} -12 & 4 \\ -18 & 20 \end{pmatrix}$

2. $\mathbf{VW} = \begin{pmatrix} -10 & -9 \\ -20 & -18 \end{pmatrix}$

$\mathbf{WV} = \begin{pmatrix} 11 & 14 & -1 \\ -25 & -30 & 5 \\ 39 & 46 & -9 \end{pmatrix}$

3. $\mathbf{VW} = (-11)$

$\mathbf{WV} = \begin{pmatrix} 4 & -10 & 18 & 4 \\ 0 & 0 & 0 & 0 \\ -6 & 15 & -27 & -6 \\ 12 & -30 & 54 & 12 \end{pmatrix}$

4. $\mathbf{VW} = \begin{pmatrix} 7 & 7 & 9 \\ -6 & -4 & -14 \end{pmatrix}$ **WV** is not possible

5. $\mathbf{VW} = \begin{pmatrix} -33 & 41 & -10 \\ 17 & -11 & -8 \end{pmatrix}$ **WV** is not possible

Exercise 18.6

1. $\mathbf{AI} = \begin{pmatrix} 2 & 1 \\ 3 & 2 \end{pmatrix}$ $\mathbf{IA} = \begin{pmatrix} 2 & 1 \\ 3 & 2 \end{pmatrix}$

2. $\mathbf{AI} = \begin{pmatrix} -2 & -4 \\ 3 & 6 \end{pmatrix}$ $\mathbf{IA} = \begin{pmatrix} -2 & -4 \\ 3 & 6 \end{pmatrix}$

3. $\mathbf{AI} = \begin{pmatrix} 4 & 8 \\ -2 & 4 \end{pmatrix}$ $\mathbf{IA} = \begin{pmatrix} 4 & 8 \\ -2 & 4 \end{pmatrix}$

4. $\mathbf{AI} = \begin{pmatrix} 3 & 2 \\ 1 & 6 \\ -2 & 5 \end{pmatrix}$ **IA** is not possible

5. $\mathbf{AI} = (-5 \ -6)$ **IA** is not possible

6. **a)** $\begin{pmatrix} 4 & -3 \\ 5 & -6 \\ 3 & 2 \\ 1 & 4 \end{pmatrix}$ **IA** is not possible

7. When **AI** exists, it is equal to **A**.

8. For a 2×2 matrix $\mathbf{AI} = \mathbf{IA} = \mathbf{A}$.

Exercise 18.7

1. **a)** 3 **b)** 3 **c)** 4 **d)** 2

2. **a)** −4 **b)** −10 **c)** −10 **d)** −1

3. **a)** 18 **b)** 14 **c)** −54 **d)** 4

4. Student's own matrices

5. Student's own matrices

6. Student's own matrices

7. **a)** 16 **b)** 131 **c)** 18 **d)** 936 **e)** −104
f) −576 **g)** 254 **h)** −576 **i)** 147

Exercise 18.8

1. **a)** $\begin{pmatrix} 4 & -5 \\ -7 & 9 \end{pmatrix}$ **b)** $\begin{pmatrix} 5 & -7 \\ -7 & 10 \end{pmatrix}$

c) $\begin{pmatrix} 1 & -1 \\ -0.8 & 1 \end{pmatrix}$ **d)** $\begin{pmatrix} -1.3 & 3 \\ -1 & 2 \end{pmatrix}$

e) $\begin{pmatrix} -1.8 & -0.8 \\ 2 & 1 \end{pmatrix}$ **f)** Not possible

2. See answers for Q.1.

3. It has no inverse as the determinant = 0.

4. A, B, D have no inverse.

5. **a)** $\begin{pmatrix} -9 & 4 \\ 7 & -3 \end{pmatrix}$ **b)** $\begin{pmatrix} -\frac{5}{4} & -\frac{3}{4} \\ 2 & 1 \end{pmatrix}$

   c) $\begin{pmatrix} 6 & -\frac{11}{4} \\ -11 & 5 \end{pmatrix}$ **d)** $\begin{pmatrix} -\frac{1}{2} & 0 \\ \frac{1}{2} & \frac{1}{11} \end{pmatrix}$

   e) $\begin{pmatrix} \frac{9}{2} & \frac{5}{2} \\ -\frac{31}{12} & -\frac{17}{12} \end{pmatrix}$ **f)** $\begin{pmatrix} -\frac{3}{4} & \frac{1}{2} \\ 2 & -1 \end{pmatrix}$

Student Assessment 1

1. **a)** $\begin{pmatrix} 4 & -1 \\ -7 & 8 \end{pmatrix}$ **b)** $\begin{pmatrix} 6 & 11 \\ -1 & -12 \\ 10 & -5 \end{pmatrix}$

   c) $\begin{pmatrix} -9 & -6 \\ -1 & 1 \end{pmatrix}$ **d)** $\begin{pmatrix} 5 & -4 \\ 1 & 5 \\ 1 & 5 \end{pmatrix}$

   e) $\begin{pmatrix} 6 & 16 & -8 \\ 2 & -12 & 14 \end{pmatrix}$ **f)** $\begin{pmatrix} 1 & -2 \\ 0 & -1.5 \end{pmatrix}$

2. **a)** $\begin{pmatrix} 18 & -17 \\ 45 & 0 \\ 30 & 15 \end{pmatrix}$ **b)** $(-83 \quad -14)$

3. **a)** 1   **b)** 6   **c)** –6   **d)** –4

4. **a)** $\begin{pmatrix} 1 & -2 \\ -\frac{3}{2} & \frac{7}{2} \end{pmatrix}$ **b)** $\begin{pmatrix} 8 & -7 \\ -9 & 8 \end{pmatrix}$

   c) $\begin{pmatrix} \frac{5}{9} & \frac{11}{18} \\ -\frac{2}{3} & \frac{5}{6} \end{pmatrix}$ **d)** $\begin{pmatrix} -\frac{1}{2} & \frac{1}{4} \\ \frac{1}{2} & -\frac{1}{12} \end{pmatrix}$

Student Assessment 2

1. **a)** $\begin{pmatrix} 3 & -1 \\ -8 & 2 \end{pmatrix}$ **b)** $\begin{pmatrix} 11 & 7 \\ 6 & -12 \\ 8 & -5 \end{pmatrix}$

   c) $\begin{pmatrix} -10 & -7 \\ -4 & 1 \end{pmatrix}$ **d)** $\begin{pmatrix} 3 & 1 \\ -4 & 1 \\ 6 & -3 \end{pmatrix}$

   e) $\begin{pmatrix} 3 & 12 & -3 \\ 6 & -12 & 15 \end{pmatrix}$ **f)** $\begin{pmatrix} 1 & -2 \\ 0 & -4 \end{pmatrix}$

2. **a)** $\begin{pmatrix} 3 & -15 \\ 42 & 10 \\ -36 & 26 \end{pmatrix}$ **b)** $(-19 \quad -32)$

3. **a)** 1   **b)** 66   **c)** 66   **d)** 18

4. **a)** $\begin{pmatrix} 9 & -8 \\ -10 & 9 \end{pmatrix}$ **b)** $\begin{pmatrix} -\frac{7}{2} & -3 \\ -\frac{9}{2} & 4 \end{pmatrix}$

   c) Not possible   **d)** $\begin{pmatrix} 67.5 & -59 \\ -75.5 & 66 \end{pmatrix}$

19 Symmetry

Exercise 19.1 page 305

1. **a)** i) Student's planes   ii) 3
   b) i) Student's planes   ii) 2
   c) i) Student's planes   ii) 4
   d) i) Student's planes   ii) 4
   e) i) Student's planes   ii) Infinite
   f) i) Student's planes   ii) Infinite
   g) i) Student's planes   ii) Infinite
   h) i) Student's planes   ii) 9

2. **a)** 2   **b)** 2   **c)** 3   **d)** 4
   e) Infinite   **f)** Infinite   **g)** Infinite
   h) 4

Exercise 19.2

1. **a)** Isosceles   **b)** Perpendicular bisector
   c) 50°   **d)** 50°   **e)** 3.6 cm (1 d.p.)
   f) 9.3 cm (1 d.p.)

2. **a)** True   **b)** True   **c)** False   **d)** True

3. **a)** False   **b)** True

Exercise 19.3

1. **a)** 70°   **b)** 72°   **c)** 21°

2. **a)** 10.9 cm (1 d.p.)   **b)** 9.3 cm (1 d.p.)
   c) 3.5 cm (1 d.p.)

Student Assessment 1

1. **a)b)c)** Student's own diagrams

2. **a)b)** Student's own diagrams

3. Proof

4. 50°

5. 11.8 mm (1 d.p.)

Student Assessment 2

1. **a)b)c)** Student's own diagrams

2. **a)b)c)** Student's own diagrams

3. Proof

4. **a)** 24°   **b)** 9.1 cm (1 d.p.)

5. 88.0 cm (1 d.p.)

20 Transformations

Core Section page 311

Exercise 20.1

1.

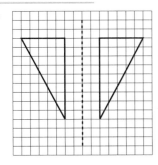

2.

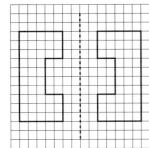

3.

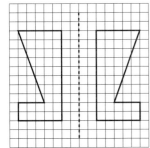

4.

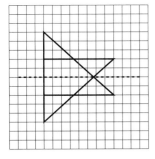

5.

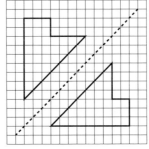

6.

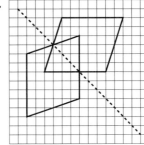

7.

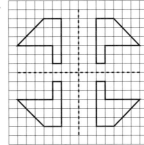

8.

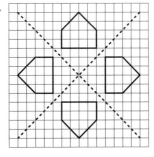

Exercise 20.2

1.

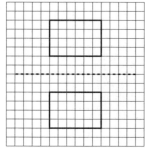

2.

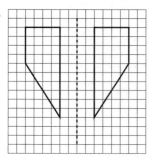

3.

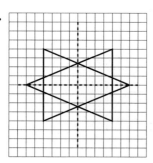

4.

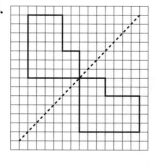

5.

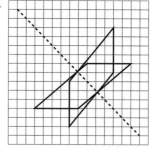

6.

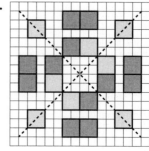

Exercise 20.3

1.

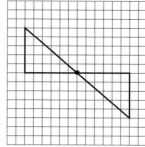

2.

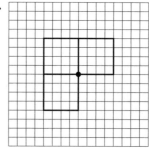

3.

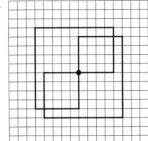

4.

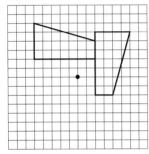

5.

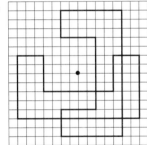

6.

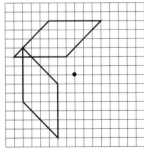

Exercise 20.4

1. **a)**

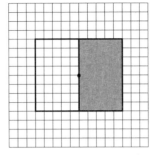

b) 180° clockwise/anti-clockwise

2. **a)**

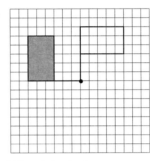

b) 90° anti-clockwise

3. **a)**

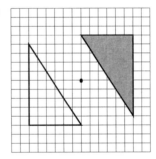

b) 180° clockwise/anti-clockwise

4. **a)**

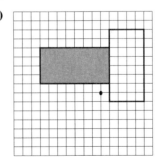

b) 90° anti-clockwise

5. **a)**

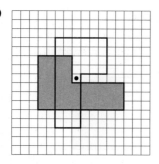

b) 90° anti-clockwise

6. **a)**

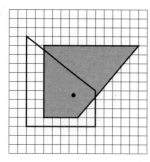

b) 90° clockwise

Exercise 20.5

1. $A \to B = \begin{pmatrix} -6 \\ 0 \end{pmatrix}$ $A \to C = \begin{pmatrix} 3 \\ 6 \end{pmatrix}$

2. $A \to B = \begin{pmatrix} 0 \\ -7 \end{pmatrix}$ $A \to C = \begin{pmatrix} -6 \\ 1 \end{pmatrix}$

3. $A \to B = \begin{pmatrix} 0 \\ 6 \end{pmatrix}$ $A \to C = \begin{pmatrix} 6 \\ -3 \end{pmatrix}$

4. $A \to B = \begin{pmatrix} 5 \\ 0 \end{pmatrix}$ $A \to C = \begin{pmatrix} -3 \\ -6 \end{pmatrix}$

Exercise 20.6

1.

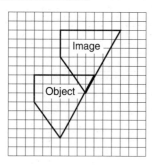

2.

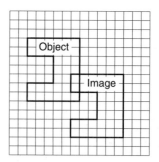

3.

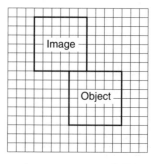

4.

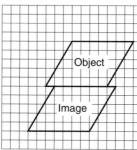

5.

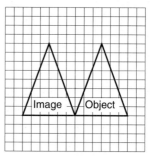

6.

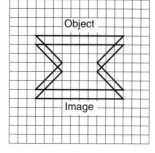

Exercise 20.7

1.

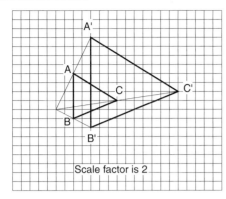

Scale factor is 2

2.

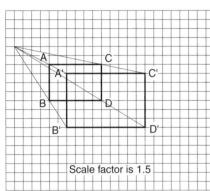

Scale factor is 1.5

3.

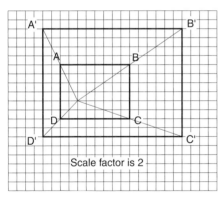

Scale factor is 2

4.

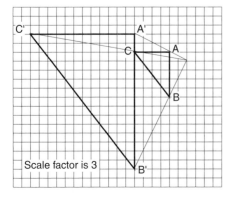

Scale factor is 3

5.

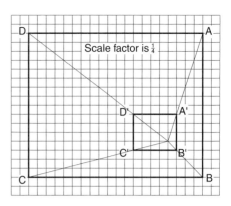

Scale factor is $\frac{1}{4}$

Exercise 20.8

1.

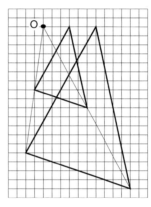

2.

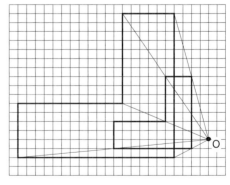

3.

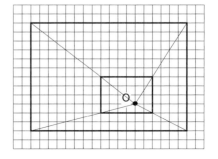

4.

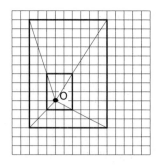

Exercise 20.9

1.

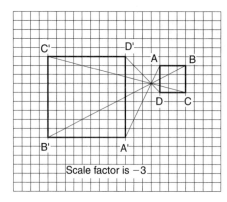

Scale factor is −3

2.

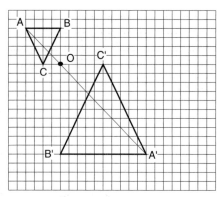

3.

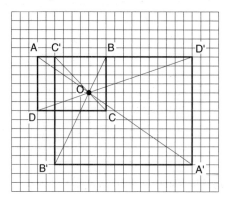

4.

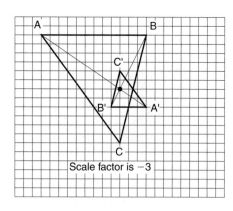

Scale factor is −3

5.

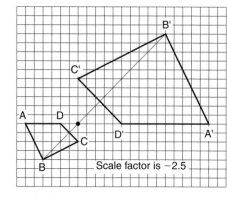

Scale factor is −2.5

6.

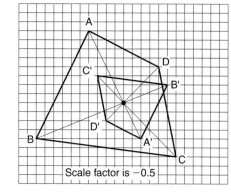

Scale factor is −0.5

Student Assessment I

1.

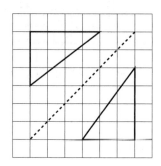

2.

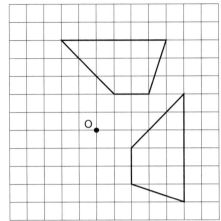

3. **a)** $\begin{pmatrix} 6 \\ 0 \end{pmatrix}$ **b)** $\begin{pmatrix} -3 \\ -5 \end{pmatrix}$

4.

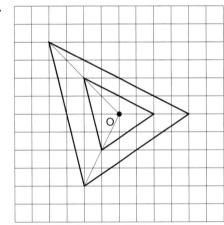

5. **a)**

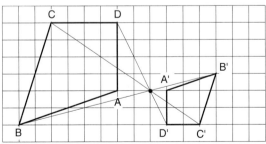

b) The scale factor of enlargement is –0.5.

Student Assessment 2

1.

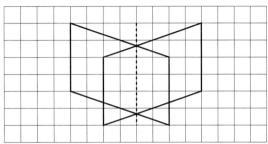

2.

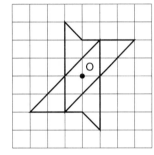

3. **a)** $\begin{pmatrix} 0 \\ -3 \end{pmatrix}$ **b)** $\begin{pmatrix} -8 \\ 2 \end{pmatrix}$

4.

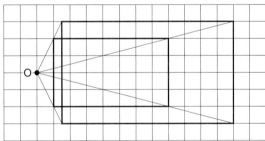

5. **a)**

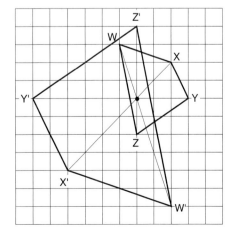

b) The scale factor of enlargement is –2.

Extended Section page 325

Exercise 20.10

1. a)

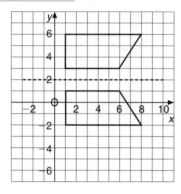

   b) $y = 2$

2. a)

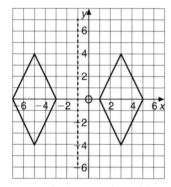

   b) $x = -1$

3. a)

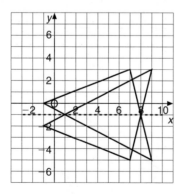

   b) $y = -1$

4. a)

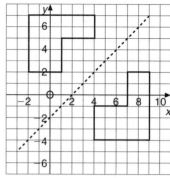

   b) $y = x - 2$

5. a)

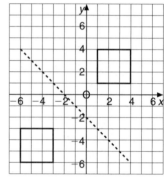

   b) $y = -x - 2$

6. a)

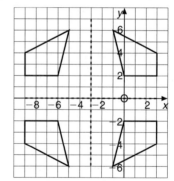

   b) $y = 0, x = -3$

7. a)

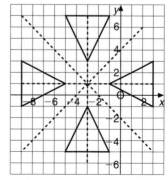

   b) $y = 1, x = -3, y = x + 4, y = -x - 2$

8. **a)**

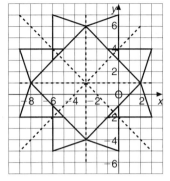

b) $y = 1$, $x = -3$, $y = x + 4$, $y = -x - 2$

Exercise 20.11

1. **a)**

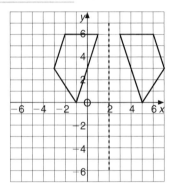

b)

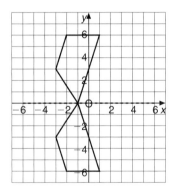

c)

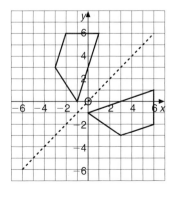

d)

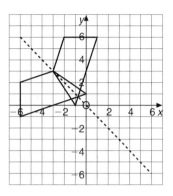

2. **a)**

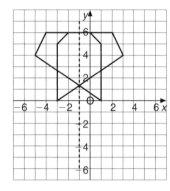

b)

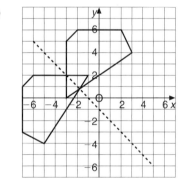

c)

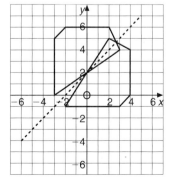

d)

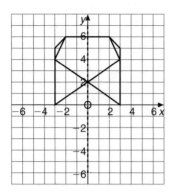

3.

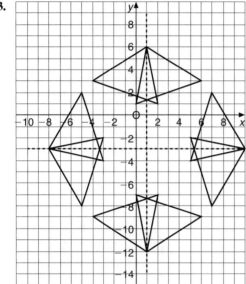

Exercise 20.12

1. Student's own diagram
2. Student's own diagram
3. Student's own diagram
4. Student's own diagram
5. Student's own diagram

Exercise 20.13

1. 1
2. 1
3. 1.5
4. 2
5. 2.5
6. 3

Exercise 20.14

1.

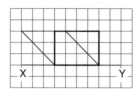

2.

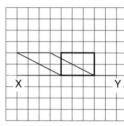

3.

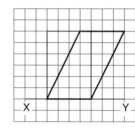

4.

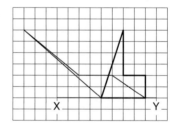

5.

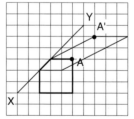

Shear factor of 2

6.

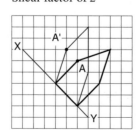

Shear factor of 0.5

7.

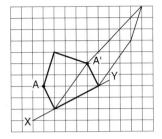

Shear factor of 2

8.

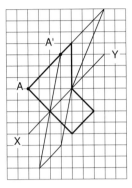

Shear factor of 1.5

Exercise 20.15

1. 2
2. 1.5
3. 3.5
4. 2
5. 1.5
6. 4

Exercise 20.16

1.

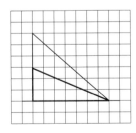

Stretch factor is 2

2.

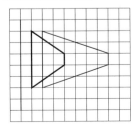

Stretch factor is 2

3.

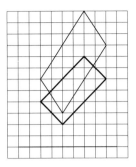

Stretch factor is 1.5

4.

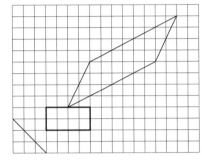

Stretch factor is 3

Exercise 20.17

1.

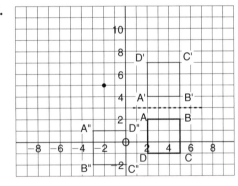

2.

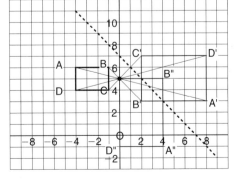

3.

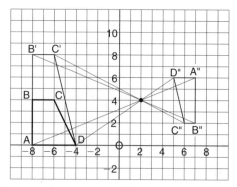

4.

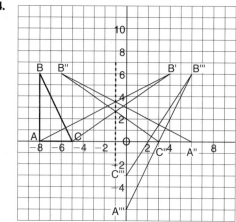

Exercise 20.18

1.

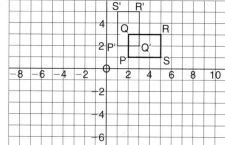

2.

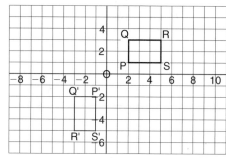

3.

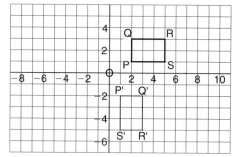

4.

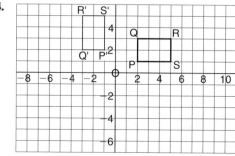

5.

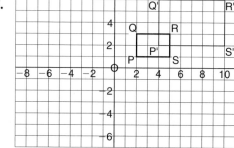

6.

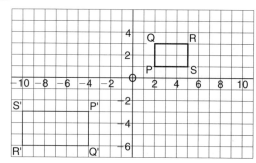

7. Q.1 represents a reflection in the line $y = x$.
Q.2 represents a reflection in the line $y = -x$.
Q.3 represents a clockwise rotation of 90°
about the origin.
Q.4 represents an anti-clockwise rotation of 90°
about the origin.

Q.5 represents an enlargement of scale factor 2 with its centre at the origin.

Q.6 represents an enlargement of scale factor –2 with its centre at the origin.

8. a)

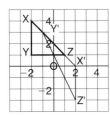

b) 4.5 units^2 **c)** 4.5 units^2 **d)** 1   **e)** 1

9. a)

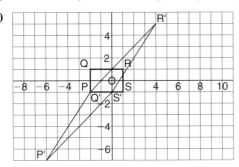

b) 6 units^2 **c)** 12 units^2 **d)** 2   **e)** 2

10. a)

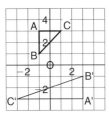

b) 2 units^2 **c)** 6 units^2 **d)** 3   **e)** 3

11. The area scale factor has the same magnitude as the determinant of the transformation matrix.

Exercise 20.19

1. d) $\begin{pmatrix} 0.5 & 0.5 \\ -1 & 0 \end{pmatrix}$

2. d) $\begin{pmatrix} 1 & 2 \\ -2 & -3 \end{pmatrix}$

3. d) $\begin{pmatrix} -1\frac{2}{3} & -\frac{2}{3} \\ -\frac{2}{3} & -\frac{2}{3} \end{pmatrix}$

4. d) $\begin{pmatrix} 0 & -\frac{2}{3} \\ -\frac{1}{2} & 0 \end{pmatrix}$

5. d) $\begin{pmatrix} \frac{1}{2} & 0 \\ \frac{2}{3} & \frac{2}{3} \end{pmatrix}$

6. f) $\begin{pmatrix} -1 & -1 \\ 2 & 1 \end{pmatrix}$ **g)** $\begin{pmatrix} -1 & 0 \\ -0.5 & -0.5 \end{pmatrix}$

h) $\begin{pmatrix} 0 & -2 \\ 1 & 2 \end{pmatrix}$ **i)** $\begin{pmatrix} 1 & 1 \\ -0.5 & 0 \end{pmatrix}$

Exercise 20.20

1. a)

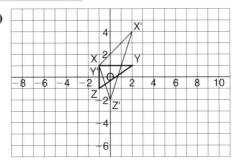

b)

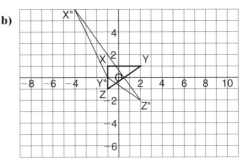

c) 6 units^2 **d)** 6 units^2

e) $\begin{pmatrix} 1 & -3 \\ -2 & 4 \end{pmatrix}$ **f)** $\begin{pmatrix} -2 & -1.5 \\ -1 & -0.5 \end{pmatrix}$

2. f) $\begin{pmatrix} -1 & 3 \\ -2.5 & 6 \end{pmatrix}$ **g)** $\begin{pmatrix} 4 & -2 \\ 1\frac{2}{3} & -\frac{2}{3} \end{pmatrix}$

3. a) $\begin{pmatrix} 1 & 0 \\ 1 & -1 \end{pmatrix}$ **b)** $\begin{pmatrix} 1 & 0 \\ 1 & -1 \end{pmatrix}$

4. a) $\begin{pmatrix} 2 & 1 \\ -1 & 2 \end{pmatrix}$ **b)** $\begin{pmatrix} 0 & -1 \\ -1 & 1 \end{pmatrix}$

c) $\begin{pmatrix} 1 & -2 \\ -3 & 1 \end{pmatrix}$ **d)** $\begin{pmatrix} -0.2 & -0.4 \\ -0.6 & -0.2 \end{pmatrix}$

e) 16 units^2 **f)** 80 units^2 **g)** 80 units^2

Student Assessment 1

1. a)

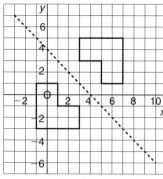

$y = x - 2$ is also a mirror line

b) $y = -x + 4$

2. a) $(3, 2)$ **b)** 90° clockwise

3. a) 1   **b)** 3.5

4. a) 1.5   **b)** 1.25

5. a) An enlargement of scale factor 2. Centre of enlargement $(3, 3)$
b) A reflection about the line $y = -x - 1$

Student Assessment 2

1.

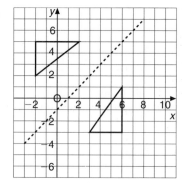

2. a)

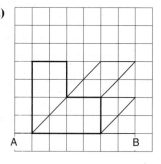

b)

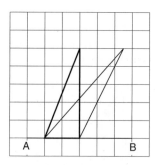

3. a) A translation of vector $\begin{pmatrix} 6 \\ 0 \end{pmatrix}$

b) An enlargement of scale factor 2. Centre of enlargement $(6, 8)$

4. a)

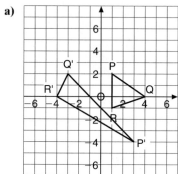

b) 2

5. a) $\begin{pmatrix} 0.5 & 1 \\ 0.5 & 2 \end{pmatrix}$ **b)** $\begin{pmatrix} 1 & 0 \\ 1 & -1 \end{pmatrix}$

Student Assessment 3

1. a) Student's own construction
b) 90° anti-clockwise

2. a) 1.5   **b)** 2

3. a) A reflection in the line $x = 0$
b) An enlargement by scale factor –0.5. Centre of enlargement $(0, -1)$

4. a) b)

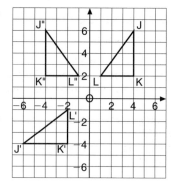

c) $\begin{pmatrix} -1 & 0 \\ 0 & 1 \end{pmatrix}$

Student Assessment 4

1. **a)** Student's own construction
   b) 180° clockwise/anti-clockwise

2. **a) b)**

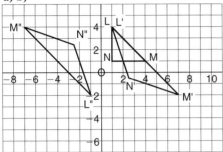

3. **a) b)**

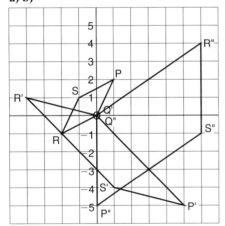

c) 3 units^2 **d)** 15 units^2 **e)** 30 units^2

f) $\begin{pmatrix} 0 & -1 \\ -0.5 & 1 \end{pmatrix}$ **g)** $\begin{pmatrix} -4 & 2 \\ -1 & -2 \end{pmatrix}$

h) $\begin{pmatrix} -0.2 & -0.2 \\ 0.1 & -0.4 \end{pmatrix}$

ICT Section page 345

3. Reflection in $y = x$

4. **a)** Reflection in $y = -x$
   b) Clockwise rotation about the origin of 90°
   c) Anti-clockwise rotation about the origin of 90°
   d) Enlargement of scale factor 2, centred at the origin.
   e) Enlargement of scale factor –2, centred at the origin.

21 Handling Data

Exercise 21.1 page 346

1. Mean = 1.7 (1 d.p.)
   Median = 1
   Mode = 1

2. Mean = 6.2
   Median = 6.5
   Mode = 7

3. Mean = 14 yrs 3 mths
   Median = 14 yrs 3 mths
   Mode = 14 yrs 3 mths

4. Mean = 26.4
   Median = 27
   Mode = 28

5. Mean = 13.9 s (1 d.p.)
   Median = 13.9 s
   Mode = 13.8 s

6. 91.1 kg

7. 103 points

Exercise 21.2

1. Mean = 3.35
   Median = 3
   Mode = 1 and 4

2. Mean = 7.03

3. **a)** Mean = 6.3 (1 d.p.)
   Median = 7
   Mode = 8
   b) The mode, as it gives the highest number of flowers per bush

Exercise 21.3

1. **a)**

Height (m)	Frequency	Mid-interval value	Frequency × mid-interval value
1.8–	2	1.85	3.7
1.9–	5	1.95	9.75
2.0–	10	2.05	20.5
2.1–	22	2.15	47.3
2.2–	7	2.25	15.75
2.3–2.4	4	2.35	9.4

b) Mean = 2.1 m (1 d.p.)
c) Modal class = 2.1–2.2 m

2. **a)** Mean = 33 h (2 s.f.)
   b) Modal class = 30–39 h

3. **a)** Mean = 6.2 cm
   b) Modal class = 6.0–6.5 cm

Exercise 21.4

1. Student's answers may differ from those given below.
   a) Possible positive correlation (strength depending on topics tested)
   b) No correlation
   c) Positive correlation (likely to be quite strong)
   d) Negative correlation (likely to be strong). Assume that motorcycles are not rare/vintage.
   e) Factors such as social class, religion and income are likely to affect results. Therefore little correlation is likely.
   f) Negative correlation (likely to be strong)
   g) 0–16 years likely to be a positive correlation
   h) Strong positive correlation

2. **a)**

Sunshine and rainfall correlation

   b) Graph shows a very weak negative correlation.

3. **a)**

Adult illiteracy and infant mortality correlation

   b) Positive correlation
   c) Student's answer. However, although there is a correlation, it doesn't imply that one variable affects the other

d)

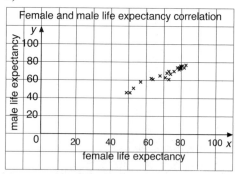

Female and male life expectancy correlation

Exercise 21.5

1.

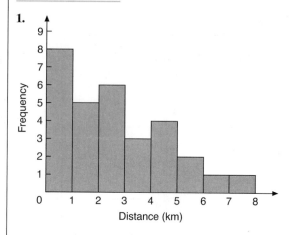

Distance (km)

2.

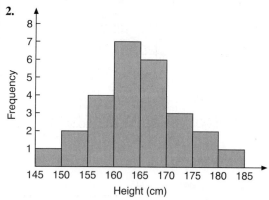

Height (cm)

Exercise 21.6

1. **a)**

Time (min)	0–	10–	15–	20–	25–	30–	40–60
Freq.	6	3	13	7	3	4	4
Freq. density	0.6	0.6	2.6	1.4	0.6	0.4	0.2

b)

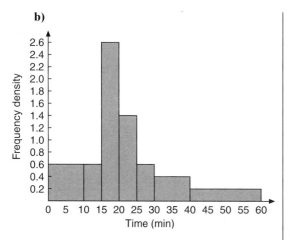

2. a)

Time (min)	Frequency	Frequency density
$0 \leqslant t < 30$	8	**0.3**
$30 \leqslant t < 45$	5	**0.3**
$45 \leqslant t < 60$	8	**0.5**
$60 \leqslant t < 75$	9	**0.6**
$75 \leqslant t < 90$	10	**0.7**
$90 \leqslant t < 120$	12	**0.4**

b)

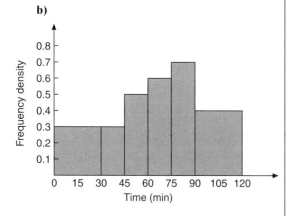

3. a)

Age (years)	0–	1–	5–	10–	20–	40–	60–90
Freq.	35	**48**	**140**	180	260	**280**	150
Freq. density	35	12	28	**18**	**13**	14	**5**

b)

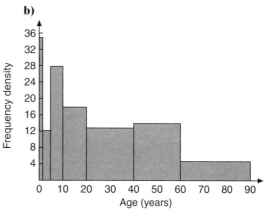

4. a)

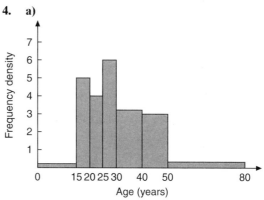

b) Student's own answers

Exercise 21.7

1. a)

Finishing time (h)	0–	0.5–	1.0–	1.5–	2.0–	2.5–	3.0–3.5
Freq.	0	0	6	34	16	3	1
Cum. freq.	0	0	6	40	56	59	60

b)

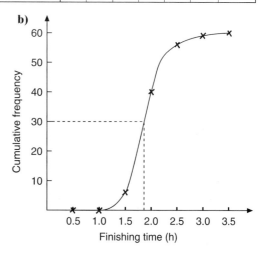

c) Median ≈ 1.8 h

d) As many runners finished before as after the median.

2. **a)**

Score	Class A Freq.	Class A Cum. freq.	Class B Freq.	Class B Cum. freq.	Class C Freq.	Class C Cum. freq.
$0 \leqslant x < 20$	1	1	0	0	1	1
$20 \leqslant x < 40$	5	6	0	0	2	3
$40 \leqslant x < 60$	6	12	4	4	2	5
$60 \leqslant x < 80$	3	15	4	8	4	9
$80 \leqslant x < 100$	3	18	4	12	8	17

b)

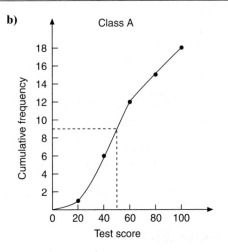

Class A

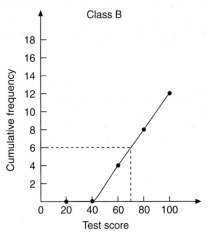

Class B

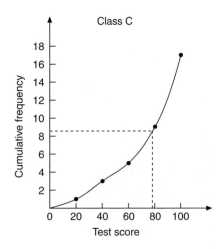

Class C

c) Class A median ≈ 50
Class B median ≈ 70
Class C median ≈ 78

d) As many students were above as below the median.

3. **a)**

Height (cm)	1996 Freq.	1996 Cum. freq.	1997 Freq.	1997 Cum. freq.	1998 Freq.	1998 Cum. freq.
150–	6	6	2	2	2	2
155–	8	14	9	11	6	8
160–	11	25	10	21	9	17
165–	4	29	4	25	8	25
170–	1	30	3	28	2	27
175–	0	30	2	30	2	29
180–185	0	30	0	30	1	30

b)

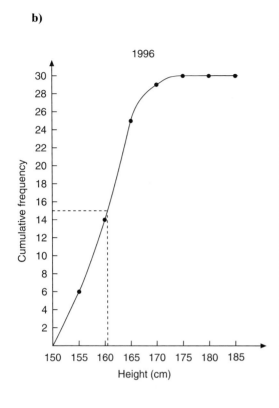

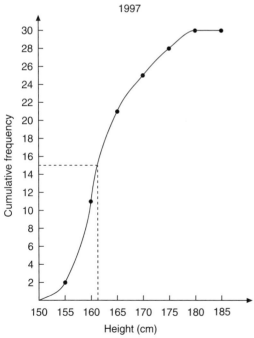

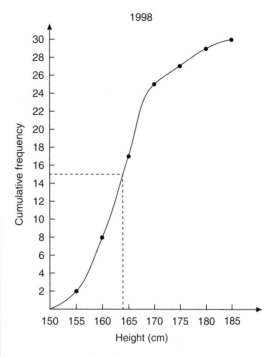

c) Median (1996) ≈ 161 cm
Median (1997) ≈ 162 cm
Median (1998) ≈ 164 cm
d) As many students are taller than the median as shorter than the median.

Exercise 21.8

1. **a)** Class A ≈ 30   Class B ≈ 30   Class C ≈ 40
b) Student's own responses
2. **a)** 1996 ≈ 7 cm   1997 ≈ 8 cm   1998 ≈ 8 cm
b) Student's own responses
3. **a)**

Distance thrown (m)	0–	20–	40–	60–	80–100
Freq.	4	9	15	10	2
Cum. freq.	4	13	28	38	40

b)

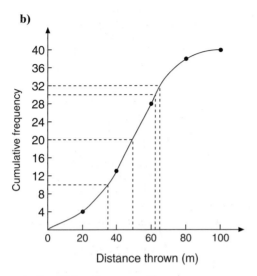

c) Qualifying distance ≈ 66 m
d) Inter-quartile range ≈ 28 m
e) Median ≈ 50 m

4. **a)**

Type A		
Mass (g)	**Frequency**	**Cum. freq.**
75–	4	4
100–	7	11
125–	15	26
150–	32	58
175–	14	72
200–	6	78
225–250	2	80

Type B		
Mass (g)	**Frequency**	**Cum. freq.**
75–	0	0
100–	16	16
125–	43	59
150–	10	69
175–	7	76
200–	4	80
225–250	0	80

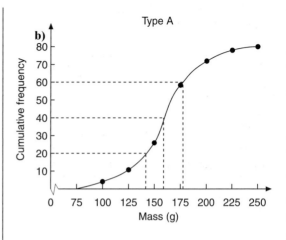

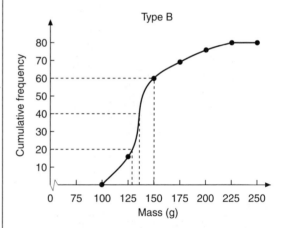

c) Median type A ≈ 157 g
Median type B ≈ 137 g
d) i) Lower quartile type A ≈ 140 g
Lower quartile type B ≈ 127 g
ii) Upper quartile type A ≈ 178 g
Upper quartile type B ≈ 150 g
iii) Inter-quartile type range type A ≈ 38 g
Inter-quartile type range type B ≈ 23 g
e) Student's own report

5. **a)** Student's own explanation
b) Student's own explanation

Student Assessment 1

1. **a)** Mean = 16
b) Median = 16.5
c) Mode = 18

2. **a)** 28
b) i) Mean 7.8 (1 d.p.)
ii) Median = 8
iii) Mode = 8

3. a) Mean = 19.1 kg (1.d.p.)
   b) Modal class is $19 \leqslant M < 20$

4. a)

Attendence	Tally	Frequency
0–3999		0
4000–7999	II	2
8000–11 999	⊬⊬⊢ II	7
12 000–15 999	⊬⊬⊢ I	6
16 000–19 000	⊬⊬⊢ ⊬⊬⊢ ⊬⊬⊢	15

b)

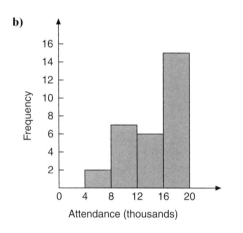

c) Mean ≈ 14 530 (4 s.f.)   **d)** Mean = 17 026

5. a)

Mark	20–	30–	40–	50–	60–	70–	80–	90–100
Freq.	2	3	5	7	6	4	2	1
Cum. freq.	2	5	10	17	23	27	29	30

b)

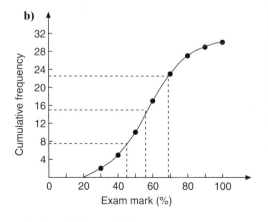

c) i) Median ≈ 57%
  ii) Lower quartile ≈ 45%
      Upper quartile ≈ 69%
  iii) Inter-quartile range ≈ 24%

6. a)

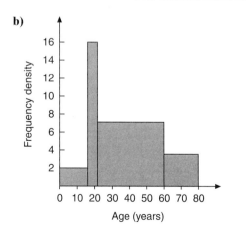

b) (Strong) positive correlation
c) It depends on their mode of transport.
d)

Distance travelled and time taken correlation

7. a)

	Age	Number	Frequency density
Juniors	0–	32	2
Intermediates	16–	80	16
Full members	21–	273	7
Seniors	60–80	70	3.5

b)

Student Assessment 2

1. Mean = 86.8 m   Median = 90.5 m
   Mode = 93 m

2. **a)** 26
   b) i) Mean = 7.7 (1 d.p.)   ii) Median = 7.5
   iii) Mode 10

3. **a)** Mean ≈ 10.0 kg (1 d.p.)
   b) Modal class is $10.0 \leqslant M < 10.1$ kg

4. **a)**

Mark (%)	Frequency	Cumulative frequency
31–40	21	21
41–50	55	76
51–60	125	201
61–70	74	275
71–80	52	327
81–90	45	372
91–100	28	400

b)

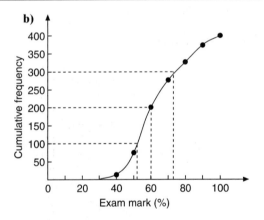

c) i) Median ≈ 60%
   ii) Lower quartile ≈ 52%
      Upper quartile ≈ 73%
   iii) Inter-quartile range ≈ 21%

5. **a)**

Mark (%)	Frequency	Cumulative frequency
1–10	10	10
11–20	30	40
21–30	40	80
31–40	50	130
41–50	70	200
51–60	100	300
61–70	240	540
71–80	160	700
81–90	70	770
91–100	30	800

b)

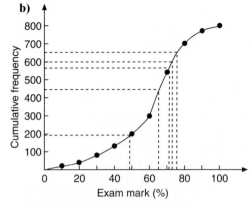

c) 'A' grade ≈ 75%
d) Lower boundary ≈ 66%
   Upper boundary ≈ 72%
e) Inter-quartile range ≈ 25%

6. **a)**

Gloves sold and outside temperature correlation

b) Positive correlation
c) Student's own answer

7. **a)**

Points	0–4	5–9	10–14	15–24	25–34	35–49
Number of games	2	3	8	9	12	3
Freq. density	0.5	0.8	2.0	1.0	1.3	0.2

b)

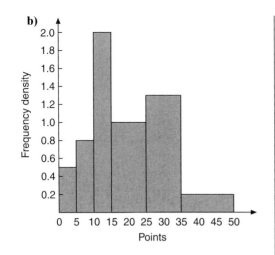

22 Probability

Exercise 22.1 page 369

1. **a)** $\frac{1}{6}$ **b)** $\frac{1}{6}$ **c)** $\frac{1}{2}$ **d)** $\frac{5}{6}$ **e)** 0 **f)** 1

2. **a)** i) $\frac{1}{7}$ ii) $\frac{6}{7}$ **b)** Total $= 1$

3. **a)** $\frac{1}{250}$ **b)** $\frac{1}{50}$ **c)** 1 **d)** 0

4. **a)** $\frac{5}{8}$ **b)** $\frac{3}{8}$

5. **a)** $\frac{1}{13}$ **b)** $\frac{5}{26}$ **c)** $\frac{21}{26}$ **d)** $\frac{3}{26}$

6. **a)** $\frac{1}{6}$

7. **a)** i) $\frac{1}{10}$ ii) $\frac{1}{4}$ **b)** i) $\frac{1}{19}$ ii) $\frac{3}{19}$

8. **a)** $\frac{1}{37}$ **b)** $\frac{18}{37}$ **c)** $\frac{18}{37}$ **d)** $\frac{1}{37}$ **e)** $\frac{21}{37}$ **f)** $\frac{12}{37}$
   g) $\frac{17}{37}$ **h)** $\frac{11}{37}$

9. **a)** RCA RAC CRA CAR ARC ACR

   b) $\frac{1}{6}$ **c)** $\frac{1}{3}$ **d)** $\frac{1}{2}$ **e)** $\frac{1}{24}$

10. **a)** $\frac{1}{4}$ **b)** $\frac{1}{2}$ **c)** $\frac{1}{13}$ **d)** $\frac{1}{26}$ **e)** $\frac{3}{13}$ **f)** $\frac{1}{52}$
    g) $\frac{5}{13}$ **h)** $\frac{4}{13}$

Exercise 22.2

1. **a)**

		Dice 1		
Dice 2		1	2	3	4
	1	1, 1	2, 1	3, 1	4, 1
	2	1, 2	2, 2	3, 2	4, 2
	3	1, 3	2, 3	3, 3	4, 3
	4	1, 4	2, 4	3, 4	4, 4

   b) $\frac{1}{4}$ **c)** $\frac{1}{4}$ **d)** $\frac{9}{16}$

2.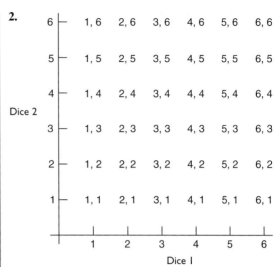

a) $\frac{1}{36}$ **b)** $\frac{1}{6}$ **c)** $\frac{1}{18}$ **d)** $\frac{1}{6}$ **e)** $\frac{1}{4}$ **f)** $\frac{3}{4}$
g) $\frac{11}{36}$ **h)** $\frac{1}{6}$ **i)** $\frac{11}{18}$

Exercise 22.3

1. **a)**

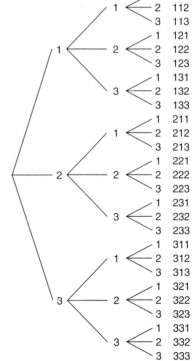

b) i) $\frac{1}{27}$ ii) $\frac{1}{3}$ iii) $\frac{1}{9}$ iv) $\frac{1}{3}$ v) $\frac{5}{9}$ vi) $\frac{1}{3}$

2. a)

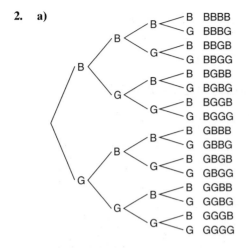

b) i) $\frac{1}{16}$ **ii)** $\frac{3}{8}$ **iii)** $\frac{15}{16}$ **iv)** $\frac{5}{16}$

3. a)

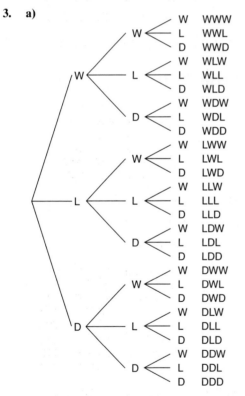

b) i) $\frac{1}{27}$ **ii)** $\frac{10}{27}$ **iii)** $\frac{19}{27}$ **iv)** $\frac{8}{27}$

4. a)

b) i) $\frac{1}{16}$ **ii)** $\frac{1}{4}$ **iii)** $\frac{1}{8}$

Exercise 22.4

1. a)

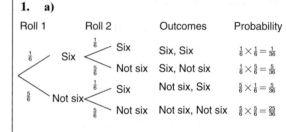

Roll 1	Roll 2	Outcomes	Probability
$\frac{1}{6}$	$\frac{1}{6}$ Six	Six, Six	$\frac{1}{6} \times \frac{1}{6} = \frac{1}{36}$
Six	$\frac{5}{6}$ Not six	Six, Not six	$\frac{1}{6} \times \frac{5}{6} = \frac{5}{36}$
$\frac{5}{6}$	$\frac{1}{6}$ Six	Not six, Six	$\frac{5}{6} \times \frac{1}{6} = \frac{5}{36}$
Not six	$\frac{5}{6}$ Not six	Not six, Not six	$\frac{5}{6} \times \frac{5}{6} = \frac{25}{36}$

b) i) $\frac{1}{6}$ **ii)** $\frac{11}{36}$ **iii)** $\frac{5}{36}$ **iv)** $\frac{125}{216}$ **v)** $\frac{91}{216}$
c) Add up to 1

2. a) $\frac{4}{25}$ **b)** $\frac{54}{125}$ **c)** $\frac{98}{125}$

3. a)

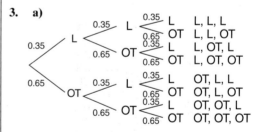

b) i) 0.275 (3 d.p.)   **ii)** 0.123 (3 d.p.)
   iii) 0.444 (3 d.p.)   **iv)** 0.718 (3 d.p)

4. a) 0.059 (3 d.p)   **b)** 0.013 (3 d.p)
   c) 0.414 (3 d.p.)   **d)** 0.586 (3 d.p.)

5. a)

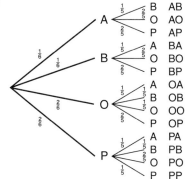

b) i) $\frac{1}{15}$ ii) $\frac{1}{15}$ iii) $\frac{3}{5}$

6. 0.027

7. a) 0.75^2 (0.56)   **b)** 0.75^3 (0.42)
   c) 0.75^{10} (0.06)

Student Assessment 1

1. a) $\frac{1}{7}$ **b)** $\frac{12}{365}$ **c)** $\frac{1}{365}$ **d)** $\frac{1}{1461}$

2. a) $\frac{1}{6}$ **b)** $\frac{1}{2}$ **c)** $\frac{2}{3}$ **d)** 0

3. a) i) $\frac{3}{5}$ ii) $\frac{1}{19}$ **b)** Infinite

4. a)

	1	2	3	4	5	6
H	1H	2H	3H	4H	5H	6H
T	1T	2T	3T	4T	5T	6T

b) i) $\frac{1}{12}$ ii) $\frac{1}{4}$ iii) $\frac{1}{4}$

5. a)

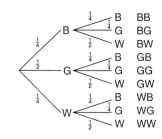

b) i) $\frac{1}{16}$ ii) $\frac{1}{8}$ iii) $\frac{1}{4}$

6. a)

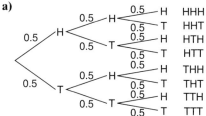

b) i) $\frac{1}{8}$ ii) $\frac{3}{8}$ iii) $\frac{1}{8}$ iv) $\frac{7}{8}$

7. a) $\frac{12}{27}$ **b)** $\frac{26}{27}$ **c)** $\frac{6}{27}$

8. a) $\frac{9}{16}$ **b)** $\frac{1}{16}$ **c)** $\frac{1}{4}$

9. 0.010 (3 d.p.)

Student Assessment 2

1. a) $\frac{1}{13}$ **b)** $\frac{1}{4}$ **c)** $\frac{1}{52}$ **f)** $\frac{3}{26}$

2. a)

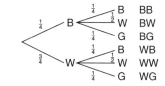

		Dice 1					
		1	2	3	4	5	6
Dice 2	1	2	3	4	5	6	7
	2	3	4	5	6	7	8
	3	4	5	6	7	8	9
	4	5	6	7	8	9	10
	5	6	7	8	9	10	11
	6	7	8	9	10	11	12

b) i) $\frac{1}{36}$ ii) $\frac{1}{6}$ iii) $\frac{1}{12}$ iv) $\frac{7}{12}$ **c)** 25

3. a) i) $\frac{1}{4}$ ii) $\frac{3}{8}$
   b) i) $\frac{3}{64}$ ii) $\frac{6}{64}$ iii) $\frac{7}{16}$ iv) $\frac{39}{64}$

4. a)

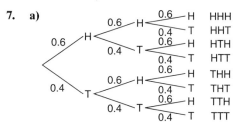

b) i) $\frac{1}{16}$ ii) $\frac{3}{8}$ iii) $\frac{3}{16}$ iv) $\frac{7}{8}$

5. a) $\frac{1}{12}$ **b)** $\frac{1}{6}$ **c)** $\frac{1}{6}$

6. a) 0.72   **b)** 0.729

7. a)

```
                          0.6  H  HHH
                   0.6 H <
              0.6 H<     0.4  T  HHT
             <     0.4 T< 0.6  H  HTH
        0.6 H<          0.4  T  HTT
       <               0.6  H  THH
        0.4 T< 0.6 H<
                   0.4  T  THT
              0.4 T< 0.6  H  TTH
                   0.4  T  TTT
```

b) i) 0.216   ii) 0.064   iii) 0.648

8. a) i) 0.8   ii) 0.7   iii) 0.3
   b) i) 0.06   ii) 0.56   iii) 0.38

ICT Section page 379

2. A possible formula for generating random
integers between 1 and 6 is:
=1+INT(RAND()*6)

3. Copy the above formula into 100 cells.

INDEX

Page numbers refer to pages where topics are described: mentions on pages devoted wholly to exercises or assessments are not indexed.